Leonid M. Blumberg

Temperature-Programmed Gas Chromatography

Further Reading

Hübschmann, H.-J.

Handbook of GC/MS

Fundamentals and Applications

2009

Hardcover

ISBN: 978-3-527-31427-0

McMaster, M.

GC/MS

A Practical User's Guide

2008

E-Book

ISBN: 978-0-470-22834-0

Rood, D.

The Troubleshooting and Maintenance Guide for Gas Chromatographers

2007

Hardcover

ISBN: 978-3-527-31373-0

Cserhati, T.

Multivariate Methods in Chromatography

A Practical Guide

2008

Hardcover

ISBN: 978-0-470-05820-6

Kuss, H.-J., Kromidas, S. (Eds.)

Quantification in LC and GC

A Practical Guide to Good Chromatographic Data

2009

Hardcover

ISBN: 978-3-527-32301-2

Fritz, J. S., Gjerde, D. T.

Ion Chromatography

4th completely revised and enlarged edition

2009

Hardcover

ISBN: 978-3-527-32052-3

Leonid M. Blumberg

Temperature-Programmed Gas Chromatography

WILEY-VCH

WILEY-VCH Verlag GmbH & Co. KGaA

The Author

Dr. Leonid M. Blumberg
Fast GC Consulting
P.O. Box 1243
Wilmington, DE 19801 19801
USA

Library of Congress Card No.: applied for

British Library Cataloguing-in-Publication Data
A catalogue record for this book is available from the British Library.

Bibliographic information published by the Deutsche Nationalbibliothek
The Deutsche Nationalbibliothek lists this publication in the Deutsche Nationalbibliografie; detailed bibliographic data are available on the Internet at http://dnb.d-nb.de.

Cover Design Adam Design, Weinheim
Typesetting Thomson Digital, Noida, India
Printing and Binding Fabulous Printers Pte Ltd

Printed in Singapore
Printed on acid-free paper

ISBN: 978-3-527-32642-6

To Irena, who made this book possible

Contents

Temperature-Programmed Gas Chromatography. Leonid M. Blumberg

ISBN: 978-3-527-32642-6

Preface

1952 was the year when James and Martin published two papers demonstrating partition gas chromatography [1, 2]. And it was also the year of Griffiths', James', and Phillips' lecture on gas chromatography (GC) [3]. The main topic of the lecture at the *First International Congress on Analytical Chemistry* (Oxford, September 4–9, 1952) was the evaluation of alternatives to a thermal-conductivity cell. The authors mentioned that, by the way, "With a mixture containing components with a wide range of boiling points, the later components tend to spread themselves out to give long bands of low concentration. This can be overcome by varying the temperature (Fig. 3)." A linear heating ramp from 0 to 50 °C in 40 min and a five-peak chromatogram were shown in Figure 3 [3]. That was all for the beginning of the temperature-programmed GC.

After 50 years, Cramers and Leclercq estimated that 80% of GC analyses are conducted with temperature programming [4].

One can distinguish two groups of studies in temperature-programmed GC. One is concerned with the prediction of retention times of particular solutes. The focus of the other is a general theory whose object is general trends (such as effect of heating rate on retention of all peaks) and the effects of those trends on performance characteristics (such as the largest possible number of resolved peaks) of GC.

Foundation of general theory of temperature-program GC can be attributed to Giddings who outlined the theory and came close to solving the problem of optimal heating rate [5]. A book entirely dedicated to general theory of temperature-programmed GC was published by Harris and Habgood in 1966 [6]. Nothing on the same subject that was as comprehensive as that book has been published since then.

This book is an attempt of an overview of the more recent state of general theory of temperature-programmed GC. The book consists of three parts. Part one is introductory. It introduces the basic terminology of GC and briefly describes the key properties of GC columns.

Part Two, *Background*, outlines mathematical and physical background of GC. The titles of its chapters – *Linear Systems, Migration of a Solid Object, Solute–Liquid Interaction in Gas Chromatography, Molecular Properties of Ideal Gas*, and *Flow of Ideal Gas* – speak for themselves. The book provides probably the broadest coverage of these topics that can be found in a single volume on GC.

Temperature-Programmed Gas Chromatography. Leonid M. Blumberg

ISBN: 978-3-527-32642-6

The core topics of the book are covered in Part Three, *Formation of Chromatogram*. Here too, the chapter titles – *Formation of Retention Times, Formation of Peak Spacing*, and *Formation of Peak Widths* – speak for themselves.

Chapter 10 on peak-width formation deserves additional comments. The chapter is based on Golay's 1958 plate height theory [7] in capillary columns. Unfortunately, widely accepted perception of the theory, while being attributed to Golay, is a significant distortion of his theory. The distortion was introduced at the same symposium (Amsterdam, May 1958) where Golay presented his theory. This time, plate height theory in packed columns published 2 years earlier by van Deemter *et al.* [8] has been distorted, but 2 years later the distortion spread to Golay's theory as well.

In the true van Deemter and Golay theories, plate height was expressed as a function of local velocity of a carrier gas. Due to gas decompression along the column, its local velocity is a coordinate-dependent quantity which is not easy to measure. It is much easier to measure the time-averaged gas velocity (briefly, average velocity). And so, for practical convenience, local gas velocity in Golay and van Deemter formulae was replaced with average velocity. The incorrectness of this distortion has been almost immediately recognized. Several leading experts in GC who adopted the distortion published the retractions. However, this was not enough. The distorted theory, being practically attractive, took on the life of its own. Since then, the pseudotheory dominated the GC literature including the textbooks and universality curricula on GC.

In majority of typical applications having weak or moderate gas decompression along the columns, the difference between actual dependence of plate height on gas average velocity and that dependence predicted from distorted theory was not alarmingly large. This was probably the key reason for the resilience of the distorted theory. However, in GC-MS with its vacuum at the column outlet, in high-resolution GC using long small-bore columns, and in other applications incurring strong gas decompression along the column, the difference is large. The fact that the accepted theory was not taken to the task of explaining experimental observations is an indication of a disregard for the theory among its own followers.

There is nothing wrong in expressing a column plate height as a function of a gas average velocity. However, the correct expressions of that kind are complex – much more so than the conveniently modified van Deemter and Golay formulae are. Also complex are the formulae for optimal average velocity. Much simpler are the formulae for the plate height as a function of a gas flow rate and the formulae for optimal flow rate. All that has been discussed in the literature. Although this simpler approach is based on sound theory and numerous experimental verifications, its acceptance lags its potential usefulness for scientifically based optimization of conventional GC and new techniques such as comprehensive multidimensional GC.

Of course the incorrect theories of GC were not the only ones. Development of correct theories by many workers never stopped. And this is what made this book possible. However, it is fair to say that the wide proliferation of the incorrect plate height theory – the theory of the most basic concept in chromatography – corrupted the trust in all theories in GC.

Recognizing that the theory presented in Chapter 10 is different from widely accepted theory, the material of the chapter is delivered differently than the material of the previous two chapters. While temperature programming is addressed from the very beginning of Chapters 8 and 9, Chapter 10 builds its case from historical perspective starting from simpler and gradually escalating to more complex issues eventually involving gas decompression and temperature programming. Both, the average gas velocity and the flow rate as independent variables in the formulae for plate height are explored. The readers who prefer to rely on the average velocity as the independent variable can see for themselves how complex the correct theory based on the average velocity is compared to its counterpart with flow rate as independent variable. Only after that, average velocity was removed from further considerations.

A special section in Chapter 10 is dedicated to the analysis of the incorrect plate height theory. This includes the motivation and evolution of the theory as well as the examples of its adoption, retractions, and criticism.

The book has a large number of mathematical formulae. Majority of them were derived from basic principles. All formulae with the exception of the simple ones were derived with the help of Mathematica® software (Wolfram Research, Champaign, IL, USA). Without Mathematica, management of this number of formula with the confidence in their correctness would be hardly possible.

A few words about myself. Educated in the USSR as electrical engineer, I immigrated to the US in 1977 and joined Avondale division of Hewlett-Packard Co. Having by the time of immigration an expertise in theory and design in low-noise electronics, I was involved for 10 years in the design of signal-processing hardware and software for GC integrators. And then began my rapid drift into GC. Ray Dandeneau, the R&D manager at that time, lured me in GC. Terry Berger was my local guide. Although I never met Golay, I consider myself to be his and Giddings' student. I met Giddings once and was lucky to have a lengthy discussion with him. I also learned a good deal from the works of Cramers and Guiochon. There was time when Cramers and I communicated on a more or less regular basis.

One of my first GC publications was a two-part series (part 1 with Berger) on theory of plate height in nonuniform time-varying chromatography [9, 10]. Although this work remains mostly unknown, I think that it was my most significant contribution to chromatographic theory (GC, LC, etc.). Thus, Giddings' formula for plate height in GC with gas decompression [11, 12] along the column and Giddings' pressure correction factor follow directly from that theory. Some results of the theory are incorporated in the text of Chapter 10.

June 2010 *Leon Blumberg*

References

1 James, A.T. and Martin, A.J.P. (1952) *Biochem. J.*, **50**, 679–690.

2 James, A.T. and Martin, A.J.P. (1952) *Analyst*, **77**, 915–932.

3 Griffiths, J., James, D., and Phillips, C.S.G. (1952) *Analyst*, **77**, 897–904.

4 Cramers, C.A. and Leclercq, P.A. (1999) *J. Chromatogr. A*, **842**, 3–13.

5 Giddings, J.C. (1962) *Gas Chromatography* (eds N. Brenner, J.E. Callen, and M.D. Weiss), Academic Press, New York, pp. 57–77.

6 Harris, W.E. and Habgood, H.W., (1966) *Programmed Temperature Gas Chromatography*, John Wiley & Sons, Inc., New York.

7 Golay, M.J.E. (1958) *Gas Chromatography 1958* (ed. D.H. Desty), Academic Press, New York, pp. 36–55.

8 van Deemter, J.J.J., Zuiderweg, F.J., and Klinkenberg, A. (1956) *Chem. Eng. Sci.*, **5**, 271–289.

9 Blumberg, L.M. and Berger, T.A. (1992) *J. Chromatogr.*, **596**, 1–13.

10 Blumberg, L.M. (1993) *J. Chromatogr.*, **637**, 119–128.

11 Stewart, G.H., Seager, S.L., and Giddings, J.C. (1959) *Anal. Chem.*, **31**, 1738.

12 Giddings, J.C., Seager, S.L., Stucki, L.R., and Stewart, G.H. (1960) *Anal. Chem.*, **32**, 867–870.

Constants, Abbreviations, Symbols

Constants:

$\mathscr{A} = 6.02214 \times 10^{23}/\text{mol}$ – Avogadro number
$e = 2.718282$ – base of natural logarithms
$p_{st} = 1$ atm – standard pressure
$\mathscr{R} = 8.31447\ \text{J K}^{-1}\ \text{mol}^{-1}$ – molar gas constant
$T_{norm} = 298.15$ K (25 °C) – normal temperature
$T_{st} = 273.15$ K (0 °C) – standard temperature
$\mathscr{V}_{st} = 22.414 \times 10^{-3}\ \text{m}^3$ – molar volume (at standard pressure and temperature)

Abbreviations:

EMG (pulse) – exponentially modified Gaussian (pulse)
GC – gas chromatography
GLC – gas–liquid chromatography
GSC – gas–solid chromatography
HETP – height equivalent to one theoretical plate
ID – internal diameter
IUPAC – International Union of Pure and Applied Chemistry
LC – liquid chromatography
mL_{norm} – milliliter at normal conditions of 1 atm and 25 °C
mL_{st} – milliliter at standard conditions of 1 atm and 0 °C
MS – mass spectrometry
OPGV – optimum practical gas velocity
OTC – open-tubular column
PLOT (column) – porous layer open tubular (column)
SFC – supercritical fluid chromatography
STP – standard temperature and pressure
WCOT (column) – wall-coated open-tubular (column)

Subscripts:

Char – characteristic parameter at fixed dimensionless film thickness ($\varphi = 0.001$), Eq. (5.37)
char – characteristic parameter at arbitrary film thickness

Temperature-Programmed Gas Chromatography. Leonid M. Blumberg

ISBN: 978-3-527-32642-6

i – inlet
i – index (1, 2, 3, ...)
init – initial conditions (at the beginning of the heating ramp)
max – maximum
min – minimum
o – outlet
opt – optimum
R – at retention time
ref – parameter measured at predetermined reference conditions
sn – significant
st – parameter measured at standard pressure ($p_{st} = 1$ atm) and standard temperature (T_{st} 273.15 K (0 °C))
thick – thick film column
thin – thin film column

Symbols (temporary symbols used within a few consecutive pagers are not included)

Symbol	Description
a	solute-specific amount (amount per length)
B_o	specific permeability, Eq. (7.13)
b	Eq. (10.39)
$C_{sol,f}$	solute concentration in stationary phase film
$C_{sol,g}$	solute concentration in gas phase
c_1	Eq. (10.40)
c_{u2}	Eq. (10.41)
D	diffusivity (of a solute in gas)
$\mathscr{D}$	(local) dispersivity, Eq. (10.55)
D_{eff}	effective diffusivity, Eq. (10.14)
D_f	solute diffusivity factor, Eq. (6.33)
D_g	gas self-diffusivity
$D_{g,st}$	D_g at $p = p_{st}$ and $T = T_{st}$
D_{pst}	D at $p = p_{st}$
D_S	solute diffusivity in stationary phase
D_{st}	D at $p = p_{st}$ and $T = T_{st}$
d	column internal diameter, Figure 2.1
d_c	internal diameter of a column tubing, Figure 2.1
d_{cx}	tube characteristic cross-sectional dimension, Eq. (7.39)
d_f	stationary phase film thickness, Figure 2.1
d_p	particle size of packing material, Figure 2.1
F	flow rate, Eq. (7.25)
f	specific flow rate, Eqs. (7.53) and (7.58)
f_R	f at retention time of a solute
G	speed gain, Eq. (8.57)
g_S	Eq. (5.10)
$\Delta\mathscr{G}$	Gibbs free energy of evaporation of solute from liquid

$\Delta\mathcal{G}_{\mathrm{H}}$	enthalpy of evaporation of solute from liquid
$\Delta\mathcal{G}_{\mathrm{S}}$	entropy of evaporation of solute from liquid
H	(apparent) plate height, Eq. (10.78)
$\mathcal{H}$	(local) plate height, Eq. (10.55)
$\mathcal{H}_{\mathrm{f}}$	stationary phase film contribution to $\mathcal{H}$, Eq. (10.43)
h	dimensionless H, Eq. (10.109)
J	energy flux (energy/area/time), Eq. (7.42)
j	James–Martin compressibility factor, Eq. (7.114)
j_{G}	Giddings compressibility factor, Eq. (10.87)
j_{H}	Halász compressibility factor, Eq. (7.89)
$\tilde{j}_{H}$	inverse Halász compressibility factor, Eq. (7.94)
K_{c}	distribution constant, Eq. (5.1)
k	retention factor, Eq. (5.4)
k_{a}	asymptotic k, Eq. (8.111)
k_{R}	k of eluting solute
$k_{\mathrm{R,a}}$	k_{a} of eluting solute, Eq. (8.116)
L	column length
L_{crit}	critical length of a tube, Eq. (7.76)
L_{ext}	vacuum extended length of a tube, Eq. (7.64)
ℓ	dimensionless length of a tube, Eq. (7.97)
ℓ_{ext}	dimensionless vacuum extended length, Eq. (7.104)
M	molar mass (mass/mole)
M_i	ith moment, Eq. (3.5)
m_i	ith normalized moment, Eq. (3.6)
$\tilde{m}_i$	ith normalized central moment, Eq. (3.7)
m_{sol}	solute amount in a column
$m_{\mathrm{sol,f}}$	solute amount in stationary phase film
$m_{\mathrm{sol,g}}$	solute amount in gas phase
N	(apparent) plate number, Eqs. (10.74) and (10.141)
$\mathcal{N}$	directly counted plate number, Eqs. (10.71) and (10.73)
P	relative pressure, Eq. (7.9)
ΔP	relative pressure drop, Eq. (7.10)
$\Delta\tilde{P}$	relative virtual pressure drop, Eq. (7.139)
p	(local) pressure
p_{a}	ambient pressure, Figure 7.2b
p_{g}	gauge pressure, Figure 7.2b
$\bar{p}$	average pressure, Eq. (7.112)
Δp	pressure drop, Eq. (7.8)
Δp_{B}	borderline pressure drop, Eq. (7.74)
$\Delta\tilde{p}$	virtual pressure, Eq. (7.133)
R_{T}	heating rate, Eq. (8.58)
$R_{\mathrm{T,norm}}$	normalized heating rate, Eq. (8.64)
r_{T}	dimensionless heating rate, Eq. (8.95)
r_{T0}	benchmark dimensionless heating rate, Eq. (8.125)
T	temperature

T_{Char}	T_{char} at $\varphi = 0.001$
T_{char}	characteristic temperature, Eq. (5.18)
T_H	thermal equivalent of enthalpy, Eq. (5.9)
T_R	solute elution temperature
$T_{R,a}$	asymptotic T_R, Eq. (8.117)
T_{ref}	reference temperature
T_{sn}	significant temperature, Eq. (10.239)
$\Delta T_{M,init}$	hold–up temperature, Eq. (8.64)
ΔT_R	thermal spacing of two peaks, Eq. (9.5)
ΔT_μ	mobilizing temperature increment, Eq. (8.76)
t	time
t_{mol}	mean time between (molecular) collisions
t_g	gas propagation time, Eq. (8.26)
$t_{g,char}$	t_g at characteristic temperature (T_{char}) of a solute
t_M	hold–up time, Eq. (7.86)
$t_{M,char}$	t_M at characteristic temperature (T_{char}) of a solute
$t_{M,ext}$	vacuum extended hold-up time, Eq. (7.100)
$t_{M,R}$	t_M at retention time of a solute
t_m	dynamic hold–up time, Eq. (8.28)
t_R	retention time
$t_{R,stat}$	Eq. (10.142)
Δt_R	temporal spacing of two peaks, Eq. (9.3)
u	gas (local) velocity, Eq. (7.2)
u_{oR}	gas outlet velocity at elution time of a solute
$\bar{u}$	average velocity of carrier gas, Eqs. (7.116) and (7.117)
v	solute (local) velocity
v_{oR}	velocity of eluting solute
w_A	peak's area-over-height width, Eq. (3.1), Figure 3.1
w_h	peak's half-height width, Figure 3.1
w_b	base width of a peak, Figure 3.1
$\langle y \rangle_x$	average of function $y(x)$, Eq. (3.15)
z	distance from column inlet
β	phase ratio, Eq. (2.5)
γ_D	empirical parameter of a gas, Eq. (6.42), Table 6.12
γ_F	Eq. (7.57)
γ_{FA}	Eq. (7.26)
γ_r	sensitivity of ΔT_R to the relative change in R_T, Eq. (9.38)
γ_T	Eq. (10.151)
γ_Ω	Eq. (7.40)
ε	packing porosity
ζ	dimensionless distance from inlet, Eq. (8.65)
η	gas viscosity
η_{st}	η at $T = T_{st}$
Θ	dimensionless temperature, Eq. (8.65)
Θ_{char}	dimensionless characteristic temperature, Eq. (8.65)

Θ_{init}	dimensionless initial temperature, Eq. (8.93)
Θ_{st}	dimensionless standard temperature, Eq. (5.50)
$\Delta\Theta_{\mu}$	dimensionless mobilizing temperature increment, Eq. (8.77)
θ_{bin}	binary thermal constant, Eq. (5.22)
θ_{Char}	θ_{char} at $\varphi = 0.001$
θ_{char}	characteristic thermal constant, Eq. (5.20)
$\theta_{Char,st}$	$\theta_{char,st}$ at $\varphi = 0.001$
$\theta_{char,st}$	characteristic thermal constant at $T = T_{st}$
θ_{T}	equivalent thermal constant, Eq. (8.88)
$\theta_{T,init}$	θ_{T} at $T = T_{init}$
λ	mean free path (of gas molecules)
μ	solute mobility factor, Eq. (8.4)
μ_{a}	asymptotic μ, Eq. (8.110)
μ_{R}	μ of eluting solute
$\mu_{R,a}$	asymptotic mobility of eluting solute, Eq. (8.115)
μ_{eff}	solute effective mobility, Eq. (8.42)
ϑ_{1}	Eqs. (10.152) and (10.153)
ϑ_{G1}	Eq. (10.25)
ϑ_{G2}	Eq. (10.26)
ξ	gas empirical parameter, Table 6.12
ρ	density
σ	width (standard deviation) of a peak
σ_{m}	unretained width of a peak, Eq. (10.80)
$\tilde{\sigma}$	width of eluting solute zone
$\tilde{\sigma}_{z}$	(local) width of a solute zone within a column
$\tilde{\sigma}_{i}$	initial width of injected solute zone, Eq. (10.86)
τ	dimensionless time, Eq. (8.54)
τ_{R}	dimensionless retention time, Eq. (8.54)
$\tilde{\tau}$	dimensionless time, Eq. (8.65)
$\tilde{\tau}_{g,char}$	dimensionless characteristic gas propagation time, Eq. (8.65)
$\tilde{\tau}_{R}$	dimensionless retention time, Eq. (8.65)
Φ	Eq. (7.144)
Φ_{1}	Eq. (7.150)
$\Delta\Phi$	Eq. (7.149)
φ	relative film thickness, Eq. (2.7)
ϕ	collision diameter (of gas molecule)
υ	average molecular speed
Ω	pneumatic resistance, Eq. (7.38)
ω	solute immobility factor, engagement factor, Eq. (8.5)
ω_{a}	asymptotic ω, Eq. (8.109)
ω_{R}	ω of eluting solute, Eq. (8.113)
$\omega_{R,a}$	ω_{a} of eluting solute, Eq. (8.114)
$\bar{\omega}_{a}$	average ω_{a}, Eq. (8.122)

Part One
Introduction

Temperature-Programmed Gas Chromatography. Leonid M. Blumberg

ISBN: 978-3-527-32642-6

1
Basic Concepts and Terms

Chromatography is a technique of separation of compounds – components of a mixture. Chromatography can be *analytical* [1], or – as in the case of *preparative chromatography* [2–4] – it can be other than analytical. The purpose of analytical chromatography is to obtain information regarding a mixture and its components rather than to make a product. Here the analytical chromatography has been considered only. How clean is the air in this room? What are the major components of the gasoline in this container? Is this pesticide present in the soil at this location, and, if yes, how much of it is there? To answer these and similar questions, one can analyze a representative sample of a mixture in question – a *test mixture*.

A chromatograph or a chromatographic instrument consists of several devices. The key device is the separation device – the one where the separation takes place. In column chromatography, the separation device is a chromatographic column – a tube that either has a special material along its inner walls (an open tubular column) or is packed with small particles (packed column) or with a porous material. The separation occurs due to different levels of nondestructive interactions of different components (analytes, species) of a test mixture with the material inside the column. As the subject of this book is the column analytical chromatography, from now on, the term chromatography will always infer that technique.

In addition to a column, a typical chromatograph includes, Figure 1.1, a sample introduction device [5–10] and a detector [5, 6, 10, 11]. A chromatograph also requires

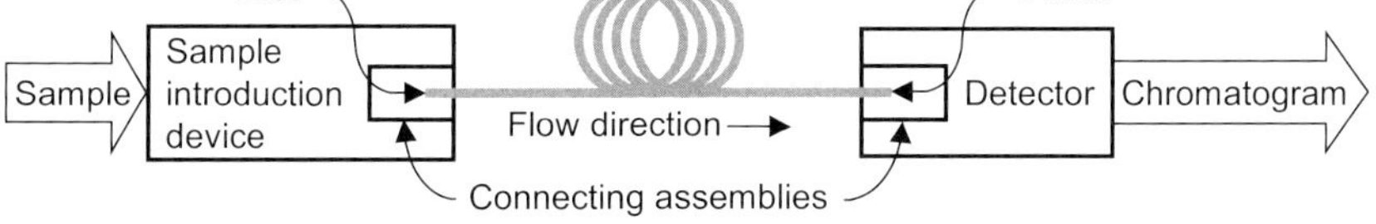

Figure 1.1 Block-diagram of a chromatograph.

Temperature-Programmed Gas Chromatography. Leonid M. Blumberg

ISBN: 978-3-527-32642-6

a fluid (eluent [6, 12], mobile phase) that flows through the column from its inlet to the outlet transporting components of a test mixture in the same direction. In gas chromatography (GC), the mobile phase is an inert carrier gas, in liquid chromatography (LC), the mobile phase is a liquid. To emphasize the fact that the components of a test mixture are soluble in the mobile phase, they are also called the *solutes*.

A set of conditions for the execution of a chromatographic analysis (a run) – the column type and temperature, the carrier gas type and its flow rate or pressure, the sample introduction device and the detector together with their operational conditions, and so forth – comprise a method of the analysis.

A chromatographic analysis starts with a quick (ideally, instantaneous) injection of a test mixture into the column inlet. While being transported through the column, different solutes differently interacting with the column interior migrate through the column with different velocities. As a result, each solute is retained (resides) in the column for different amount of time, known as the retention time or the residence time. Different retention times cause the solutes to elute – to pass through the column outlet – separately from each other constituting the separation of the solutes.

When an eluite [6, 13] (a solute eluting from a column) mixed with effluent [12] (mobile phase leaving a column) passes through a detector, the latter generates a response indicative of the presence of the solute in the detector. Ideally, a detector response to each solute should be proportional to the solute amount or concentration.

A way to observe the separation result is through a chromatogram representing a plot of a detector signal – the detector response as a function of time elapsed since the injection of a test mixture. A simple chromatogram resulted from analysis of a two-component mixture is shown in Figure 1.2. Ideally, it should be a line chromatogram [14–16] shown in Figure 1.2a. Unfortunately, no matter how quick was the injection, each solute migrating along the column occupies a zone (a band) whose width gradually increases with time. As a result, each eluite and a corresponding peak have nonzero width as shown in Figure 1.2b.

Usually, a chromatographic analysis does not end with the generation of a chromatogram. A contemporary *chromatographic system* might include a chromatograph and a data analysis subsystem. The latter might quantify and identify the peaks

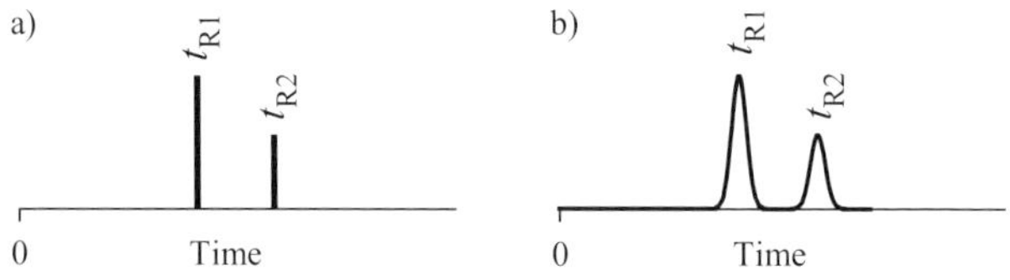

Figure 1.2 Chromatograms of a two-component mixture. The markers, t_{R1} and t_{R2}, are the retention times of respective components. (a) A *line chromatogram* that would occur if there were no broadening of the solute zones during the solute migration along the column. (b) A realistic two-peak chromatogram resulted from the separation and the broadening of the solute zones.

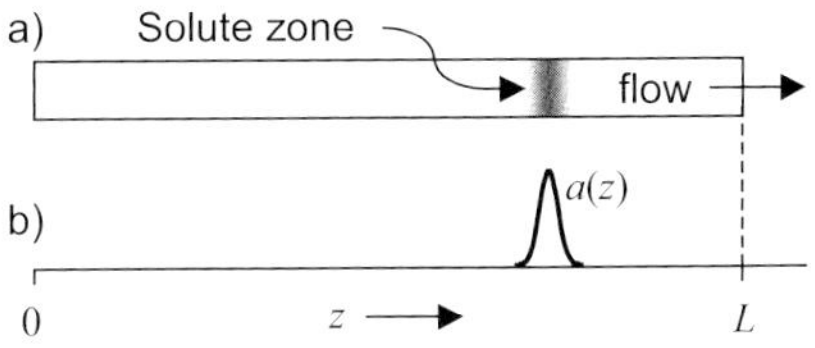

Figure 1.3 Solute zone within a *L*-long column (a), and the distribution, $a(z)$, of its amount along the *z*-axis (b). The proportions of the column and the zone shown here are not typical. Typically, a column is several orders of magnitude longer than its internal diameter, and the longitudinal width of a zone – its spread along the *z*-axis – is hundreds of times larger than the diameter.

and report retention time, width, area, height, amount, concentration, and other information regarding each peak.

Two different concepts – a solute zone and a peak – have already been mentioned in the preceding text. A zone, Figure 1.3, is a space occupied by a solute migrating in a column. The distribution of a solute zone along a column can be described (Figure 1.3b) by the solute's specific amount, $a(z)$, – the solute amount (mass, mole, and so forth) per unit of length. The width of a zone is measured in units of length along the column. A peak, on the other hand, can be a zone elution rate [6], a detector signal in response to elution of the zone, or a portion of chromatogram (Figure 1.2b) representing that signal. In either case, the width of a peak is measured in units of time. The distinction between the terms a zone and a peak is recognized throughout the book. Typically, both a zone and a peak representing the zone have similar pulse-like shape with a clearly identifiable maximum.

Throughout the book, the width of a pulse (a peak or a zone) is identified with its standard deviation defined later. For now, it is sufficient to assume that standard deviation of a pulse (denoted as σ for a peak and $\tilde{\sigma}$ for a zone) is equal to about 40% of its half-height width [12].

Quality of separation of two peaks depends on their spacing [15, 17, 18] – the difference between the peak retention times – and widths. For the peaks of the same width, Figure 1.4, the further they are apart from each other the better they are separated. On the other hand, for the same spacing, Figure 1.5, the narrower are the peaks the better is their separation.

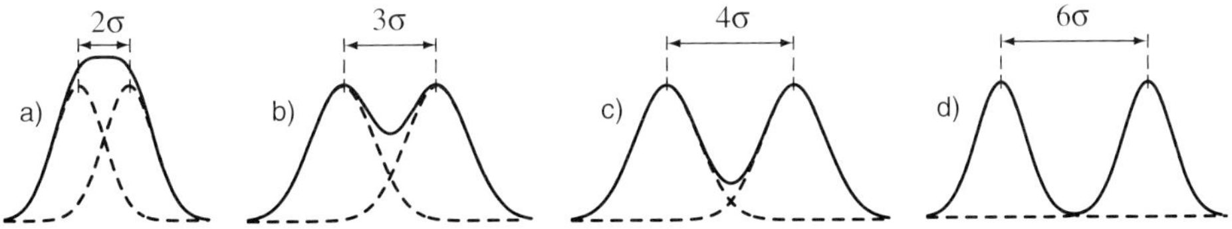

Figure 1.4 Four pairs of peaks. The peaks (dashed lines) in each pair have the same width and shape, but different distances between their apexes. The distances increase from (a) to (c) and so does the appearance of the peak separation in each pair.

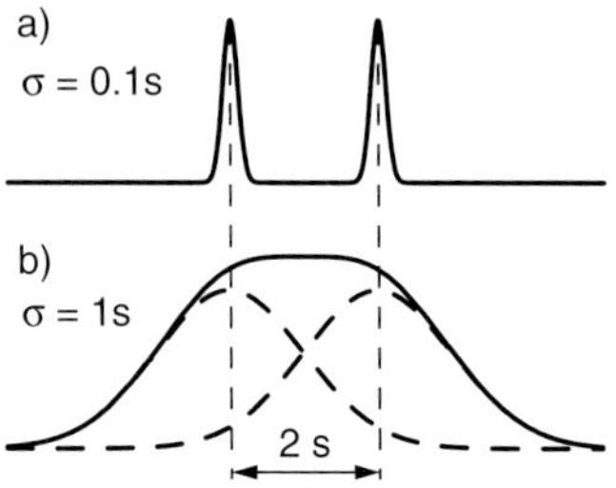

Figure 1.5 Pairs, (a) and (b), of peaks. In both pairs, the distance between the peaks (dashed curves, – – –, in pair (b)) is the same. However, the two peaks in pair (a) are 10 times narrower and appear to be more separate than the peaks in pair (b).

References

1 Giddings, J.C. (1991) *Unified Separation Science*, John Wiley & Sons, Inc., New York.
2 Biblingmeyer, B.A. (1987) *Preparative Liquid Chromatography*, Elsevier, Amsterdam.
3 Guiochon, G., Shirazi, S.G., and Katti, A.M. (1994) *Fundamentals of Preparative and Nonlinear Chromatography*, Academic Press, San Diego.
4 Schmidt-Traub, H. (2005) *Preparative Chromatography of Fine Chemicals and Pharmaceutical Agents*, Wiley-VCH, Weinheim.
5 Lee, M.L., Yang, F.J., and Bartle, K.D. (1984) *Open Tubular Gas Chromatography*, John Wiley & Sons, Inc., New York.
6 Guiochon, G. and Guillemin, C.L. (1988) *Quantitative Gas Chromatography for Laboratory Analysis and On-Line Control*, Elsevier, Amsterdam.
7 Klee, M.S. (1990) *GC Inlets – An Introduction*, Hewlett-Packard Co., USA.
8 Jennings, W., Mittlefehldt, E., and Stremple, P. (1997) *Analytical Gas Chromatography*, 2nd edn, Academic Press, San Diego.
9 Grob, K. (2001) *Split and Splitless Injection for Quantitative Gas Chromatography – Concepts, Processes, Practical Guidelines, Sources of Error*, Wiley-VCH, Weinheim.
10 Poole, C.F. (2003) *The Essence of Chromatography*, Elsevier, Amsterdam.
11 McNair, H.M. and Miller, J.M. (1998) *Basic Gas Chromatography*, John Wiley & Sons, Inc., New York.
12 Ettre, L.S. (1993) *Pure Appl. Chem.*, **65**, 819–872.
13 Haidacher, D., Vailaya, A., and Horváth, S. (1996) *Proc. Natl. Acad. Sci. USA*, **93**, 2290–2295.
14 Davis, J.M. and Giddings, J.C. (1985) *Anal. Chem.*, **57**, 2168–2177.
15 Giddings, J.C. (1990) in *Multidimensional Chromatography Techniques and Applications* (ed. H.J. Cortes), Marcel Dekker, New York and Basel, pp. 1–27.
16 Giddings, J.C. (1995) *J. Chromatogr.*, **703**, 3–15.
17 Dolan, J.W. and Snyder, L.R. (1998) *J. Chromatogr. A*, **799**, 21–34.
18 Dolan, J.W. (2003) *LC-GC*, **21**, 350–354.

2
A Column

In this chapter, the most prominent structures, retention mechanisms, and operational modes of GC columns are discussed with the main purpose of introducing the relevant terminology. A more detailed information can be found in many sources [1–13].

2.1
Retention Mechanisms

As mentioned earlier, the root cause of separation of solutes – components of a test mixture – in chromatography is the different levels of *nondestructive interaction* of the solutes with the *column interior*. In GC, that interaction is the sorption (absorption or adsorption) of the solutes by stationary (not moving) sorbent – a column stationary inner material also known as the *stationary phase*. The sorption is balanced by the opposing process of a solute desorption into the inert carrier gas – the *mobile phase* or the *eluent*. At their sorption/desorption equilibrium, different solutes become differently distributed between the mobile and the stationary phase. As a result, different solutes are carried by the carrier gas toward the column outlet with different net velocities. This causes different time of retention (residence) of the solutes in the column, and their subsequent separation.

Two types of stationary phase sorbents are used in GC.

Liquid organic polymers act as absorbents of the solutes. The absorption/desorption of the solutes can also be viewed as their solvation in the stationary phase and evaporation into the mobile phase. GC based on this principle has been proposed by Martin and Synge [14–16] in 1941 and demonstrated by James and Martin [17, 18] in 1952. This separation technique is known as partition GC or as gas–liquid chromatography (GLC). The strength of the solute–liquid interaction in partition GC substantially depends on a solute and a liquid polarity [8, 12, 19]. A *polar solute* is the one whose molecules have a distinct dipole-like electric field. On the other end of a solute polarity scale are the *apolar solutes* – the ones, like the *n*-alkanes, with the molecules whose electric fields have no clear direction. Polar liquids tend to better absorb polar solutes, while apolar ones tend to most strongly absorb apolar solutes. In

Temperature-Programmed Gas Chromatography. Leonid M. Blumberg

ISBN: 978-3-527-32642-6

partition GC analysis, the apolar solutes tend to elute in the order of increase in their boiling points, as described in *Trouton's rule* [3, 6, 20]. The development of metrics of polarity is still in progress [8, 19].

An adsorption on solid surface is another type of interaction of components of a test mixture with the stationary phases in GC. GC based on this principle is known as gas–solid chromatography (GSC) or as adsorption GC. In most cases, adsorption GC is used for the separation of gases and other low boiling species that, at room temperature, are in the gaseous state.

2.2 Structures

A column can be designed either as an open tube – an open-tubular column (OTC) – or as a tube packed with particles of more or less similar shape and size, Figure 2.1. The more regularly shaped and the more identical to each other are the particles in a packed column, the more compact can be the packing. The ideally regular packing cannot be achieved in practice.

The first columns in GC were the packed ones – a natural evolution of *packed distillation columns* (*distillation towers*) for oil refinery [14, 21–28] and packed separation tubes in other fields [29]. In 1956, Golay invented [30, 31], and shortly after that publicly described [32, 33], an open-tubular (capillary) column.

The length, L, of a typical commercial capillary column can be up to 100 m while the internal diameter (ID), d_c, of its tubing typically ranges between 0.1 and 0.53 mm. A typical commercial packed column is several meters long and has ID of several millimeters. The particle size, d_p, Figure 2.1b, of a packing material is typically measured in hundreds of microns.

A liquid organic polymer acting as a liquid stationary phase in GLC, also known as partition GC [17, 18], can be bonded to a column packing material in the case of a packed column, or to a column inner walls as a film – typically 0.1 μ to several microns in thickness, d_f, Figure 2.1a – in the case of a *wall-coated open tubular* (WCOT) column. These columns also known as simply the *capillary columns* are utilized in overwhelming majority of GC analyses, and are the subject of main attention in this book.

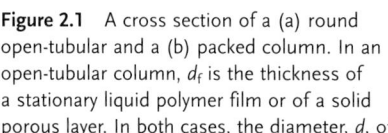

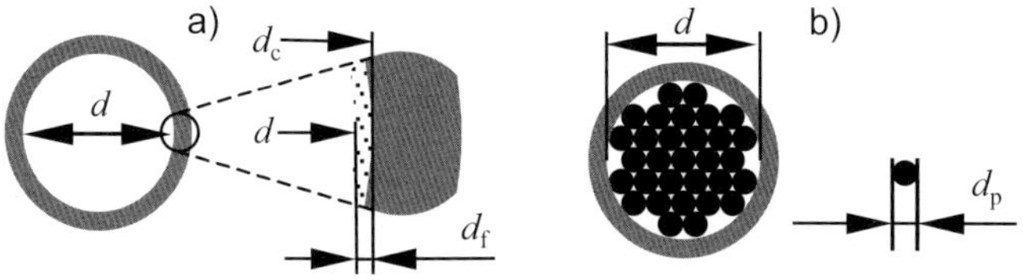

Figure 2.1 A cross section of a (a) round open-tubular and a (b) packed column. In an open-tubular column, d_f is the thickness of a stationary liquid polymer film or of a solid porous layer. In both cases, the diameter, d, of the column internal open space is smaller by $2d_f$ than the internal diameter, d_c, of a column tubing. In a packed column, d_p is the particle size of a packing material.

In adsorption GC, a stationary adsorbent can be either a surface of a *solid porous packing material* (packed columns), or a surface of a *solid porous layer* – typically ranging in thickness d_f, Figure 2.1a, from several microns to several tens of microns – bonded to the inner walls of a *porous layer open tubular* (PLOT) column. As a separation technique, gas adsorption chromatography has been pioneered in the early 1940s by Cremer [34] and her school. A potential advantage of PLOT columns over their wall-coated counterparts has been originally described by Golay in 1958 [30, 33, 35], who envisioned such a layer as "an inert porous material with a very large surface coated with a . . . thinner layer of the retentive material." The first experimental PLOT columns of this type for GC were described by Halász and Horváth in 1963 [30, 36, 37]. Currently, the stationary phases in commercial PLOT columns are almost exclusively the *solid adsorbents*.

Typically, vendors of capillary columns specify ID, d_c, of a column tubing [38], and film thickness, d_f. The diameter, d, Figure 2.1a, of an open internal space of a capillary column can be found as

$$d = d_c - 2d_f \tag{2.1}$$

Typically, the difference between d and d_c is negligible, that is,

$$d_f \ll d \approx d_c \tag{2.2}$$

The ratio

$$\beta = \frac{V_g}{V_f} \tag{2.3}$$

where V_g and V_f are the volumes of, respectively, the gas and the stationary phase in a column is known as the *phase ratio* [38]. For a WCOT column, Figure 2.1a,

$$\beta = \frac{\pi d^2}{4} \times \frac{4}{\pi (d_c^2 - d^2)} = \frac{d^2}{d_c^2 - d^2} = \frac{(d_c - 2d_f)^2}{d_c^2 - (d_c - 2d_f)^2} \tag{2.4}$$

In view of Eq. (2.2), this becomes

$$\beta = \frac{V_g}{V_f} \approx \frac{d_c}{4d_f} \approx \frac{d}{4d_f} \tag{2.5}$$

Phase ratio is one way to describe the film thickness in relative terms. *Dimensionless film thickness*,

$$\varphi = \frac{V_f}{4V_g} = \frac{1}{4\beta} \tag{2.6}$$

is another way of doing so. While β is inversely proportional to the film thickness (d_f) quantity φ is proportional to d_f. Indeed, due to Eq. (2.2), φ can be expressed as

$$\varphi = \frac{V_f}{4V_g} = \frac{1}{4\beta} \approx \frac{d_f}{d_c} \approx \frac{d_f}{d} \tag{2.7}$$

It also follows from Eq. (2.2) that, typically,

$$\varphi \ll 1 \tag{2.8}$$

2.3
Operational Modes

In the simplest static [39] (steady-state [40], time-invariant [39, 41–45]) mode of operation, all column parameters, including its temperature, inlet and outlet pressure, and others remain constant during chromatographic analysis. Static analysis of a complex mixture consisting of hundreds or thousands of components might take prohibitively long time. Substantial reduction in the analysis time without sacrificing potential number of adequately separated peaks can be achieved by gradually increasing the column temperature during the analysis according to a predetermined temperature program. This technique has been originally described in 1952 by Griffith, James, and Phillips [46–48].

In a temperature-programmed analysis, column temperature, T, is typically a piece-wise linear function, $T(t)$, of time, t. The function consists of a combination of rising linear heating ramps (briefly, heating ramp) where T increases with constant heating rate, and temperature plateaus where T remains constant for a predetermined time period.

The first GC instruments utilizing the temperature programming allowed only the *isobaric* (constant pressure) mode of operation. During isobaric temperature-programmed analysis, the carrier gas flow rate gradually declines due to increase in the gas viscosity caused by the temperature increase. Virtually all contemporary commercial GC instruments are equipped with electronic *pressure programming* along with or without the temperature programming. A mode of a column operation where its temperature and/or pressure changes with time is a *dynamic* [39] (non-steady-state [40], time-varying [39, 41–45]) one. An important application of the pressure programming is maintaining a fixed flow rate during a temperature-programmed analysis. The *isorheic* (constant flow) mode of a temperature-programmed analysis has been originally described by Costa Neto *et al.* [49] in 1964. This operational mode has several advantages [50, 51] over the isobaric mode. The most important of them is providing a stable operational environment for the detectors.

2.4
Specific and General Properties of a Column

One can study specific or general properties of GC analysis and a column. Specific properties are the ones that are relevant to prediction of retention times of particular solutes. General properties, on the other hand, are the ones that represent the trends (such as the effect of the heating rate on retention times, elution order, width, and other parameters of *all* solutes) and their effect on general performance (analysis time, number of peaks that can be adequately separated, and so forth) of analysis. Only the general properties of GC analyses are considered below.

2.5 Boundaries

Conditions of GC analysis incorporate method parameters, solute parameters, and descriptive properties of the analysis. Method parameters include column parameters (dimensions, stationary phase type, and so forth) and operational parameters (carrier gas type, pressure, column temperature, and so forth). Solute parameters include its interaction with stationary phase, diffusivity in the carrier gas, and so forth. Descriptive properties of a method include absence or presence of gas decompression along a column, absence or presence of temperature programming, and so forth.

Theoretical prediction of a column operation might be based on idealizing assumptions (carrier gas behaves as an ideal gas, temperature does not change column dimensions, and so forth). Although a predicted operation might be different from the actual one, the idealization might be useful for the simplicity of the prediction. In a properly chosen idealizations, the prediction error can be either practically undetectable and, therefore, irrelevant, or, even if the error is measurable, it can be acceptable due to its minor effect on a column operations. Thus, a 50% error in prediction of optimal flow rate in a static analysis lowers a would-be optimal separation of two solutes by less than 4%. If this error is acceptable then one can assume that, if the flow rate is optimal for one solute pair, then it is optimal for all solutes although actually this might not be the case. Many factors might affect a column operation. The net error in prediction of a column operation results from accumulation of specific errors. This makes it important to reduce the contributing errors whenever possible.

Following are the boundaries of what is considered throughout the book to be regular conditions of a GC analysis. These regular conditions are always assumed to exist unless the contrary is explicitly stated.

Column cross section. A column is made of a round tubing with uniform – the same everywhere along the column – cross section. All characteristics of a stationary phase (dimensions, composition, and so forth) in an open-tubular column as well the characteristics of a packing material and its packing in a packed column are also uniform. Column dimensions do not change during the analysis.

Column length. A column has a fixed length, L, and is sufficiently long so that

$$L \gg d \qquad (2.9)$$

where, Figure 2.1, d is the column ID.

Capillary column. It is assumed that the column is a capillary (wall-coated open tubular) one. As mentioned earlier, the special attention to these columns is based on their use in overwhelming majority of analytical applications.

Thickness, d_f, of a stationary phase film in a capillary column is negligible compared to ID, d_c, of a column tubing so that Eqs. (2.5) and (2.7) are always valid.

Column temperature, T. It stays within the following range:

$$300\ \text{K} \leq T \leq 600\ \text{K (roughly } 25\ ^\circ\text{C} \leq T \leq 325\ ^\circ\text{C)} \tag{2.10}$$

The amount of experimental thermodynamic data regarding chromatographic properties of liquid stationary phases is sufficiently large (Chapter 5) within the regular range outlined in Eq. (2.10), and is more limited outside that range.

Uniform temperature. At any time, temperature is the same at any point along and across a column.

Carrier gas. Carrier gas can be helium, hydrogen, nitrogen, or argon.

Inlet pressure, p_i. In a column it is higher than outlet pressure, p_o, and does not exceed 30 atm, that is,

$$p_o \leq p_i \leq 30\ \text{atm} \tag{2.11}$$

Ideal gas (Chapter 6). The state of a carrier gas – the relationship between its pressure, temperature, and volume – is governed by the ideal gas law. The gas viscosity is pressure-independent, and its self-diffusivity is inversely proportional to pressure.

Inert gas. A gas does not chemically interact with any material.

Flow of a carrier gas is laminar (Appendix 7.A), mass-conserving (Chapter 7), and has no inertia.

Linearity and pressure-independence of a solute interaction with stationary phase. For any fixed temperature, the distribution of any solute between the stationary phase and the carrier gas in a column does not depend on the amount of that or any other solute in the column (linear isotherm), and on the gas pressure.

No inertia. Like the flow of a carrier gas, the flow of the solutes dissolved in it has no inertia.

n-Alkanes. Whenever *n*-alkanes are mentioned without further qualification,

$$n\text{-alkanes } C_5 \text{ through } C_{40} \tag{2.12}$$

are assumed. The *n*-alkanes, having a simple regular molecular structure, allow for the most specific empirical description of their properties (interaction with liquid phases, diffusion in carrier gases, and so forth) that are relevant to operation of a GC column. On the other hand, statistics of interaction of a large number of solutes with several typical liquid phases (Chapter 5) allow one to conclude that those properties of a solute–liquid interaction that are relevant to general operation of analyses of complex mixtures are well represented by the respective properties of *n*-alkanes. In a typical temperature-programmed analysis covering the temperature range in Eq. (2.10), *n*-alkanes C_5–C_{40} elute with practically acceptable retention.

In addition to the listed here regular conditions of a GC analysis, typical conditions are frequently mentioned in the text. These are those regular conditions that are most frequently utilized.

References

1 Keulemans, A.I.M. (1959) *Gas Chromatography*, 2nd edn, Reinhold Publishing Corp., New York.
2 Dal Nogare, S. and Juvet, R.S. (1962) *Gas-Liquid Chromatography. Theory and Practice*, John Wiley & Sons, Inc., New York.
3 Purnell, J.H. (1962) *Gas Chromatography*, John Wiley & Sons, Inc., New York.
4 Halász, I., Hartmann, K., and Heine, E. (1965) in *Gas Chromatography 1964* (ed. A. Goldup), The Institute of Petroleum, London, pp. 38–61.
5 Harris, W.E. and Habgood, H.W. (1966) *Programmed Temperature Gas Chromatography*, John Wiley & Sons, Inc., New York.
6 Littlewood, A.B. (1970) *Gas Chromatography: Principles, Techniques, and Applications*, 2nd edn, Academic Press, New York.
7 Lee, M.L., Yang, F.J., and Bartle, K.D. (1984) *Open Tubular Gas Chromatography*, John Wiley & Sons, Inc., New York.
8 Guiochon, G. and Guillemin, C.L. (1988) *Quantitative Gas Chromatography for Laboratory Analysis and On-Line Control*, Elsevier, Amsterdam.
9 Giddings, J.C. (1991) *Unified Separation Science*, John Wiley & Sons, Inc., New York.
10 Ettre, L.S. and Hinshaw, J.V. (1993) *Basic Relations of Gas Chromatography*, Advanstar, Cleveland, OH.
11 Grob, R.L. (1995) *Modern Practice of Gas Chromatography*, 3rd edn, John Wiley & Sons, Inc., New York.
12 Jennings, W., Mittlefehldt, E., and Stremple, P. (1997) *Analytical Gas Chromatography*, 2nd edn, Academic Press, San Diego.
13 Poole, C.F. (2003) *The Essence of Chromatography*, Elsevier, Amsterdam.
14 Martin, A.J.P. and Synge, R.L.M. (1941) *Biochem. J.*, **35**, 1358–1368.
15 Bayer, E. (1961) *Gas Chromatography*, Elsevier, Amsterdam.
16 Smolková-Keulemansová, E. (2000) *J. High Resolut. Chromatogr.*, **23**, 497–501.
17 James, A.T. and Martin, A.J.P. (1952) *Analyst*, **77**, 915–932.
18 James, A.T. and Martin, A.J.P. (1952) *Biochem. J.*, **50**, 679–690.
19 Rohrschneider, L. (2001) *J. Sep. Sci.*, **24**, 3–9.
20 Giddings, J.C. (1962) *J. Chem. Educ.*, **39**, 569–573.
21 Peters, W.A. (1922) *J. Ind. Eng. Chem.*, **14**, 476–479.
22 Bragg, L.B. (1939) *Ind. Eng. Chem. Anal. Ed.*, **11**, 283–287.
23 Podbielniak, W.J. (1931) *Ind. Eng. Chem. Anal. Ed.*, **3**, 177–188.
24 Westhaver, J.W. (1942) *Ind. Eng. Chem.*, **34**, 126–130.
25 Forbes, R.J. (1948) *Short History of the Art of Distillation*, Brill, Leiden.
26 Billet, R. (1979) *Distillation Engineering*, Chemical Publishing Co., New York.
27 Kister, H.Z. (1992) *Distillation Design*, McGraw-Hill, New York.
28 Stichlmair, J.G. and Fair, J.R. (1998) *Distillation Principles and Practices*, Wiley-VCH, New York.
29 Westhaver, J.W. (1947) *J. Res. Nat. Bur. Stand.*, **38**, 169–183.
30 Ettre, L.S. (1987) *J. High Resolut. Chromatogr.*, **10**, 221–230.
31 Ettre, L.S. (2002) *Milestones in the Evolution of Chromatography*, ChromSource, Inc., Franklin, TN.
32 Golay, M.J.E. (1958) in *Gas Chromatography* (eds V.J. Coates, H.J. Noebels, and I.S. Fagerson), Academic Press, New York, pp. 1–13.
33 Golay, M.J.E. (1958) in *Gas Chromatography 1958* (ed. D.H. Desty), Academic Press, New York, pp. 36–55.
34 Ettre, L.S. (2008) *LC-GC N. Amer.*, **26**, 48–60.
35 Golay, M.J.E. (1979) in *75 Years of Chromatography – A Historical Dialogue* (eds L.S. Ettre and A. Zlatkis), Elsevier, Amsterdam, pp. 109–113.
36 Halász, I. and Horváth, S. (1963) *Anal. Chem.*, **35**, 499–505.
37 Guiochon, G. (1969) in *Advances in Chromatography*, vol. 8 (eds J.C. Giddings and R.A. Keller), Marcel Dekker, New York, pp. 179–270.
38 Ettre, L.S. (1993) *Pure Appl. Chem.*, **65**, 819–872.

39 Blumberg, L.M. (1994) *Chromatographia*, **39**, 719–728.
40 Ohline, R.W. and DeFord, D.D. (1963) *Anal. Chem.*, **35**, 227–234.
41 Kailath, T. (1980) *Linear Systems*, Prentice-Hall, Englewood Cliffs, NJ.
42 Reid, G.J. (1983) *Linear System Fundamentals*, McGraw-Hill Book Company, New York.
43 Tsakalis, K.S. and Ioannou, P.A. (1993) *Linear Time-Varying Systems*, Prentice-Hall, Englewood Cliffs, NJ.
44 Blumberg, L.M. and Berger, T.A. (1992) *J. Chromatogr.*, **596**, 1–13.
45 Blumberg, L.M. (1993) *J. Chromatogr.*, **637**, 119–128.
46 Griffiths, J., James, D., and Phillips, C.S.G. (1952) *Analyst*, **77**, 897–904.
47 Drew, C.M. and McNesby, J.R. (1957) in *Vapour Phase Chromatography* (ed. D.H. Desty), Academic Press, New York, pp. 213–221.
48 Ettre, L.S. (2003) *LC-GC*, **21**, 144–149, 167.
49 Costa Neto, C., Köffer, J.T., and De Alencar, J.W. (1964) *J. Chromatogr.*, **15**, 301–313.
50 Blumberg, L.M., Berger, T.A., and Klee, M.S. (1995) *J. High Resolut. Chromatogr.*, **18**, 378–380.
51 Blumberg, L.M., Wilson, W.H., and Klee, M.S. (1999) *J. Chromatogr. A*, **842** (1–2), 15–28.

Part Two
Background

Temperature-Programmed Gas Chromatography. Leonid M. Blumberg

ISBN: 978-3-527-32642-6

3
Linear Systems

3.1
Problem Review: Metrics for Peak Retention Time and Width

The key parameters of a chromatographic peak are its retention time and width.

Ideally, a zone of a test mixture injected in a column should have the infinitesimal width, and neither the detector nor the connecting assemblies, Figure 1.1, should affect retention or shape of any peak in a chromatogram. In practice, however, these ideal conditions do not exist. The question then becomes, what is necessary to know about the actual shape of the injected zone of a test mixture, about the properties of the column, the detector, and the connecting assemblies in order to predict the properties of the chromatogram. This suggests that although the main subject of the studies in the book is a column, there is a need to describe the properties of a column in such a way that can be useful in the accounting for the *extracolumn* contributions [1] to a chromatogram from other devices in a chromatograph.

Suppose that a pulse-like function (briefly, a pulse) in Figure 3.1 is a chromatographic peak. Intuitively, retention time of a peak can be identified with the time coordinate of its apex. Unfortunately, there is a fundamental problem with this approach. If just one device in a chromatograph can cause a sufficient asymmetry of some peaks, then it would be difficult to come-up with a simple way of predicting the retention times. This would be true even if all parameters of all devices (the injector, the column, the detector, and others) in a chromatograph, Figure 1.1, which might affect the peak retention times were known.

When it comes to the peak width measurement, the half-height width [2], w_h, Figure 3.1, of a peak appears to be a simple and straightforward metric. One needs only a "ruler and a pencil" to measure w_h. A little less straightforward, but still conceptually simple, are the base width [2, 3], w_b, and the area-over-height width, w_A, Figure 3.1, of a peak. The latter can be found as [4–11]

$$w_A = \frac{\text{peak area}}{\text{peak height}} \tag{3.1}$$

and is convenient for computer calculations.

Temperature-Programmed Gas Chromatography. Leonid M. Blumberg

ISBN: 978-3-527-32642-6

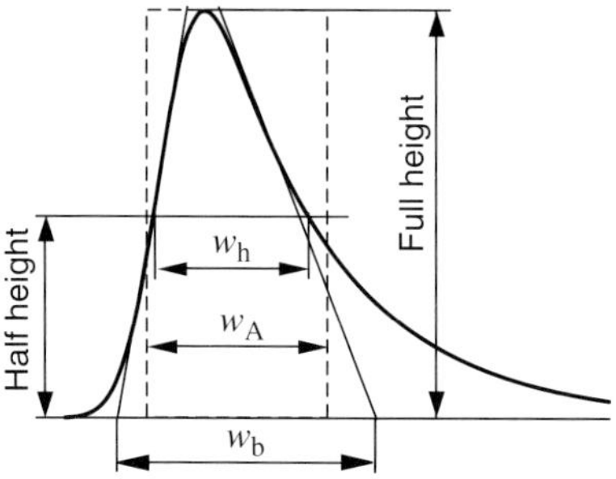

Figure 3.1 Base, w_b, half-height, w_h, and area-over-height, w_A, widths of a pulse. The base width is measured as the "segment of the peak base intercepted by the tangents drawn to the inflection points on either side of the peak" [2]. The dashed box is the equivalent rectangle that has the same area and the same height as those of the pulse.

As with the retention time, there are fundamental problems with the above-mentioned peak width metrics. Unless it was known a priori that all peaks would not sufficiently depart from the *Gaussian* shape (see below), it would be difficult to predict any of those metrics from the parameters of the devices comprising a chromatograph. As in the case of the retention time, the difficulty would exist even if all parameters of all devices in a chromatograph that might affect the widths of all peaks were known.

There are the solutions for the above-mentioned problems. They are based on identifying the peak parameters, including their retention times and the widths, with the mathematical moments of the peaks.

3.2
Chromatograph as an Information Processing System

A chromatograph can be viewed as an information processing system, Figures 3.2a, and b, that takes an arbitrary test mixture as its input, I, applies a certain separating operation, O, to the input, and returns a chromatogram as a response, $O(I)$. A system might consist of several serially connected, Figure 3.2c, devices or subsystems. Such subsystems of a typical chromatograph are, Figure 1.1, a sample introduction device, a column, a detector, and connecting assemblies. On the other hand, a chromatograph itself can be a subsystem of a larger system that includes, for example, a chromatograph, a data analysis, and other subsystems.

Generally, a prediction of a system response to a given input can be a difficult or an impossible task. A response is typically easier to predict for a linear system.

A system described by an operation O is linear [7, 12–16] if the sum, $I_1 + I_2$, of two arbitrary inputs, I_1 and I_2, is also a valid input for the system, and the response, $O(I_1 + I_2)$, to $I_1 + I_2$ is the same as the sum, $O(I_1) + O(I_2)$, of the responses, $O(I_1)$ and $O(I_2)$, to inputs I_1 and I_2 applied separately. In other words, a system described by an operation O is linear if, for two arbitrary inputs, I_1 and I_2,

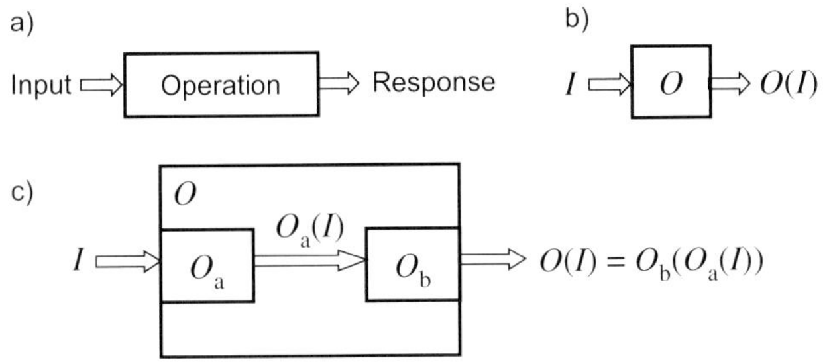

Figure 3.2 Block-diagrams of an information processing systems. Symbols *I*, *O*, and *O*(*I*) in the block-diagram (b) denote the input, the system operation, and the response, respectively, shown in the block-diagram (a). The serially connected devices O_a and O_b in (c) are subsystems of the system *O*.

$$O(I_1 + I_2) = O(I_1) + O(I_2) \tag{3.2}$$

It follows directly from this definition that

Statement 3.1

Any number of serially connected linear subsystems is a linear system.

Thus, for a system in Figure 3.2c,

$$O_b(O_a(I_1 + I_2)) = O_b(O_a(I_1) + O_a(I_2)) = O_b(O_a(I_1)) + O_b(O_a(I_2)) \tag{3.3}$$

Equation (3.2) also implies that in order to be a linear system, a chromatograph must have the following two properties:

1) *No interaction between the solutes.* A peak corresponding to a given solute should be the same regardless of the presence of other solutes in the test mixture.
2) *No overloading in a system.* A change in the amount of any solute in a test mixture should only cause a proportional change in the height of the corresponding peak, and neither the shape nor the retention time of the peak should depend on that amount.

Strictly speaking, one should not expect that a real chromatographic system is a linear one. Some degree of nonlinearity can always exist. However, when the amount of the test mixture is sufficiently small, the nonlinearity can have a negligible effect. In that sense, the no-overloading requirement can be satisfied only up to a certain maximal amount of a test mixture. Further increase in the amount might lead to an unacceptably disproportional change in the response. This is known as the *overloading*. It might be a column overloading [16], a detector overloading, and so forth. A system can be viewed as a linear one only if the amount of a test mixture does not exceed the maximal nonoverloading amount. Frequently, there might be no clearly recognizable maximal nonoverloading amount of a test mixture for a particular chromatograph. The first noticeable sign of a column overloading might appear as a slight distortion of a peak shape [16]. Further increase in the amount can lead to more pronounced and then to unacceptable distortions. A choice of the maximal non-overloading amount might be different for different test mixtures and different analyses. In all cases, however, the limit to the amount of the test mixture always exists, and it might be up to the analyst to define its level.

Typically, in order to make it more predictable, an effort is made in chromatography to assure that all devices in a chromatograph behave in a linear manner. From now on, unless otherwise specifically indicated, the *linearity* of all relevant systems and devices is always assumed. As demonstrated by Eq. (3.3), linearity of all subsystems implies the linearity of the entire system.

3.3 Properties of Linear Systems

Properties of linear systems significantly simplify prediction of system responses to arbitrary inputs.

An operation performed by a linear system on an arbitrary input is completely described by the system's *impulse response* [7, 12–15]. An infinitely narrow positive pulse, δ, whose area is equal to unity, is known as *Dirac's delta function*. Impulse response, r_δ, of a linear system is its response to a Dirac's delta function. To find r_δ experimentally, one simply needs to apply δ as the system input and observe the system response, r_δ.

In chromatography, the impulse response of a column to a given solute can be the rate of elution of the solute resulted from a sufficiently sharp injection of the solute into the column. The impulse response of a detector to a solute can be the detector response to a sufficiently sharp injection of the solute directly into the detector. Any nonoverloading amount, A, of a solute can be injected for these measurements. The impulse response, r_δ, can be found from an actual response, r, to a sharp injection as $r_\delta = r/A$. From this point of view, an ideal chromatogram is no more than a sum of the scaled impulse responses of a column to the solutes in the test mixture. As shown below (Section 3.5.1), practical requirement to the sharpness of the injection can be significantly relaxed compared to the Dirac's delta function required under ideal test conditions.

Impulse response of a linear system as a function of absolute time (clock time), t_{abs}, depends on absolute time, t_{in}, of application of Dirac's delta function to the system's input. In other words, r_δ as a function, $r_\delta\,(t_{abs}, t_{in})$, of t_{abs} depends on t_{in}. In a simple case of a static [17] (steady-state [18], time-invariant [14, 17,19–22]) system like isothermal and isobaric GC, $r_\delta\,(t_{abs}, t_{in})$ is a function of the difference $t_{abs} - t_{in}$ regardless of the absolute values of t_{abs} and t_{in}. As a result, a response of a static system to an arbitrary input is also a function of only the difference $t_{abs} - t_{in}$. On the other hand, in a pressure- and/or temperature-programmed column representing a dynamic [17] (non-steady-state [18], time-varying [14, 17,19–22]) system, a shift in t_{in} relative to the program can significantly change the shapes of all peaks and their retention times relative to t_{in}.

In reality, time programs in chromatography are always synchronized with injection of a sample, and t_{in} is always treated as time zero. In other words, it is always assumed that analysis time, t, and retention times, t_R, of all peaks are defined as $t = t_{abs} - t_{in}$, $t_R = t_{R,abs} - t_{in}$.

So far, it was assumed that inputs to chromatographic systems and their responses were functions of time, or temporal functions. In other words, behavior of chromatographic systems in time domain was considered. To understand the process of formation and separation of chromatographic peaks, one might also need to consider the shapes of the solute zones migrating through a column. These shapes can be expressed as functions of distance, z, from the column inlet, that is, as the pulses in the distance-domain of spatial pulses. To accommodate the terms time-invariant and time-varying to this case, more general terms *shift-invariant* and *shift-varying*, respectively, could be used. A column is a shift-invariant system in the distance-domain if conditions along the column are uniform [17, 21, 22] (the same at any z). Otherwise, a column is a shift-varying system in the distance domain. Gas decompression along a column is the key source of the nonuniform [17, 21, 22] (and, therefore, shift-varying in distance domain) conditions in GC.

In general, a shift-invariant system is the one for which a shift in the system's input causes the same shift (and only the shift) in the system's response, that is, if $r(x)$ is the system's response to input $I(x)$ then $r(x + \Delta x)$ is its response to input $I(x + \Delta x)$.

Statement 3.2

Any number of serially connected shift-invariant subsystems is a shift-invariant system.

An operation,

$$y(t) = \int_{-\infty}^{\infty} y_1(t-x)\, y_2(x)\, dx \tag{3.4}$$

and its result, $y(t)$, are known as *convolution* [1, 7, 12–15] of functions $y_1(t)$ and $y_2(t)$.

Statement 3.3

A response of a linear, shift-invariant system to an arbitrary input is convolution of the input and the system's impulse response.

This together with Statements 3.1 and 3.2 implies that

Statement 3.4

Impulse response of a linear, shift-invariant system is the convolution of the impulse responses of all its serially connected subsystems.

One can conclude that the impulse response of a linear shift-invariant system or its subsystems is all that one needs to know about the system in order to find its response to any input.

Although the impulse response of a linear shift-invariant system allows predicting a complete description of its response to any input, a description of the response might be too complex for a practical use. What is important, however, is that the facts described in Statements 3.3 and 3.4 provide a solid ground for further substantial simplifications described below.

3.4
Mathematical Moments of Functions

A function can be described by several parameters in such a way that the parameters of a function resulted from the convolution of two other functions can be easily calculated from the corresponding parameters of the convolved functions. The parameters in question are the *mathematical moments* of the function. They provide an important and, sometimes, complete information about the function [7, 12, 23].

The mathematical concept of the moments is widely known from the fields of statistics and the probability theory [12, 13, 24]. However, some aspects of the terminology related to statistical moments are different from the terminology adopted in a general theory of mathematical moments [7]. In chromatography, both aspects – the general and the statistical – of the moments are typically referred to as the *statistical moments* [15,25–28]. This appears to be a source of some counterintuitive or even confusing terminology that, in turn, creates a challenge for the attempts to clearly describe some chromatographic concepts using the conventional terms.

The main subject of this and the following sections is a general concept of the mathematical moments, and particularly, the *normalized moments* [7, 27, 28] of pulses. These moments can be used as metrics of a peak retention time, location of the solute zones, as well as the width and other parameters of the peaks and the solute zones. In this context, in spite of the widely accepted tradition [15,25–28], referring to the moments as to the statistical ones will be avoided.

Note 3.1

Because chromatography is a measurement technique, it needs to deal with the statistical treatment of measurement errors [29]. For that, it utilizes such statistical concepts as the average error and the standard deviation of the error that are based on the statistical moments of the errors. Typically, these concepts are not treated as an integral part of the linear system theory – the subject of this chapter. Nevertheless, it appears to be useful for this chapter to provide, where appropriate, several remarks regarding the statistical moments in order to better identify their place within the general concept of the mathematical moments and, by that, to minimize the possible confusions. □

The following definitions combine a general mathematical treatment [7] of the moments with the one adopted in chromatography [15, 23, 27, 28].

Consider a function $y(x)$. Variable x can be time, t, when $y(t)$ is a temporal function or distance, z, when $y(z)$ is a spatial function. The quantity

$$M_i = \int_{-\infty}^{\infty} x^i y(x)\,dx, \quad i = 0, 1, 2, \ldots \tag{3.5}$$

is the ith *moment* of $y(x)$. The quantity

$$m_i = \frac{M_i}{M_0} = \frac{1}{M_0}\int_{-\infty}^{\infty} x^i y(x)\,dx, \quad i = 0, 1, 2, \ldots \tag{3.6}$$

is the ith *normalized moment* of $y(x)$. The quantity

$$\tilde{m}_i = \frac{1}{M_0}\int_{-\infty}^{\infty} (x - m_1)^i y(x)\,dx, \quad i = 0, 1, 2, \ldots \tag{3.7}$$

is the ith *normalized central moment* of $y(x)$. It follows directly from these definitions that

$$m_0 = \tilde{m}_0 = 1, \quad \tilde{m}_1 = 0 \tag{3.8}$$

The following are the interpretations of some moments.

The 0th moment, M_0, of $y(x)$ is the area of $y(x)$.

Example 3.1

In chromatography, the 0th moment

$$M_0 = \int_0^L a(z)\,dz \tag{3.9}$$

of the specific amount, $a(z)$, of a solute distributed along the column, Figure 1.3, is the total amount of the solute residing in the column. For a solute migrating along the column, the amount of its part residing in the column begins to decline when the solute begins to elute from the column. If this amount is calculated before the beginning of the elution of the solute, then Eq. (3.9) gives the total amount injected in the column. Eventually, the entire amount of the solute leaves the column. That amount can also be found as the 0th moment

$$M_0 = \int_0^{\infty} F_{\mathrm{sol,o}}(t)\,dt \tag{3.10}$$

of the solute elution rate, $F_{\mathrm{sol,o}}(t)$.

Note 3.2

In probability theory and in statistics, a random variable, x, can be described [12, 13, 24] by its probability density function (PDF) – a non-negative function, $y(x)$, whose area (the 0th mathematical moment, Eq. (3.5)) is equal to one, that is $M_0 = 1$. Equations (3.6) and (3.7) for the first and the higher normalized moments of $y(x)$ become

$$m_i = \int_{-\infty}^{\infty} x^i y(x)\,\mathrm{d}x, \quad i = 1, 2, 3, \ldots \tag{3.11}$$

$$\tilde{m}_i = \int_{-\infty}^{\infty} (x - m_1)^i\, y(x)\,\mathrm{d}x, \quad i = 1, 2, 3, \ldots \tag{3.12}$$

So far, speaking of the moments, we used the same terminology as the one that was used for the mathematical moments in the main text. The terminology actually used in probability theory and in statistics is different.

Let us recall that, from general point of view of the main text, the moments that were described in Eqs. (3.11) and (3.12) would be the moments of the PDF $y(x)$. In probability theory and in statistics, the quantities described in Eqs. (3.11) and (3.12) are treated as the moments of the random variable x. This subtle difference in the terminology is, in some cases, the source of a not so subtle difference in the treatment of some moments in the two contexts. Unfortunately, in other cases (chromatography included), when the terminology developed for the probability theory and for the statistics is used in conjunction with the moments of the functions rather than of the random variables, the difference in the basic terminology becomes a source of unintuitive terms or even of the annoying confusions. □

It is worth noticing that the most general definition, Eq. (3.5), of moments requires no special properties of $y(x)$ except for the existence of the respective integrals. The definitions, Eqs. (3.6) and (3.7), of the normalized moments are just a bit more restrictive. They indicate that $y(x)$ must have a nonzero area ($M_0 \neq 0$) for these moments to exist. From now on, we will assume that, when necessary, $M_0 \neq 0$, meaning that a function $y(x)$ can always be normalized. Furthermore speaking of the first and the higher moments of a function, we will always infer the normalized moments unless otherwise is specifically stated, and will frequently omit the adjective "normalized."

3.4.1
The First and Higher Moments of a Pulse

The requirements for the interpretation of the moments to be meaningful in chromatography are more stringent than the general ones. Those more stringent requirements are satisfied in the *pulses*. A pulse, Figure 3.1, is a single-sign function that has one and only one extremum – only the maximum or only the minimum. The

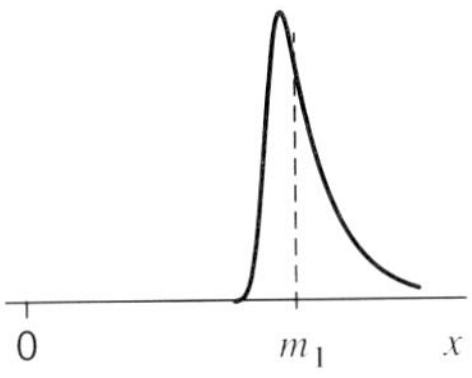

Figure 3.3 Centroid, m_1, of a pulse.

property of being a single-sign function, $y(x)$, means that $y(x)$ can only be a *positive pulse*, that is, $y(x) \geq 0$, or a *negative pulse*, that is $y(x) \leq 0$. Normalization $y(x)/M_0$ of a pulse, $y(x)$, converts it into a positive pulse.

The first moment, m_1, of a pulse, $y(x)$, is known as its *centroid* [7, 27, 28] (center of gravity [15, 26]). This quantity can be positive or negative; it is measured in the same units as the units of x. It represents the location of $y(x)$ on the x-axis, and it can be interpreted as the distance of $y(x)$ from the origin of x, Figure 3.3. If $y(x)$ is a symmetric pulse, then m_1 coincides with the abscissa (the x-coordinate) of the extremum (minimum or maximum) of $y(x)$. Otherwise, m_1 can be different from the abscissa of the extremum of $y(x)$.

Example 3.2

In chromatography, the first moment of a specific amount, $a(z)$, of a solute in a column is measured in units of length and represents the distance of the solute from the origin of the z-axis (typically, a column inlet) [21, 22, 27, 28]. The first moment of a peak in a chromatogram is the peak's retention time – the time past since the injection of the solute that produced the peak.

Note 3.3

The first moment, m_1, of a random variable (probability theory, statistics) is known as the expected value (mean, average) of that random variable [12, 13, 24]. Frequently, an average of a random variable x is denoted as $\bar{x}$. According to Eq. (3.11) for $i = 1$,

$$\bar{x} = \int_{-\infty}^{\infty} x\, y\,(x) \mathrm{d}x \tag{3.13}$$

which is very similar to the first moment of a normalized pulse $y(x)$. In view of that, one might choose to treat the first moment of a chromatographic peak as its statistical average. However, it would not be the average of the peak itself (its amount, concentration, and so forth), but the average, $t_{\mathrm{R}} = \overline{t_{\mathrm{R,mol}}}$ of retention times, $t_{\mathrm{R,mol}}$, of all molecules of the solute that produced the peak. Similarly, the first mathematical moment of a solute zone can be treated as the average, $z = \overline{z_{\mathrm{mol}}}$, of the distances, z_{mol}, from the inlet of all molecules composing the zone. This was the approach in some earlier theoretical studies of chromatography (Klinkenberg and Sjenitzer [30], for example, treated retention time – "holding time" was the actual term – of a peak as the mean of the "holding time distributions"), but this is also an illustration of a possible confusion that can result from mixing the terminologies associated with the concepts of mathematical and statistical moments. □

Note 3.4

Similar to the concept of the average of a random variable is a concept of the *average*

$$\bar{x} = \frac{1}{n}(x_1 + x_2 + x_3 + \cdots + x_n) \tag{3.14}$$

of n numbers, $x_1, x_2, x_3, \ldots, x_n$, or the average

$$\bar{y} = \langle y \rangle_x = \frac{1}{x_2 - x_1}\int_{x_1}^{x_2} y(x)\,dx \tag{3.15}$$

of a function $y(x)$.

If quantities $x_1, x_2, x_3, \ldots, x_n$ are the instances (the realizations) of a random variable x, then $\bar{x}$ in Eq. (3.14) is a statistical estimate of the "true" average of x. (If, for example, $x_1, x_2, x_3, \ldots, x_n$ are the errors found in n measurements of some physical quantity by some instrument, then $\bar{x}$ in Eq. (3.14) is a statistical estimate [12, 29] of the average error.) □

The second central moment, $\tilde{m}_2$, of $y(x)$ is a positive number known as the variance of $y(x)$. The variance is measured in the same units as the units of x^2 and is typically denoted by σ^2, that is,

$$\sigma = \sqrt{\tilde{m}_2} \tag{3.16}$$

From Eq. (3.7), one has

$$\sigma^2 = \frac{1}{M_0}\int_{-\infty}^{\infty} (x - m_1)^2 y(x)\,dx \tag{3.17}$$

The quantity

$$\sigma = \sqrt{\frac{1}{M_0}\int_{-\infty}^{\infty} (x - m_1)^2 y(x)\,dx} \tag{3.18}$$

is known as the *standard deviation* of $y(x)$. To reveal the meaning of this quantity, notice that it is measured in the same units as the units of x. Furthermore, consider two functions $y_1(x)$ and $y_1(x/x_o)$, Figure 3.4, where x_o is a positive number. Both functions have the same shape, but the latter is x_o times wider than the former. On the other hand, it follows directly from Eq. (3.18) that if σ_1 is the standard deviation of $y_1(x)$, then the standard deviation of $y_1(x/x_o)$ is $x_o\sigma_1$. This means that the standard

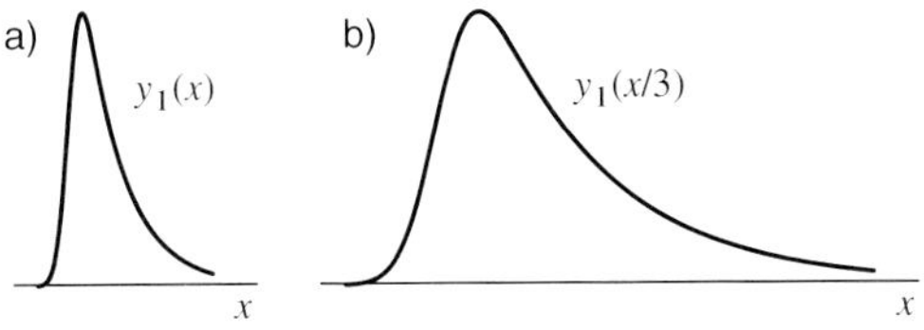

Figure 3.4 Two pulses of the same shape. The pulse in (b) is three times wider and has three times larger standard deviation than the one in (a).

deviation of a pulse is proportional to its width and, therefore, can be used as a measure of the width of the pulse.

One property of standard deviation, σ, sets it apart from many other pulse width metrics. The quantity σ provides a definitive information about the distribution of the area of the pulse. Let $y(x)$ be a positive pulse, and

$$A\,(x_0) = \int_{-\infty}^{m_1 - x_0} y(x)\,\mathrm{d}x + \int_{m_1 + x_0}^{\infty} y(x)\,\mathrm{d}x \tag{3.19}$$

be a part of its area located outside of a $2x_0$-wide segment $(m_1 - x_0,\ m_1 + x_0)$, centered around the centroid, m_1, of the pulse, and $A = A(0)$ be the net area of the pulse. According to *Chebyshev's inequality* [12]:

$$\frac{A(x_0)}{A} \le \left(\frac{\sigma}{x_0}\right)^2 \tag{3.20}$$

In the case of a symmetric pulse, a tighter inequality applies:

$$\frac{A(x_0)}{A} \le \left(\frac{2\sigma}{3x_0}\right)^2 \tag{3.21}$$

It suggests, for example, that the area outside of a $(-3\sigma,\ 3\sigma)$ interval around a centroid of a symmetric positive pulse of any shape is smaller that 5% (2.5% on each side).

Note 3.5

It is very unfortunate that while σ in Eq. (3.18) represents a measure of a width of a pulse, the term *standard deviation of a pulse* assigned to this quantity does not have an intuitive association with the concept of the width of a pulse. Intuitively, the term standard deviation of a pulse seems to infer a deviation of a pulse from something. Actually, no inference to such deviation was deliberately intended for the term within the general context of the mathematical moments. The nonintuitive terminology might be one of the reasons for a somewhat mysterious view of the standard deviation as a measure of the width of a peak in chromatography and for the limited acceptance of that measure in chromatographic practice. □

Note 3.6

The term standard deviation comes from the probability theory and the statistics where quantity

$$\sigma = \sqrt{\int_{-\infty}^{\infty} (x-\bar{x})^2 y(x)\, dx} \qquad (3.22)$$

represents the width of the PDF, $y(x)$, of a random variable x. In the context of the probability theory and the statistics, σ in Eq. (3.22) is defined as the standard deviation of a random variable x (rather than as the standard deviation of the PDF, $y(x)$, as it would be defined in the context of the general mathematical moments). In the statistical context, the term standard deviation appeals to an intuitive meaning of the notion of the deviation (the departure) of x from its average $\bar{x}$. The larger is the standard deviation, σ, of x, the wider is the distribution, $y(x)$, of x around its average $\bar{x}$, and, hence, the larger are the likely deviations of x from $\bar{x}$. This treatment of standard deviation was characteristic for some earlier theoretical studies of chromatography. Klinkenberg and Sjenitzer [30], for example, expressed the width of a chromatographic peak via the standard deviation (from the mean holding time) of the "holding time distributions" (holding time is the same as retention time in current terminology). □

Note 3.7

Sometimes, it can be desirable to know the standard deviation, σ, of a random error, x, in a measurement of some physical quantity by some instrument. PDF of x is frequently unknown in practice and, hence, Eq. (3.22) cannot be used for the calculation of σ. A statistical estimate, σ_{est}, of the standard deviation can be found from an appropriately organized series of a large number, n, of measurements. If x_1, x_2, x_3, ..., x_n are n experimentally found errors, then [12, 29]

$$\sigma_{est} = \sqrt{\frac{1}{n-1}\left((x_1-\bar{x})^2 + (x_2-\bar{x})^2 + \cdots + (x_n-\bar{x})^2\right)} \qquad (3.23)$$

Strictly speaking, so found σ_{est} is the (unbiased) estimate of the standard deviation [12] of x, but not the standard deviation itself. However, in nonmathematical literature, that distinction is typically ignored [29]. In this book, that tradition is honored as well. □

The third central moment, $\tilde{m}_3$, of $y(x)$ represents the asymmetry of $y(x)$, and is equal to zero if $y(x)$ is a symmetric pulse. A dimensionless quantity,

$$c_{skew} = \frac{\tilde{m}_3}{\sigma^3} \qquad (3.24)$$

where σ^2 and $\tilde{m}_3$ are, respectively, the second and the third central moments of $y(x)$, is the *coefficient of skewness* of $y(x)$ [12].

The fourth central moment, $\tilde{m}_4$, of $y(x)$, represents the *sharpness* of $y(x)$. A dimensionless quantity

$$c_{\text{excess}} = \frac{\tilde{m}_4}{\sigma^4 - 3} \tag{3.25}$$

where σ^2 and $\tilde{m}_4$ are the second and the fourth central moments of $y(x)$, c_{excess} is the *coefficient of excess* of $y(x)$ [12] that describes the sharpness of $y(x)$ in comparison with the sharpness of a *Gaussian* pulse (described below) whose coefficient of excess is set to zero.

3.5 Properties of Mathematical Moments

The moments, M_o, m_1, $\tilde{m}_2$, $\tilde{m}_3$ and $\tilde{m}_4$, of a pulse resulted from the convolution of arbitrarily shaped pulses "a," "b," and so forth can be found as follows:

$$M_o = M_{o,a} \times M_{o,b} \times \cdots \tag{3.26}$$

$$m_1 = m_{1,a} + m_{1,b} + \cdots \tag{3.27}$$

$$\tilde{m}_2 = \tilde{m}_{2,a} + \tilde{m}_{2,b} + \cdots \tag{3.28}$$

$$\tilde{m}_3 = \tilde{m}_{3,a} + \tilde{m}_{3,b} + \cdots \tag{3.29}$$

$$\tilde{m}_4 = \tilde{m}_{4,a} + 6\tilde{m}_{2,a}\tilde{m}_{2,b} + \tilde{m}_{4,b} \tag{3.30}$$

Equation (3.26) means that the area of a system impulse response can be found as a product of the areas of the input pulse and of the impulse responses of all serially connected subsystems of the system. With an appropriate scaling, an area of an impulse response of a linear device can be interpreted as the *sensitivity* or as the *gain* of the device. In that case, Eq. (3.26) becomes a statement of the fact that the sensitivity or the gain of a system is a product of the sensitivities or the gains, respectively, of its serially connected subsystems.

Generally, if a metric of a pulse resulted from convolution of several pulses can be found from similar metrics of the convolved pulses, then the metric can be said to have a property of the *closure* [12] under the convolution. Equations (3.26)–(3.29) show that each of the moments M_o, m_1, $\tilde{m}_2$, and $\tilde{m}_3$ is closed under the convolution, meaning, for example, that in order to find the area, M_o, of the convolution of several pulses, one only needs to know the areas of the convolved pulses. Similarly, in order to find the centroid, m_1, of the convolution of several pulses, one only needs to know the centroids of the convolved pulses, and so forth. Equations (3.27)–(3.29) also show that the property of closure of the moments m_1, $\tilde{m}_2$, and $\tilde{m}_3$ under the convolution has a simple form of *additivity* meaning that each of these metrics of a pulse resulted from the convolution of several pulses is a sum of the same metrics of the convolved pulses.

Not all moments are closed under the convolution. A rarely used fourth central moment, $\tilde{m}_4$, is an example of such moments. Indeed, according to Eq. (3.30), the knowledge of the fourth central moments, $\tilde{m}_{4,a}$ and $\tilde{m}_{4,b}$, of two pulses is insufficient for finding $\tilde{m}_4$ of their convolution.

The property of the closure in general and the additivity in particular are extremely important properties of the first three moments, m_1, $\tilde{m}_2$, and $\tilde{m}_3$. Their additivity substantially simplifies prediction of these moments and related metrics of the responses of a linear, shift-invariant system.

3.5.1 Standard Deviation of Convolution of Two Pulses

It has been shown earlier that standard deviation of a pulse can be used as a measure of its width. Due to Eqs. (3.16) and (3.28), the standard deviation, σ, of the convolution of two pulses can be found as (Figure 3.5)

$$\sigma = \sqrt{\sigma_1^2 + \sigma_2^2} \tag{3.31}$$

where σ_1 and σ_2 are the standard deviations of the convolved pulses.

Let the second pulse be significantly narrower than the first one, that is, $\sigma_2 \ll \sigma_1$. Equation (3.31) can be approximated as

$$\sigma = \sqrt{\sigma_1^2 + \sigma_2^2} \approx \sigma_1 \times \left(1 + \frac{1}{2}\left(\frac{\sigma_2}{\sigma_1}\right)^2\right) \quad (\text{when } \sigma_2 \ll \sigma_1) \tag{3.32}$$

The relative difference, $|\sigma - \sigma_1|/\sigma_1$, between σ and the standard deviation, σ_1, of the wider pulse can be estimated as

$$\frac{|\sigma - \sigma_1|}{\sigma_1} \approx \frac{1}{2}\left(\frac{\sigma_2}{\sigma_1}\right)^2 \quad (\text{when } \sigma_2 \ll \sigma_1) \tag{3.33}$$

indicating that the narrower is the narrowest of the two convolved pulses, the disproportionably smaller is its contribution to the width of the result.

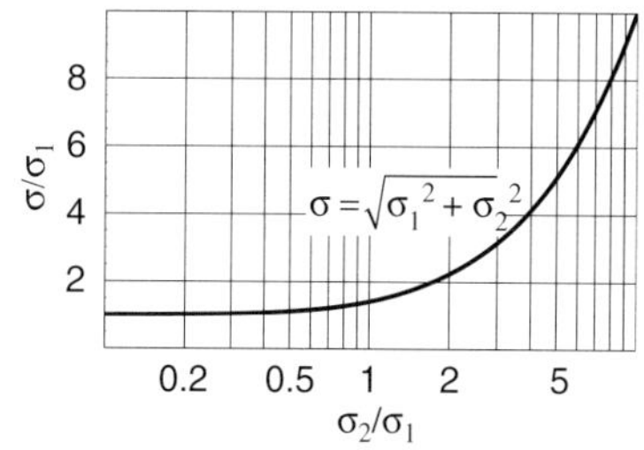

Figure 3.5 Standard deviation, σ, of convolution of two pulses with standard deviations σ_1 and σ_2. The largest standard deviation disproportionably dominates the result.

Example 3.3

If σ_2 is 30% of σ_1, then σ is less than 5% larger than σ_1, that is $\sigma \approx 1.05\sigma_1$. If σ_2 is 10% of σ_1, then σ is less than 0.5% larger than σ_1, that is $\sigma \approx 1.005\sigma_1$.

It is also interesting to point out that not only the width but the shape of the widest pulse dominates the shape of the system response.

These observations have several practically important implications for linear shift-invariant systems.

1) To measure a system impulse response, one can use a rather relaxed requirements to the width of an input pulse. As long as the input pulse is several times narrower than the impulse response, the width and the shape of the former have a minor effect on the latter.
2) In a system consisting of serially connected devices, the ones that have relatively narrow impulse responses make almost no effect on the width of the system impulse response.

3.6 Pulses

Many pulse-like functions can model the inputs, outputs, and the impulse responses of linear systems [1, 7, 13, 14]. A review of the application of several specific pulses in chromatography is known from Sternberg [1]. Some of these pulses are as follows, Figure 3.6, Table 3.1:

- A Gaussian pulse

$$y(x) = \frac{1}{\sigma\sqrt{2\pi}} \exp\left(-\frac{x^2}{2\sigma^2}\right), \quad \sigma > 0 \tag{3.34}$$

- An exponential pulse (exponential decay pulse)

$$y(x) = \begin{cases} 0 & \text{when } x < 0 \\ \sigma^{-1} e^{-x/\sigma} & \text{when } x \geq 0 \end{cases}, \quad \sigma > 0 \tag{3.35}$$

- A rectangular pulse

$$y(x) = \begin{cases} 0 & \text{when } x < 0 \text{ or } x > 2\,\sigma\sqrt{3} \\ 1/(2\sigma\sqrt{3}) & \text{when } 0 \leq x \leq 2\,\sigma\sqrt{3} \end{cases}, \quad \sigma > 0 \tag{3.36}$$

- An exponentially modified Gaussian (EMG) pulse [1, 31–33] – a convolution of exponential and Gaussian pulses. The shape of an EMG pulse is closer to the shape of its widest component, and can be nearly Gaussian or nearly exponential, and everything in between.

All pulses in Eqs. (3.34–3.36) are scaled in such a way that the area of each pulse is equal to 1 and its standard deviation is equal to the quantity σ in the defining expression. It is evident from Eqs. (3.34) to (3.36) that the width of each pulse is proportional to σ. This confirms the earlier made observation that the standard deviation of a pulse can indeed be used as a measure of its width.

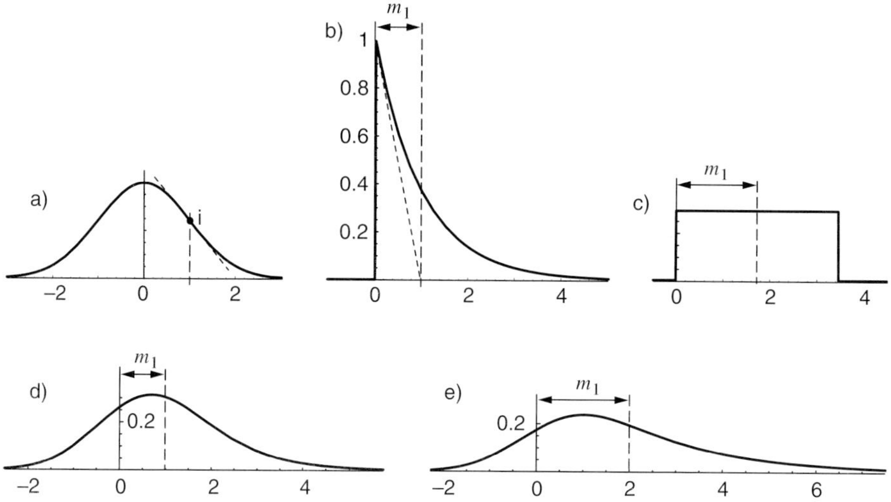

Figure 3.6 Gaussian (a), exponential (b), rectangular (c), and EMG pulses (d) and (e). Each pulse has the unity area. The abscissa of each vertical dashed line is the centroid, m_1, of a respective pulse. The first three pulses have the unity standard deviation, σ. The Gaussian and the exponential components in the EMG pulses have the following standard deviations, σ_G and σ_E, respectively: (d) $\sigma_G = \sigma_E = 1$; (e) $\sigma_G = 1$, $\sigma_E = 2$. Graphically, σ for a Gaussian pulse can be found as the length of the segment in the horizontal axis between its origin and abscissa of the inflection point "i." For an exponential pulse, σ can be constructed as the length of the segment in the horizontal axis between its origin and the intersection with the tangent line through the highest point of exponential curve. For a rectangular pulse, $\sigma^2 = 1/12$ and, therefore, σ is about 0.346 of its total width.

Table 3.1 Moments, Eqs. (3.6) and (3.7), and the coefficients of skewness, c_{skew}, Eq. (3.24) and excess, c_{excess}, Eq. (3.25), of the pulses described in Eqs. (3.34)–(3.36).

Pulse type	m_1	$\tilde{m}_2$	$\tilde{m}_3$	$\tilde{m}_4$	c_{skew}	c_{excess}
Gaussian	0	σ^2	0	$3\sigma^4$	0	0
Exponential	σ	σ^2	$2\sigma^3$	$9\sigma^4$	2	6
Rectangular	$3\sqrt{\sigma}$	σ^2	0	$1.8\sigma^4$	0	−1.2

Note 3.8

The pulses described in Eqs. (3.34)–(3.36) are typically used to model the following devices.

The shape of a solute zone injected into a column can be modeled by the rectangular pulse [1]. The pulse approaches an ideal delta function, δ, when its standard deviation, σ, approaches zero. The rectangular pulse is also suitable [34] for modeling the impulse response of a cell of a *thermal-conductivity detector* (TCD) [5, 16].

Ideally, a solute injected as a delta function migrates along the column as a Gaussian zone (in distance) [1, 5, 26, 35, 36] and elutes from the column as a nearly Gaussian peak (in time) [15, 25, 37, 38]. In other words, both the spatial and the temporal impulse responses of a column have a Gaussian or nearly Gaussian shape. This is why the chromatographic peaks are ideally expected to be Gaussian.

An exponential pulse can be used [1, 14] to model the impulse responses of electronic front-ends of some detectors. It can also model the impulse responses of the so-called "dead-volumes" [39] such as those that can be caused by the traps in the connecting assemblies, Figure 1.1, and other devices. □

References

1 Sternberg, J.C. (1966) in *Advances in Chromatography*, vol. **2** (eds J.C. Giddings and R.A. Keller), Marcel Dekker, New York, pp. 205–270.

2 Ettre, L.S. (1993) *Pure Appl. Chem.*, **65**, 819–872.

3 van Deemter, J.J.J., Zuiderweg, F.J., and Klinkenberg, A. (1956) *Chem. Eng. Sci.*, **5**, 271–289.

4 James, A.T. and Martin, A.J.P. (1952) *Analyst*, **77**, 915–932.

5 Littlewood, A.B. (1970) *Gas Chromatography: Principles, Techniques, and Applications*, 2nd edn, Academic Press, New York.

6 Blumberg, L.M. (1984) *Anal. Chem.*, **56**, 1726–1729.

7 Bracewell, R.N. (1986) *The Fourier Transform and its Application*, 2nd edn, revised, McGraw-Hill Book Company, New York.

8 Dose, E.V. (1987) *Anal. Chem.*, **59**, 2414–2419.

9 Blumberg, L.M. and Berger, T.A. (1993) *Anal. Chem.*, **65**, 2686–2689.

10 Yan, B., Zhao, J., Brown, J.S., Blackwell, J., and Carr, P.W. (2000) *Anal. Chem.*, **72**, 1253–1262.

11 Blumberg, L.M. and Klee, M.S. (2001) *J. Chromatogr. A*, **933**, 1–11.

12 Korn, G.A. and Korn, T.M. (1968) *Mathematical Handbook for Scientists and Engineers*, McGraw-Hill Book Company, New York.

13 Papoulis, A. (1965) *Probability, Random Variables, and Stochastic Processes*, McGraw-Hill Book Company, New York.

14 Kailath, T. (1980) *Linear Systems*, Prentice-Hall, Englewood Cliffs, NJ.

15 Jönsson, J.A. (1987) in *Chromatographic Theory and Basic Principles* (ed. J.A. Jönsson), Marcel Dekker, New York, pp. 27–102.

16 Guiochon, G. and Guillemin, C.L. (1988) *Quantitative Gas Chromatography for Laboratory Analysis and On-Line Control*, Elsevier, Amsterdam.

17 Blumberg, L.M. (1994) *Chromatographia*, **39**, 719–728.

18 Ohline, R.W. and DeFord, D.D. (1963) *Anal. Chem.*, **35**, 227–234.

19 Reid, G.J. (1983) *Linear System Fundamentals*, McGraw-Hill Book Company, New York.

20 Tsakalis, K.S. and Ioannou, P.A. (1993) *Linear Time-Varying Systems*, Prentice-Hall, Englewood Cliffs, NJ.

21 Blumberg, L.M. and Berger, T.A. (1992) *J. Chromatogr.*, **596**, 1–13.

22 Blumberg, L.M. (1993) *J. Chromatogr.*, **637**, 119–128.

23 Grushka, E., Myers, M.N., Schettler, P.D., and Giddings, J.C. (1969) *Anal. Chem.*, **41**, 889.

24 Law, A.M. and Kelton, W.D. (1982) *Simulation Modeling and Analysis*, McGraw-Hill Book Company, New York.

25 Grubner, O. (1968) in *Advances in Chromatography*, vol. 6 (eds J.C. Giddings and R.A. Keller), Marcel Dekker, New York, pp. 173–209.

26 Giddings, J.C. (1991) *Unified Separation Science*, John Wiley & Sons, Inc., New York.

27 Lan, K. and Jorgenson, J.W. (2000) *Anal. Chem.*, **72**, 1555–1563.
28 Lan, K. and Jorgenson, J.W. (2001) *J. Chromatogr. A*, **905**, 47–57.
29 Miller, J.C. and Miller, J.N. (1988) *Statistics for Analytical Chemistry*, 2nd edn, Ellis Horwood Limited, Chichester.
30 Klinkenberg, A. and Sjenitzer, F. (1956) *Chem. Eng. Sci.*, **5**, 258–270.
31 Foley, J.P. and Dorsey, J.G. (1984) *J. Chromatogr. Sci.*, **22**, 40–46.
32 Hanggi, D. and Carr, P.W. (1985) *Anal. Chem.*, **57**, 2394–2395.
33 Lan, K. and Jorgenson, J.W. (2001) *J. Chromatogr. A*, **915**, 1–13.
34 Blumberg, L.M. and Dandeneau, R.D. (1995) *J. High Resolut. Chromatogr.*, **18**, 235–242.
35 Giddings, J.C. (1965) *Dynamics of Chromatography*, Marcel Dekker, New York.
36 Jennings, W. (1987) *Analytical Gas Chromatography*, Academic Press, London.
37 Kučera, E. (1965) *J. Chromatogr.*, **19**, 237–248.
38 Grushka, E. (1972) *J. Phys. Chem.*, **76**, 2586–2593.
39 Maynard, V.R. and Grushka, E. (1972) *Anal. Chem.*, **44**, 1427–1434.

4
Migration of a Solid Object

In many ways, description of migration of a solute along a chromatographic column is similar to the description of *one-dimensional migration* (*movement*) of a solid object along a predetermined migration path.

4.1
Velocity of an Object

Let z be the distance traveled by an object during time t assuming that at the starting point, $z = 0$ and $t = 0$, that is,

$$z = 0 \text{ implies } t = 0, \; t = 0 \text{ implies } z = 0 \tag{4.1}$$

In chromatography, the starting point of migration of all solutes is a column inlet, and the starting event is injection of a solute mixture in the inlet.

The quantities z and t describing migration of an object are mutually dependent variables, that is, z can be expressed as a function, $z = z(t)$, of t, or t can be expressed as a function, $t = t(z)$, of z. The velocity, v, of an object is defined as

$$v = \frac{dz}{dt} \tag{4.2}$$

It is assumed throughout this chapter that v is a positive quantity, that is,

$$v > 0 \tag{4.3}$$

4.2
Parameters of Migration Path

If a moving object is, for example, a car traveling from one city to another, then the driver might have a significant freedom to choose the car velocity. Even in that case, however, there could be some restrictions to the car velocity dictated by the route: different maximal velocity might be allowed along different stretches of the route,

Temperature-Programmed Gas Chromatography. Leonid M. Blumberg

ISBN: 978-3-527-32642-6

traffic lights might program the movement across certain spots along the route, and so forth. In GC (and in other chromatographic techniques), the velocity of each solute is totally controlled by the conditions along the column. The velocity depends on the type and the amount of stationary phase, on column temperature, on carrier gas velocity, and so forth. A method of a GC analysis uniquely prescribes velocity of a solute for each particular location along the column at each particular time.

Reflecting the conditions that exist in chromatography, let us assume that migration velocity of an object is a known function, $v = v(z, t)$, of its coordinate, z, and time, t. Typically, it is much easier to describe function $v(z, t)$, for an arbitrary z and t regardless of whether or not the object can actually be at a particular location, z, at a particular time, t. In other words, when describing a path rather than an object migrating in it, it is convenient to treat z and t as mutually independent variables. In this case, z is an arbitrary coordinate along the path, and t is an arbitrary time.

Example 4.1

Let v be expressed as a product

$$v = v(z,t) = \frac{1}{1 + (\mu_i^{-1} - 1)e^{-\gamma_t t}} \frac{u_i}{\sqrt{1 - \gamma_z z}} \tag{4.4}$$

where the first term describes dependence of v on t at any z, while the second term the dependence of v on z at any t. This formula is meaningful in GC, but, for now, it is sufficient to treat it just as a formula where u_i, γ_t, γ_z, and μ_i are known non-negative quantities. □

When z and t are treated as mutually independent variables, their values $z = 0$ and $t = 0$ still represent the coordinate of a starting point, and the starting event – the time when the object migration started. However, $v(z, 0)$ is meaningful for any z and, vice versa, $v(0, t)$ is meaningful for any t. In other words, when z and t are independent variables, the implications described in Eq. (4.1) do not apply.

Sometimes, not only the velocity, $v(z, t)$, of an object migrating along a path, but other quantities – parameters of the path – can be of interest in addition to or instead of $v(z, t)$. In GC, such parameters could be column temperature, pressure, and so forth. Let $y = y(z, t)$ be $v(z, t)$ or any other parameter of the path.

Depending on the values of its partial derivatives, $\partial y/\partial z$ and $\partial y/\partial t$, quantity y can be uniform or nonuniform, it can also be static (steady state, time-invariant) or dynamic (time-varying) [1–5]. The quantity y is uniform if $\partial y/\partial z = 0$ at any z. Otherwise, y is nonuniform. If y is a uniform quantity, then its local values $y(z_1, t)$ and $y(z_2, t)$ at any two arbitrary locations, z_1 and z_2, are equal to each other at any given time, t. Otherwise, there should exist different locations, z_1 and z_2, where, sometimes or always, $y(z_1, t) \neq y(z_2, t)$. The quantity y is static if $\partial y/\partial t = 0$ at any t; otherwise, it is dynamic. If y is a static quantity, then its instant values $y(z, t_1)$ and $y(z, t_2)$ at any two arbitrary time instants, t_1 and t_2, are equal to each other at any location, z. Any combination of properties of being uniform or nonuniform and static or dynamic could exist. For example, a static parameter could be nonuniform (different at different locations along the path). Similarly, a uniform parameter can be dynamic. It is convenient to assign the properties of velocity, $v(z, t)$, of being uniform or nonuniform, and static or dynamic to the migration path itself. In this case, a migration path is uniform if $v(z, t)$ is uniform, a migration path is dynamic if $v(z, t)$ is dynamic, and so forth.

4.3 Relations between Path Parameters and Object Parameters

If dependence, $v = v(z, t)$, of v on arbitrary z and t is known for a path, then v for an object migrating along that path, that is, v as a function, $v = v(z)$, of location, z, of the object or as a function $v = v(t)$ of its migration time, t, can be found as

$$v = v(z) = v(z, t(z)), \quad v = v(t) = v(z(t)t) \tag{4.5}$$

In both cases, one needs first to find $t = t(z)$ or $z = z(t)$. Due to Eq. (4.3), both are monotonically increasing functions that can be found by solving Eq. (4.2).

Example 4.2

Equation (4.2) with v from Eq. (4.4) is the first-order ordinary differential equation. Its solution with initial conditions described in Eq. (4.1) can be expressed as

$$\frac{1-(1-\gamma_z z)^{3/2}}{3u_i\gamma_z} = \frac{\ln(1+(e^{\gamma_t t}-1)\mu_i)}{2\gamma_t} \tag{4.6}$$

The solutions of this equation for $t = t(z)$ and for $z = z(t)$ are

$$t = t(z) = \frac{1}{\gamma_t}\ln\left(1+\mu_i^{-1}\left(\exp\left(\frac{2\gamma_t}{3u_i\gamma_z}\left(1-(1-\gamma_z z)^{3/2}\right)\right)-1\right)\right) \tag{4.7}$$

$$z = z(t) = \frac{1}{\gamma_z}\left(1-\left(1-\frac{3u_i\gamma_z}{2\gamma_t}\ln(1+(e^{\gamma_t t}-1)\mu_i)\right)^{2/3}\right) \tag{4.8}$$

Once functions $t = t(z)$ and $z = z(t)$ for the object are known, its velocity as an explicit function of the object location, z, or as a function of its migration time, t, can be found from the substitution of one of these functions in Eq. (4.4). That yields

$$v = v(z) = v(z, t(z)) = \frac{u_i}{\sqrt{1-\gamma_z z}}\left(1-(1-\mu_i)\exp\left(-\frac{2\gamma_t}{3u_i\gamma_z}\left(1-(1-\gamma_z z)^{3/2}\right)\right)\right) \tag{4.9}$$

$$v = v(t) = v(z(t), t) = \frac{u_i}{1+(\mu_i^{-1}-1)e^{-\gamma_t t}}\left(1-\frac{3u_i\gamma_z}{2\gamma_t}\ln(1+(e^{\gamma_t t}-1)\mu_i)\right)^{-1/3} \tag{4.10}$$

It is not unusual for the expressions, such as Eqs. (4.9) and (4.10), explicitly describing the velocity of an object as functions $v = v(z)$ and $v = v(t)$ of the object's location, z, or of its migration time, t, to be more complex and less transparent than the equation $v = v(z, t)$, such as Eq. (4.4), for v at an arbitrary z and t. □

If expressions $t = t(z)$ or $z = z(t)$ are known, one can find $v = v(z)$ and $v = v(t)$ directly from Eq. (4.2) as $v(t) = \mathrm{d}z(t)/\mathrm{d}t$ or $v(z) = 1/(\mathrm{d}t(z)/\mathrm{d}z)$. However, the approach described in Eq. (4.5) is suitable not only for the object's velocity, but for any parameter, y, describing the object's migration path.

Not only the velocity, v, can be a parameter of a path (described as a function $v(z, t)$ of arbitrary z and t) and a parameter of an object migrating along the path (described as a function, $v(z)$, of the object's location, z, or as a function, $v(t)$, of the object's migration time, t). Any parameter, $y = y(z, t)$, of a path can give rise to functions,

$$y = y(z) = y(z, t(z)), \quad y = y(t) = y(z(t), t) \tag{4.11}$$

that represent parameter y of the path at the time, t, when an object migrating along the path is located at z. The parameter y described in Eq. (4.11) can be viewed as a parameter of the object itself. Equation (4.11) represents generalization of Eq. (4.5).

Ordinary derivatives, $\mathrm{d}y/\mathrm{d}z$ and $\mathrm{d}y/\mathrm{d}t$, of parameter y of an object describe, respectively, the *spatial* and the *temporal rate* of increase in that parameter. In order to find these rates, consider an elementary increment, $\mathrm{d}y$, in y due to migration of an object from location z_1 to $z_1 + \mathrm{d}z$ during the time $\mathrm{d}t$, Figure 4.1. Two factors contribute to the value of $\mathrm{d}y$. First, if y is nonuniform then the displacement of the object by $\mathrm{d}z$ causes the change $(\partial y/\partial z)\mathrm{d}z$ in y. Second, if y is dynamic then the passage of time $\mathrm{d}t$ causes y to change by $(\partial y/\partial t)\mathrm{d}t$. The net change,

$$\mathrm{d}y = \frac{\partial y}{\partial z}\mathrm{d}z + \frac{\partial y}{\partial t}\mathrm{d}t \tag{4.12}$$

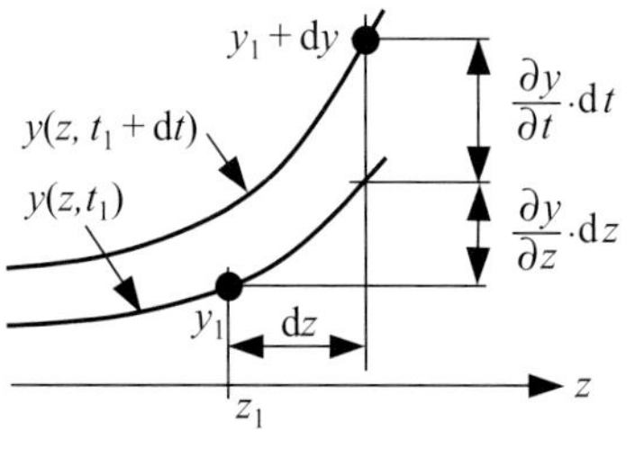

Figure 4.1 Increments in the value of parameter y of a migrating object. When the object migrates from location z_1, where it resided at the time t_1, to location $z_1 + \mathrm{d}z$ at time $t_1 + \mathrm{d}t$, the value of y changes from y_1 to $y_1 + \mathrm{d}y$. The net elementary increment, $\mathrm{d}y$, in y is described in Eq. (4.12).

in y is the sum, Figure 4.1, of the changes caused by both factors. This, in view of Eq. (4.2), yields

$$\frac{dy}{dt} = \frac{dy(t)}{dt} = \frac{\partial y}{\partial z}\frac{dz}{dt} + \frac{\partial y}{\partial t} = \left(v(z,t)\frac{\partial y(z,t)}{\partial z} + \frac{\partial y(z,t)}{\partial t} \right)\bigg|_{z=z(t)} \tag{4.13}$$

$$\frac{dy}{dz} = \frac{dy(z)}{dz} = \frac{\partial y}{\partial z} + \frac{1}{v}\frac{\partial y}{\partial t} = \left(\frac{\partial y(z,t)}{\partial z} + \frac{1}{v(z,t)}\frac{\partial y(z,t)}{\partial t} \right)\bigg|_{t=t(z)} \tag{4.14}$$

In this chapter and in several forthcoming chapters, dual use of variables z and t is employed. They are used as mutually independent variables describing a migration path, and as mutually dependent variables describing the coordinate, z, of a migrating object at the time, t, since the start of its migration. This dual use can be confusing. To avoid a possible confusion, one can use different notations [2, 3, 6, 7] for the distance and the time variables in the case when they are mutually independent and in the case when they represent mutually dependent location and migration time of an object. Unfortunately, this approach has its own shortcomings. One of them is the lack of flexibility forcing a choice between the notations even when the difference between them is irrelevant, or when the meaning of a variable is clear from the context, and so forth.

Reliance on usual mathematical conventions and on additional care when necessary allows one to avoid possible confusions without resorting to additional notations. Thus, the ordinary derivative, dy/dt, representing all changes in y caused by the change in t and by the subsequent change in z, naturally represents the rate of the change in parameter y of an object migrating along a path having property $y(z, t)$. Similar observations are true for the ordinary derivative, dy/dz. On the other hand, partial derivatives $\partial y/\partial t$ and $\partial y/\partial z$ naturally represent a path with parameter $y(z, t)$ being a function of mutually independent variables t and z.

Not always the ordinary and the partial derivatives are different from each other. Equation (4.12) implies that

Statement 4.1

If parameter y of a path is uniform (i.e., $\partial y/\partial z = 0$), then $dy/dt = \partial y/\partial t$. Similarly, $dy/dz = \partial y/\partial z$ if y is static (i.e., $\partial y/\partial t = 0$).

When a parameter is uniform and/or static, there is also no need to distinguish between its roles as a parameter of a path or as a parameter of a migrating object. Indeed, a static y does not change with t, and, therefore, its description as $y = y(z)$ is complete whether z is an independent variable or a coordinate of a migrating object. Similarly, a uniform y does not change with z, and, therefore, its description as $y = y(t)$ is complete whether t is an independent variable or it is an object migration time.

References

1 Giddings, J.C. (1963) *Anal. Chem.*, **35**, 353–356.
2 Blumberg, L.M. and Berger, T.A. (1992) *J. Chromatogr.*, **596**, 1–13.
3 Blumberg, L.M. (1993) *J. Chromatogr.*, **637**, 119–128.
4 Blumberg, L.M. (1994) *Chromatographia*, **39**, 719–728.
5 Blumberg, L.M. (1995) *Chromatographia*, **40**, 218.
6 Lan, K. and Jorgenson, J.W. (2000) *Anal. Chem.*, **72**, 1555–1563.
7 Lan, K. and Jorgenson, J.W. (2001) *J. Chromatogr. A*, **905**, 47–57.

5
Solute–Liquid Interaction in Gas Chromatography

Solute separation in GC results from the difference in their migration velocities in a column. That difference, in turn, is a result of the difference in the solute interaction with the column's stationary phase. The stronger is the interaction, the lower is the migration velocity. Depending on the stationary phase type – solid or liquid – a solute can be adsorbed on the solid surface, or it can be absorbed by the liquid (dissolved in it). This chapter focuses on the GC-related parameters of interaction of solutes with liquid stationary phases.

5.1
Distribution Constant and Retention Factor

Consider a layer of a liquid (the liquid phase) and a layer of an inert gas (the gas phase) that are in contact with each other, Figure 5.1. In this chapter, the layers represent organic liquid polymer film and a carrier gas in a capillary (open tubular) column, respectively, Figure 2.1a. Suppose that a solute was added to the gas, and the entire three-component system was sealed. A fraction of the solute added to the gas can be absorbed by the liquid (dissolved in it). Of the primary concern for this chapter is a chromatographically meaningful description of equilibrium of a solute solvation–evaporation (solvation in the liquid and evaporation into the gas) or, in broader terms, equilibrium of a solute–liquid interaction. The equilibrium takes place at a certain distribution of a solute between the gas and the liquid. The gas is inert and has no effect on the equilibrium distribution of a solute between the two phases. This means that replacement of one gas with another has no effect on the equilibrium.

The equilibrium distribution of a solute between a gas and a liquid, Figure 5.1, can be described by the *distribution constant* [1] (*partition coefficient* [2–8]), K_c, defined as [2–7, 9]

$$K_c = \frac{C_{sol,f}}{C_{sol,g}} \tag{5.1}$$

where $C_{sol,g}$ and $C_{sol,f}$ are concentrations (mass per volume, mole per volume, and so forth) of the solute in a liquid film and a gas, respectively.

Temperature-Programmed Gas Chromatography. Leonid M. Blumberg

ISBN: 978-3-527-32642-6

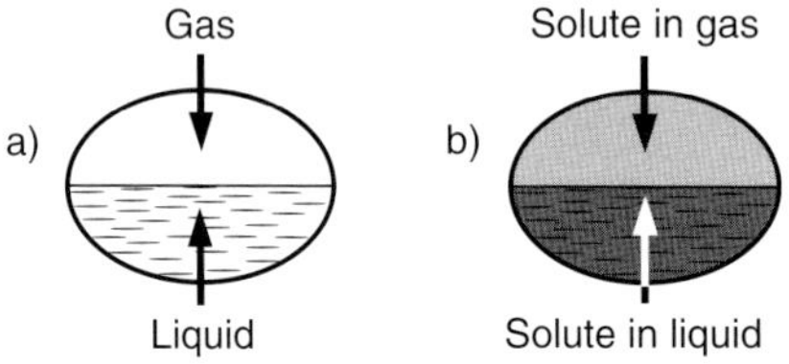

Figure 5.1 A piece of a cut through a boundary between the layers of a liquid and a gas. Both layers are pure in (a). A solute is distributed between the layers in (b). At the equilibrium in a typical GC analysis, the ratio, K_c, Eq. (5.1), of the solute concentration in the liquid to the concentration in the gas is much larger than unity.

The value of K_c depends on the properties of the solute and the liquid. It also depends on their temperature, T. Generally, K_c also depends on the solute concentration, $C_{sol,f}$, in the liquid. However, when $C_{sol,f}$ is not very high, K_c at a fixed T is nearly independent of $C_{sol,f}$. In that case, K_c can be expressed as [6, 10]

$$K_c = \exp\frac{\Delta\mathcal{G}}{\mathcal{R}T}, \quad \Delta\mathcal{G} = \Delta\mathcal{G}_H - \Delta\mathcal{G}_S T \tag{5.2}$$

where $\mathcal{R} = 8.314510\,\mathrm{J\,K^{-1}\,mol^{-1}}$ is the molar gas constant, and quantities $\Delta\mathcal{G}$, $\Delta\mathcal{G}_S$, and $\Delta\mathcal{G}_H$ are, respectively, the Gibbs free energy, the entropy, and the enthalpy of evaporation of the solute from the liquid. As an example, the values of $\Delta\mathcal{G}_S$ and $\Delta\mathcal{G}_H$ for several n-alkanes are listed in Table 5.1.

Generally, parameters $\Delta\mathcal{G}_S$ and $\Delta\mathcal{G}_H$ themselves can change with temperature [8, 12–20]. However, within a relatively narrow temperature range (under 100 °C or so) around its elution temperature in a typical temperature-programmed analysis, the solute parameters $\Delta\mathcal{G}_S$ and $\Delta\mathcal{G}_H$ are essentially temperature independent. A model in Eq. (5.2) in which parameters $\Delta\mathcal{G}_S$ and $\Delta\mathcal{G}_H$ are fixed quantities (different for different solute–liquid pairs) is treated here as the ideal thermodynamic model of a solute–liquid interaction. Later, we will consider other models. To distinguish one model from others, its type will be explicitly marked in-line with the formula describing the model.

It follows from Eq. (5.2) that, ideally, $\ln K_c$ is a linear function, Figure 5.2,

$$\ln K_c = -\frac{\Delta\mathcal{G}_S}{\mathcal{R}} + \frac{\Delta\mathcal{G}_H}{\mathcal{R}T} \quad \text{(ideal thermodynamic model)} \tag{5.3}$$

Table 5.1 Entropy, $\Delta\mathcal{G}_S$, and enthalpy, $\Delta\mathcal{G}_H$, of evaporation of n-alkanes (C_{10} through C_{32}) from (5%-phenyl)-dimethyl polysiloxane film (reconstructed from the data in the literature [11]).

	C_{10}	C_{12}	C_{14}	C_{16}	C_{18}	C_{20}	C_{22}	C_{24}	C_{26}	C_{28}	C_{30}	C_{32}
$\Delta\mathcal{G}_S$ (J/mol/K)	74.3	76.8	79.3	81.7	85.1	88.4	91.7	95	99.2	103.4	108.4	112.5
$\Delta\mathcal{G}_H$ (kJ/mol)	47	51.6	56.2	60.8	65.4	70.1	74.6	79.3	83.9	88.5	93.1	97.7

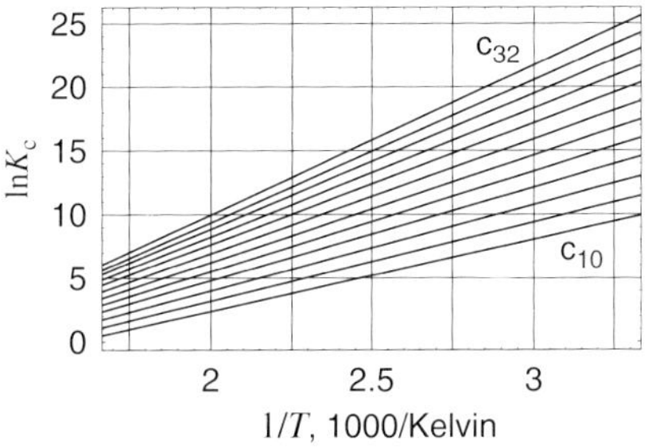

Figure 5.2 Logarithms of distribution constants, K_c, Eqs. (5.2) and (5.3), of *n*-alkanes in Table 5.1 as functions of inverse temperature, $1/T$.

of inverse temperature, $1/T$. In that function, quantities $-\Delta\mathscr{G}_S/R$ and $\Delta\mathscr{G}_H/R$ are the intercept and the slope, respectively.

In chromatography, it is more important not how the concentrations, $C_{sol,g}$ and $C_{sol,f}$, of a solute in a gas and a liquid relate to each other but how the solute amounts, $m_{sol,g}$ and $m_{sol,f}$, in gas and stationary phase film are distributed between the phases. The ratio

$$k = \frac{m_{sol,f}}{m_{sol,g}} \tag{5.4}$$

is the retention factor [1] (partition ratio [3, 4, 7, 9, 21, 22], capacity ratio [1], and capacity factor [5, 6, 23–29]) of a solute.

The amounts $m_{sol,g}$ and $m_{sol,f}$ can be found as

$$m_{sol,g} = C_{sol,g}V_g, \qquad m_{sol,f} = C_{sol,f}V_f \tag{5.5}$$

where V_g and V_f are the volumes of the gas and the liquid film, respectively. This allows one to express k as, Figure 5.3,

$$k = \frac{K_c}{\beta} = 4\varphi K_c \tag{5.6}$$

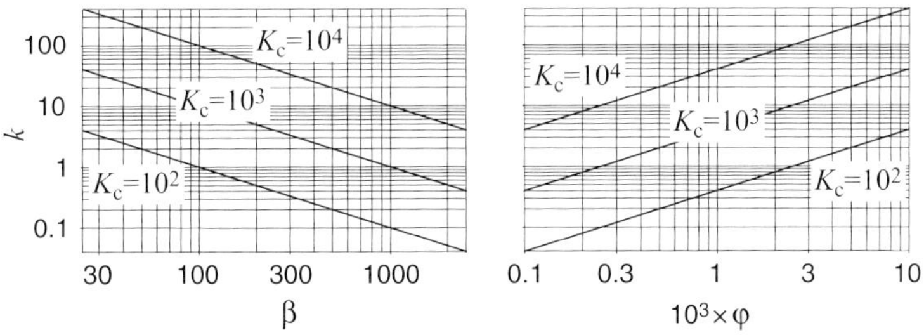

Figure 5.3 Retention factors, k, Eq. (5.6), as functions of phase ratio, β, and dimensionless film thickness, φ.

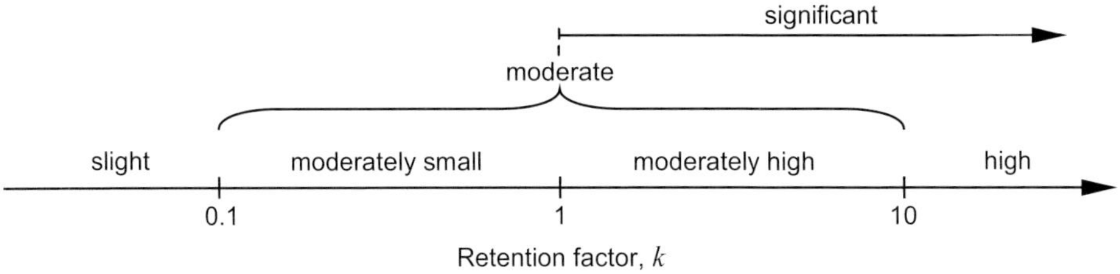

Figure 5.4 Levels of a solute retention – slight, moderate, high, and so forth.

where $\beta = V_g/V_f \approx d/(4d_f)$, Eqs. (2.3) and (2.5), is the phase ratio, and $\varphi = 1/(4\beta) \approx d_f/d$, Eq. (2.7), is the dimensionless film thickness. The value of k can be any nonnegative number from zero for unretained solutes to thousands for highly retained ones. Several levels of solute retention (slight, moderate, high, and so forth) are identified in Figure 5.4 for future references.

Film thickness significantly affects many aspects of a solute–liquid interaction and a column performance itself. In many cases – Eq. (5.6) is one of them – it is not the absolute film thickness, d_f, but its relation to the column diameter, d, that provides an easily identifiable effect of the film thickness on other column parameters. For that reason, it is convenient to express the film thickness via relative parameters like β or φ.

Widely known metric of the relative film thickness in chromatography is the phase ratio, β. Unfortunately, this metric has significant shortcoming. Being inversely proportional, Eqs. (2.3) and (2.5), to the thickness, d_f, and to the amount, V_f, of the stationary phase film, quantity β obscures interpretations and illustration of the effects of the film thickness on column parameters. Thus, the dependence of k on β is represented in Figure 5.3 by the declining lines obscuring the fact that k increases when the film thickness increases.

Dimensionless film thickness, φ, is proportional, Eq. (2.7), to the thickness and the amount of stationary phase film. As a result, it does not suffer from the aforementioned shortcomings of β. Thus, the dependence of k on φ is represented in Figure 5.3 by the rising lines directly reflecting the fact that k increases when the film thickness increases. In the future, dimensionless film thickness, φ, would be typically used as a measure of the relative amount of the film. If desired, Eq. (2.6) can be used for replacing the quantity φ with quantity β in any formula.

Equations (5.2) and (5.6) allow one to express k as

$$k = 4\varphi \cdot \exp\frac{\Delta\mathcal{G}}{\mathcal{R}T} = 4\varphi \cdot \exp\left(-\frac{\Delta\mathcal{G}_S}{\mathcal{R}} + \frac{\Delta\mathcal{G}_H}{\mathcal{R}T}\right) \quad \text{(ideal thermodynamic model)} \tag{5.7}$$

$$\ln k = \ln(4\varphi) - \frac{\Delta\mathcal{G}_S}{\mathcal{R}} + \frac{\Delta\mathcal{G}_H}{\mathcal{R}T} \quad \text{(ideal thermodynamic model)} \tag{5.8}$$

As does Eq. (5.3), these formulae meaningfully describe thermodynamics of a solute–liquid interaction and a solute retention in a given column. However, they have several considerable shortcomings from chromatographic point of view. Thermodynamic parameters $\mathscr{R}$, $\Delta\mathscr{G}_H$, and $\Delta\mathscr{G}_S$ in these formulae have no direct chromatographic meaning and, therefore, do not expose characteristics of a solute behavior in a chromatographic analysis. What, for example, do parameters $\Delta\mathscr{G}_S = 88.4$ J/mol/K and $\Delta\mathscr{G}_H = 70\ 100$ J/mol of C_{20} in Table 5.1 tell about its chromatographic behavior? What would be the solute elution temperature – the temperature at the time of the solute elution – in a typical temperature-programmed analysis? How would the elution temperature be affected by heating rate – the rate of the temperature change?

As a first step toward expressing Eqs. (5.7) and (5.8) via chromatographically meaningful parameters is the reduction in the number of parameters in these formulae. Equations (5.7) and (5.8) express the dependence of k on T via four parameters, $\mathscr{R}$, $\Delta\mathscr{G}_H$, $\Delta\mathscr{G}_S$, and φ. The quantity $\ln k$ in Eq. (5.8) is a linear function of $1/T$ (as is quantity $\ln K_c$ in Eq. (5.3)). Only two parameters – the slope,

$$T_H = \Delta\mathscr{G}_H/\mathscr{R} \tag{5.9}$$

and the intercept $\ln(4\varphi) - \Delta\mathscr{G}_S/\mathscr{R}$ – are necessary for a complete representation of a linear function. Because typically $\varphi \ll 1$, the intercept is a negative quantity. Using its positive value,

$$g_S = \Delta\mathscr{G}_S/\mathscr{R} - \ln(4\varphi) \tag{5.10}$$

allows one to express Eqs. (5.7) and (5.8) as

$$k = \exp\left(-g_S + \frac{T_H}{T}\right) \quad \text{(ideal thermodynamic model)} \tag{5.11}$$

$$\ln k = -g_S + \frac{T_H}{T} \quad \text{(ideal thermodynamic model)} \tag{5.12}$$

It follows from the definitions in Eqs. (5.9) and (5.10) that g_S is a dimensionless quantity and T_H is measured in units of temperature. The latter can be interpreted as [11] the thermal equivalent of enthalpy of a solute–liquid interaction. For a typical film thickness ($\varphi = 0.001$), quantities g_S and T_H for n-alkanes of Table 5.1 are shown in Table 5.2. The data in Table 5.2 suggest that parameters g_S and T_H have no better chromatographic meaning than that of parameters $\Delta\mathscr{G}_H$ and $\Delta\mathscr{G}_S$. However, parameters g_S and T_H provide a convenient bridge to a set of two chromatographically meaningful parameters of a solute–liquid interaction in a column.

Table 5.2 Parameters T_H and g_S, Eqs. (5.9) and (5.10), for n-alkanes in Table 5.1 at $\varphi = 0.001$.

	C_{10}	C_{12}	C_{14}	C_{16}	C_{18}	C_{20}	C_{22}	C_{24}	C_{26}	C_{28}	C_{30}	C_{32}
T_H (K)	5655	6209	6763	7317	7871	8425	8979	9533	10 087	10 641	11 195	11 749
g_S	14.7	15.0	15.3	15.6	16.0	16.4	16.8	17.2	17.7	18.2	18.8	19.3

5.2 Chromatographic Parameters of Solute–Liquid Interaction

Temperature, T, is a chromatographically meaningful variable, Eq. (5.12). The same cannot be said about variable $1/T$. On the other hand, the quantity $\ln k$ in Eq. (5.12) is a linear function of $1/T$, but it is not a linear function, Figure 5.5, of T.

Let $\ln k(T)$ be an arbitrary (not necessarily describing the ideal thermodynamic model of a solute–liquid interaction) monotonically declining function of T, and Figure 5.5 be just a general illustration of such function. Consider a linear approximation to $\ln k(T)$ by linear Taylor series expansion of $\ln k(T)$ at $k = k_0$, where k_0 is some predetermined value of k. This approximation is a tangent line to $\ln k(T)$ at $k = k_0$, Figure 5.5.

The linear Taylor series expansion of $\ln k(T)$ at $k = k_0$ can be described as

$$\ln k = \ln k_0 + \left(\left. \frac{\mathrm{d}(\ln k(T))}{\mathrm{d}T} \right|_{T=T_0} \right)(T - T_0) \quad \text{(linear approximation to } \ln k(T)) \tag{5.13}$$

As a Taylor series expansion, this formula specifies the derivative at a predetermined temperature T_0 rather than at a predetermined retention factor k_0, as we need. However, a monotonic function, $\ln k(T)$, of T can be reversed allowing to express T as a monotonic function, $T(k)$, of k. As a result, if $\ln k(T)$ is a known function of T, then T can be found for any k. Particularly,

$$T_0 = T|_{k=k_0} \tag{5.14}$$

Because $\ln k(T)$ declines with T, its slope and the slope of its linear approximation are negative quantities. It is convenient to express Eq. (5.13) in the form

$$\ln k = \ln k_0 - \frac{T - T_0}{\theta} \quad \text{(linear approximation to } \ln k(T)) \tag{5.15}$$

where a positive quantity θ is an inverse of the absolute value of the slope of the linear approximation in Eqs. (5.13) and (5.15). The quantity θ is measured in units of temperature, and is conceptually analogous to the time constant known in several

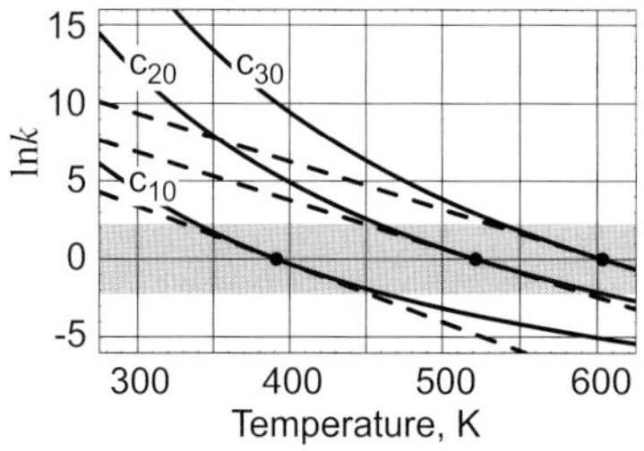

Figure 5.5 Functions $\ln k$ of T, Eq. (5.12), for n-alkanes C_{10}, C_{20}, and C_{30} in Table 5.2 at $\varphi = 0.001$ (solid lines), and their tangents (dashed lines) at $k = 1$. Within the shaded area where $0.1 \leq k \leq 10$, the difference between the solid and the dashed lines for the same solute is small.

technical fields. This suggests that the quantity θ can be called as the *thermal constant* of a solute–liquid interaction in a column. Comparing Eqs. (5.13) and (5.15), one can conclude that

$$\theta = -\left(\left.\frac{\mathrm{d}(\ln k)}{\mathrm{d}T}\right|_{T=T_0}\right)^{-1} = -\left(\left.\left(\frac{1}{k}\frac{\mathrm{d}k}{\mathrm{d}T}\right)\right|_{T=T_0}\right)^{-1} \tag{5.16}$$

Recognizing that T_0 is a unique function, Eq. (5.14), of k_0, the last formula can be extended as

$$\theta = -\left(\left.\frac{\mathrm{d}(\ln k)}{\mathrm{d}T}\right|_{k=k_0}\right)^{-1} = -\left(\left.\frac{1}{k_0}\frac{\mathrm{d}k}{\mathrm{d}T}\right|_{k=k_0}\right)^{-1} = -k_0\left.\frac{\mathrm{d}T}{\mathrm{d}k}\right|_{k=k_0} \tag{5.17}$$

The last formula can be further simplified if k_0 is chosen as $k_0 = 1$ (the solute material is evenly distributed between stationary and mobile phases). The parameters T_0 and θ corresponding to this special condition will be called as [11, 30–32] *characteristic temperature*,

$$T_{\mathrm{char}} = T_0|_{k_0=1} = T|_{k=1} \tag{5.18}$$

and characteristic thermal constant,

$$\theta_{\mathrm{char}} = \theta|_{k_0=1} \tag{5.19}$$

of a solute–liquid interaction in a column. It follows from Eq. (5.17) that

$$\theta_{\mathrm{char}} = \theta|_{k_0=1} = -\left(\left.\frac{\mathrm{d}(\ln k)}{\mathrm{d}T}\right|_{k=1}\right)^{-1} = -\left(\left.\frac{\mathrm{d}k}{\mathrm{d}T}\right|_{k=1}\right)^{-1} = -\left.\frac{\mathrm{d}T}{\mathrm{d}k}\right|_{k=1} \tag{5.20}$$

Equation (5.15) becomes

$$\ln k = -\frac{T - T_{\mathrm{char}}}{\theta_{\mathrm{char}}} \quad \text{(linear approximation to } \ln k(T)\text{)} \tag{5.21}$$

As an example, characteristic parameters of *n*-alkanes of Table 5.2 are listed in Table 5.3.

The parameters T_0 and θ of a linear approximation to $\ln k(T)$, as well as their special cases T_{char} and θ_{char}, have straightforward chromatographic meaning. Furthermore,

Table 5.3 Characteristic temperatures, T_{char}, and characteristic thermal constants, θ_{char}, Eqs. (5.18) and (5.20), of *n*-alkanes in Table 5.2 at $\varphi = 0.001$ (boiling temperatures [33], T_b, are provided for comparison).

	C_{10}	C_{12}	C_{14}	C_{16}	C_{18}	C_{20}	C_{22}	C_{24}	C_{26}	C_{28}	C_{30}	C_{32}
T_{char} (K)	385	414	442	469	492	514	534	554	570	585	595	609
T_{char} (°C)	112	141	169	196	219	241	261	281	297	312	322	336
T_b (°C)	174	216	254	287	316	343	369	391	412	432	450	467
θ_{char} (K or °C)	26.2	27.6	28.9	30	30.7	31.3	31.8	32.2	32.2	32.1	31.7	31.5

typically, like it is in Figure 5.5, $\ln k(T)$ is a smooth (mildly curved) function of T. As a result, $\ln k(T)$ in vicinity of $T = T_{char}$ is almost equal to its linear approximation in Eq. (5.21), and parameters of that approximation well represent function $\ln k(T)$ itself.

Consider parameters θ and θ_{char}. As mentioned earlier, thermal constant, θ, and characteristic thermal constant, θ_{char}, in particular are inversions of absolute value of a slope of a function $\ln k(T)$. Specifically, the quantity θ_{char} is an inverse of the absolute value of the slope of $\ln k(T)$ at $k = 1$ and, therefore, at $T = T_{char}$. As a result, the quantity θ_{char} is approximately equal to an increase in T that reduces $\ln k(T)$ in vicinity of $T = T_{char}$ by one (say, from $\ln k(T) \approx 1.5$ to $\ln k(T + \theta_{char}) \approx 0.5$). Conversely, lowering T by θ_{char} in vicinity of $T = T_{char}$ increases $k(T)$ by one. As for the retention factor, k, itself, raising T by θ_{char} in vicinity of $T = T_{char}$ causes approximately e-fold reduction ($e \approx 2.72$) in $k(T)$ (say, from $k(T) \approx 5$ to $k(T + \theta_{char}) \approx 1.6$). Conversely, lowering T by θ_{char} in vicinity of $T = T_{char}$ makes $k(T)$ approximately e-fold larger. For n-alkanes in Table 5.3, θ_{char} is confined between the values of 26.2 and 31.5 °C. Suppose that $\theta_{char} = 30\,°C$. Then 30 °C increase in T in vicinity of $k = 1$ causes about 2.72-fold reduction in k.

The quantities $1/\theta$ or $1/\theta_{char}$ are absolute values of the slopes of function $\ln k(T)$ at predetermined values of k (at $k = 1$ in the case of θ_{char}). These slopes are measured in the units of inverse temperature (such as 1/°C) and describe the rate of reduction in $\ln k$ due to an increase in T. The quantities $1/\theta$ and $1/\theta_{char}$ also describe a relative decrease in k per unit of increase in T. For example, if $\theta_{char} = 30\,°C$ then $1/\theta_{char} = 0.033/°C$. This means that $\ln k$ gets lower by 0.033 per each 1 °C increase in a column temperature. This also means that k gets lower by a factor of 1.033 per each 1 °C increase in a column temperature. Thus, if $k = 2.000$ at 100 °C, then it falls to 1.935 at 101 °C.

These observations can be summarized as follows:.

Statement 5.1

Thermal constant, θ, of a solute–liquid pair is the temperature increase that causes e-fold ($\approx$2.72-fold) reduction in retention factor, k, in vicinity of its predetermined value, k_0. The quantity $1/\theta$ is the relative reduction in k per unit of increase in T.

Statement 5.2

The characteristic thermal constant, θ_{char}, of a solute–liquid pair is the temperature increase that causes e-fold ($\approx$2.72-fold) reduction in retention factor, k, in vicinity of $k = 1$. The quantity $1/\theta_{char}$ is the relative reduction in k per unit of increase in T.

Similar to the characteristic thermal constant, θ_{char}, is what can be called as *binary thermal constant*, θ_{bin}, – the temperature increment [34–36] that causes twofold reduction in a solute retention factor, k. The quantities θ_{char} and θ_{bin} are related as

$$\theta_{bin} \approx \theta_{char} \ln 2 \approx 0.7\theta_{char} \tag{5.22}$$

The quantity θ_{bin} might be more convenient for quick evaluations. On the other hand, quantity θ_{char} yields simpler mathematical formulae, and it is quantity $1/\theta_{char}$ rather than $1/\theta_{bin}$ that is the rate of the relative decline in k per 1 °C increase in T.

Direct or implicit estimates [34–37] of the values of θ_{bin} are known as $20\,°C \le \theta_{bin} \le 40\,°C$ (Giddings [34, 37]), $\theta_{bin} \approx 20\,°C$ (Fialkov, Gordin *et al.* [35]), and so forth. These estimates are in general agreement with more specific data in Table 5.3 (and Table 5.6). The difference between the estimates of θ_{bin} known from the literature and the values of θ_{bin} obtained from applying Eq. (5.22) to reported data for θ_{char} can be attributed to several factors. First, θ_{char} is measured at $k = 1$. Second, as shown below, there is correlation between T_{char} and θ_{char}. The solutes eluting with $k = 1$ or so at higher temperature tend to have larger θ_{char}. Third, θ_{char} depends on film thickness. All these factors affect both θ_{bin} and θ_{char}, but they were not taken into account in the known estimates of θ_{bin}.

The solute parameters T_0 and T_{char} are the column temperatures at which the solute has a predetermined retention factor, k_0. Thus,

Statement 5.3

The characteristic temperature, T_{char}, of a solute–liquid pair is the temperature at which the solute retention factor is equal to one (the solute is evenly distributed between the liquid and the gas).

During a long heating ramp in a temperature-programmed analysis, all solutes elute with approximately equal retention factors (Chapter 8). Suppose that the heating rate is such that the elution retention factors, k_R, of all peaks are close to unity ($k_R \approx 1$). In this case, characteristic temperature of a solute closely represents its elution temperature. It also means that, generally, the solutes that have higher characteristic temperatures are longer retained in a column, and the solutes elute in the order of increase in their characteristic temperatures.

Actually, at optimal heating rate, retention factors of eluting solutes are closer to two than to unity [32, 38, 39]. This means that, under optimal conditions, elution temperature of each solute is some 20–30 °C lower than its characteristic temperature. Even in this case, characteristic temperature of a solute is a good indicator of a solute retention in the column, its elution temperature, and its elution order during a realistic temperature program.

In isothermal analysis, the solutes elute in the order of increase in retention factors at the actual column temperature, T. Observing the curves in Figure 5.5, one can conclude that the solutes that require higher temperature to obtain $k = 1$ ($\ln k = 0$), generally, are more retained at a specific temperature, T. This means that, in isothermal analysis, the solutes also elute generally in the order of increase in their characteristic temperatures.

It has been emphasized in the previous observations that the solutes elute only generally in the order of increase in their characteristic temperatures. Elution order of solutes with close characteristic temperatures (different by less than 10 °C or so, Chapter 9) depends on relationship between their characteristic thermal constants and the heating rate in temperature-programmed GC or on column temperature in isothermal GC. This means that relationship between characteristic temperatures and characteristic thermal constants of two solutes determines the possibility and the extent of reversal of their elution order under specific conditions of analysis.

All in all, parameters T_0 and θ as well as their special cases T_{char} and θ_{char} have a clear chromatographic meaning.

5.3
Alternative Expressions of Ideal Retention Model

As quantities T_0, θ, T_{char}, and θ_{char} have straightforward chromatographic meaning, it is desirable to use them as parameters of solute–liquid interaction in a column.

The same ideal model of a solute–liquid interaction in a column (ideal retention model in it) can be described in several ways shown in Eqs. (5.7), (5.8), (5.11) and (5.12). Solving together Eqs. (5.11), (5.12), (5.15), and (5.17) allows one to express parameters g_S and T_H of Eqs. (5.11) and (5.12) via parameters T_0 and θ as

$$g_S = \frac{T_0}{\theta} - \ln k_0 \tag{5.23}$$

$$T_H = \frac{T_0^2}{\theta} \tag{5.24}$$

Substitution of these two formulae in Eq. (5.12) yields

$$\ln k = \ln k_0 + \frac{T_0}{\theta}\left(\frac{T_0}{T} - 1\right) \quad \text{(ideal thermodynamic model)} \tag{5.25}$$

The last formula represents the same function k of T as the one in Eqs. (5.7), (5.8), (5.11), and (5.12) except that, in this case, the function is expressed via parameters T_0 and θ instead of parameters $\Delta\mathcal{G}_H$, $\Delta\mathcal{G}_S$, $\mathcal{R}$, and φ in Eqs. (5.7) and (5.8) or parameters g_S and T_H in Eqs. (5.11) and (5.12). If necessary, Eqs. (5.23) and (5.24) could be inversed to express parameters T_0 and θ via parameters g_S and T_H as

$$T_0 = \frac{T_H}{g_S + \ln k_0} \tag{5.26}$$

$$\theta = \frac{T_H}{(g_S + \ln k_0)^2} \tag{5.27}$$

If necessary, these formulae could be used to transform Eq. (5.25) back into the form of Eqs. (5.11) and (5.12) that, by using Eqs. (5.9) and (5.10), could be further transformed back into the conventional form of Eqs. (5.7) and (5.8).

At $k_0 = 1$, parameters T_0 and θ become characteristic parameters T_{char} and θ_{char}, respectively, and the last formulae become

$$g_S = \frac{T_{char}}{\theta_{char}} \tag{5.28}$$

$$T_H = \frac{T_{char}^2}{\theta_{char}} \tag{5.29}$$

$$\ln k = \frac{T_{char}}{\theta_{char}}\left(\frac{T_{char}}{T} - 1\right) \quad \text{(ideal thermodynamic model)} \tag{5.30}$$

$$T_{char} = \frac{T_H}{g_S} \tag{5.31}$$

$$\theta_{char} = \frac{T_H}{g_S^2} \tag{5.32}$$

Equation (5.30) represents the same ideal thermodynamic model of a solute retention as Eqs. (5.7), (5.8), (5.11), (5.12), and (5.25) do, except that, in this case, the model is expressed via characteristic parameters T_{char} and θ_{char} rather than parameters $\Delta\mathcal{G}_H$, $\Delta\mathcal{G}_S$, $\mathcal{R}$, and φ in Eqs. (5.7) and (5.8), parameters g_S and T_H in Eqs. (5.11) and (5.12), and parameters T_0 and θ in Eq. (5.25).

Unlike other expressions of the same model, Eqs. (5.25) and (5.30) describe that model via the pair of chromatographically meaningful parameters T_0 and θ in the case of Eq. (5.25) and via the pair of chromatographically meaningful parameters T_{char} and θ_{char} in the case of Eq. (5.30). This means that previously discussed properties of parameters T_0, θ, T_{char}, and θ_{char} follow directly from Eqs. (5.25) and (5.30). Of these two formulae, the latter has fewer components and, therefore, is simpler than the former. Generally, formulae utilizing characteristic parameters T_{char} and θ_{char} are simpler than the ones utilizing less-specific parameters T_0 and θ. Out of these two parameter pairs, only the pair T_{char} and θ_{char} will be used in the future. Equation (5.30) can be also expressed as

$$k = \exp\left(\frac{T_{char}}{\theta_{char}}\left(\frac{T_{char}}{T} - 1\right)\right) \quad \text{(ideal thermodynamic model)} \tag{5.33}$$

Sometimes, it might be necessary to find a column temperature that yields a predetermined retention factor for a solute with given characteristic parameters. Inversion of Eq. (5.33) yields

$$T = \left(1 - \frac{\theta_{char} \ln k}{T_{char} + \theta_{char} \ln k}\right) T_{char} \quad \text{(ideal thermodynamic model)} \tag{5.34}$$

5.4
Linearized Retention Model

Equations (5.30) and (5.33) express ideal thermodynamic model of a solute retention in a column via chromatographically meaningful characteristic parameters T_{char} and θ_{char}. It makes these formulae easier to interpret than formulae in Eqs. (5.7), (5.8), (5.11), and (5.12), but it does not make them mathematically simpler. All six formulae describe the same model illustrated in Figure 5.5 where $\ln k$ is not a linear function of T and, therefore, k is not an exponent of a linear function of T. This nonlinearity is a source of significant complications in mathematical treatment of temperature programming in GC [9, 13, 40–42] which, in Giddings' words [34], "always leads to rather formidable integrals whose solutions are not easily obtained."

Consider again the curves in Figure 5.5 and their linear approximations at $k = 1$. As mentioned earlier, for the k-values within the range $0.1 \leq k \leq 10$ (shaded area in Figure 5.5), each curve itself and its approximation are close to each other. This means that, in some cases, linear approximation in Eq. (5.21) can be treated as an alternative, linearized model

$$\ln k = -\frac{T - T_{\text{char}}}{\theta_{\text{char}}} \quad \text{(linearized model)} \tag{5.35}$$

$$k = \exp\left(-\frac{T - T_{\text{char}}}{\theta_{\text{char}}}\right) \quad \text{(linearized model)} \tag{5.36}$$

of a solute retention in a column. This model is similar to the linear solvent strength model of gradient elution LC [25, 43–50], where relative composition of a solvent plays the role of temperature in GC. Linearization of function $\ln k(T)$ was also used for computer simulations of solute migration in GC [25–28].

The linearized model of a solute retention provides a simple alternative to ideal thermodynamic model. In many cases, only the linearized model leads to close form solutions. It is shown later in the book that numerical solutions obtained from the linearized model are sufficiently close to numerical solutions obtained from ideal thermodynamic model.

The following were additional empirical factors that influenced acceptance of the linearized model.

As mentioned earlier, the goal of this book is the evaluation of general properties of GC analyses rather than, say, accurate prediction of retention times of some or all peaks. The issues that will be addressed later in the book are the ones that can eventually help to answer such question as how well and fast a column can analyze a test mixture of a given complexity rather than, say, what would be the retention times of certain components of the mixture. Such approach allows for more relaxed requirements to the accuracy of a retention model in favor of its simplicity.

Numerical values of parameters of ideal thermodynamic model gradually change if T changes within a wide temperature range [8, 12–20]. This means that the ideal thermodynamic model (fixed parameters in Eqs. (5.7), (5.8), (5.11), (5.12), (5.25), and (5.30)) gives a good approximation to reality only within a relatively narrow temperature range. And so does the linearized model. This means that, although the ideal thermodynamic model provides better insight into thermodynamics of a solute retention, it is not necessarily more accurate than the linearized one, especially within a moderate temperature range where both models are close to each other.

Furthermore, majority of known numerical values of thermodynamic parameters of each solute–liquid pair were obtained from the measurements of the solute retention in the columns with respective liquid phases [11, 51–72]. The isothermal analyses that were usually employed for such measurements had several limitations. On the one hand, it was difficult to accurately measure very low retentions. On the other hand, measurements involving high retention were time consuming. As a result, the measurements were made at the temperatures yielding more or less moderate retentions of the solutes of interest – exactly in the region where numerical difference between the two models is the smallest. This measurement approach also matches the reality of a solute migration during a typical temperature program where each solute migrates through the major portion of a column at more or less moderate retention (Chapter 8) for which numerical difference between the models is small.

All models based on known experimental parameters are suitable only for qualitative, but not for quantitative evaluations of a column performance at very high retention factors.

5.5 Relations Between Characteristic Parameters

Generally, θ_{char} tends to increase [30] with the increase in T_{char}. In this section, the correlation of θ_{char} with T_{char} is first evaluated at a fixed dimensionless film thickness, and then the film thickness effect on the correlation is considered.

5.5.1 Fixed Dimensionless Film Thickness

Dimensionless film thickness $\varphi = 0.001$ has already been used in several examples. This value of φ is also used below in the first step of evaluation of the correlation of θ_{char} with T_{char}. To a certain degree, the choice of $\varphi = 0.001$ is arbitrary. It is based on the following considerations. The number 0.001 is a simple decimal number. The dimensionless film thickness of 0.001 is more or less in the middle of practically available options. The choice of $\varphi = 0.001$ for the study of relationship between θ_{char} and T_{char} at a fixed φ has no effect on the final results accounting for variable φ.

Temperature of 273.15 K (0 °C) is typically used (together with 1 atm pressure) as a component of the standard temperature and pressure condition for measurement of gas volumes, densities, and so forth. Standard temperature, $T_{st} = 273.15$ K, is a convenient reference temperature in many evaluations.

Let us denote throughout the rest of this chapter T_{Char} and θ_{Char} (capital "C" in the superscript) as

$$T_{Char} = T_{char} \text{ at } \varphi = 0.001, \qquad \theta_{Char} = \theta_{char} \text{ at } \varphi = 0.001 \tag{5.37}$$

It has been shown in the literature [30] that relation between θ_{Char} and T_{Char} follows a trend that can be described as

$$\theta_{Char} = \left(\frac{T_{Char}}{T_{st}}\right)^{0.7} \theta_{Char,st} \tag{5.38}$$

where $\theta_{Char,st}$ is the characteristic thermal constant of a solute whose T_{Char} is equal to T_{st}. In the original report [30], the correlation model in Eq. (5.38) was demonstrated with experimentally found (T_{Char},θ_{Char}) pairs for about 200 solutes and two types of stationary phase polymers [11, 64, 71]. A much larger and more comprehensive database was used for the evaluation described below.

A large database of thermodynamic parameters of solute–liquid interaction in GC is compiled in commercial *Pro ezGC* software [73] covering more than 3000 solutes – each in several liquid polymers widely used for making stationary phases in GC columns [74]. The database was compiled from published thermodynamic data [51–62, 64–66] and from experimental results developed by the software vendor [74]. All the following evaluations come from retention times generated by the Pro ezGC software. The final results were also cross-checked against the previously developed ones [11] that were based on other sources [11, 64, 71].

Figure 5.6 covers 2412 solute–liquid pairs representing interactions of about 2000 solutes with one or two of six widely used liquid polymers, Table 5.4, of different polarity [5, 7, 75] – apolar, intermediate polarity, and polar. It has been found that

Statement 5.4

There exists a single value

$$\theta_{Char,st} = 22\,^{\circ}\text{C} \tag{5.39}$$

of quantity $\theta_{Char,st}$ in Eq. (5.38) – a typical characteristic thermal constant at 273.15 K and $\varphi = 0.001$ ($\beta = 250$) – that is suitable for evaluated (T_{Char},θ_{Char}) pairs in all columns.

Due to Eq. (5.39), Eq. (5.38) can be extended as

$$\theta_{Char} = \left(\frac{T_{Char}}{T_{st}}\right)^{0.7} \theta_{Char,st} = \left(\frac{T_{Char}}{273.15\text{ K}}\right)^{0.7} 22\,^{\circ}\text{C} \tag{5.40}$$

The standard deviation of the departures of actual θ_{Char} values in Figure 5.6 at the same T_{Char} from the value in Eq. (5.38) is about 6.5%. Overwhelming majority of all (T_{Char},θ_{char}) pairs in Figure 5.6 is contained within a stripe, Figure 5.6j, whose absolute width,

$$|\Delta\theta_{Char}| = |\theta_{Char,2} - \theta_{Char,1}| \leq 0.3\theta_{Char}, \quad \text{(at the same } T_{Char}\text{)} \tag{5.41}$$

gradually increases with increase in T_{Char}, so that its relative width, $\Delta\theta_{Char}/\theta_{Char}$, remains the same at any T_{Char}. As an example, characteristic parameters of several pesticides in Figure 5.6e are listed in Table 5.5.

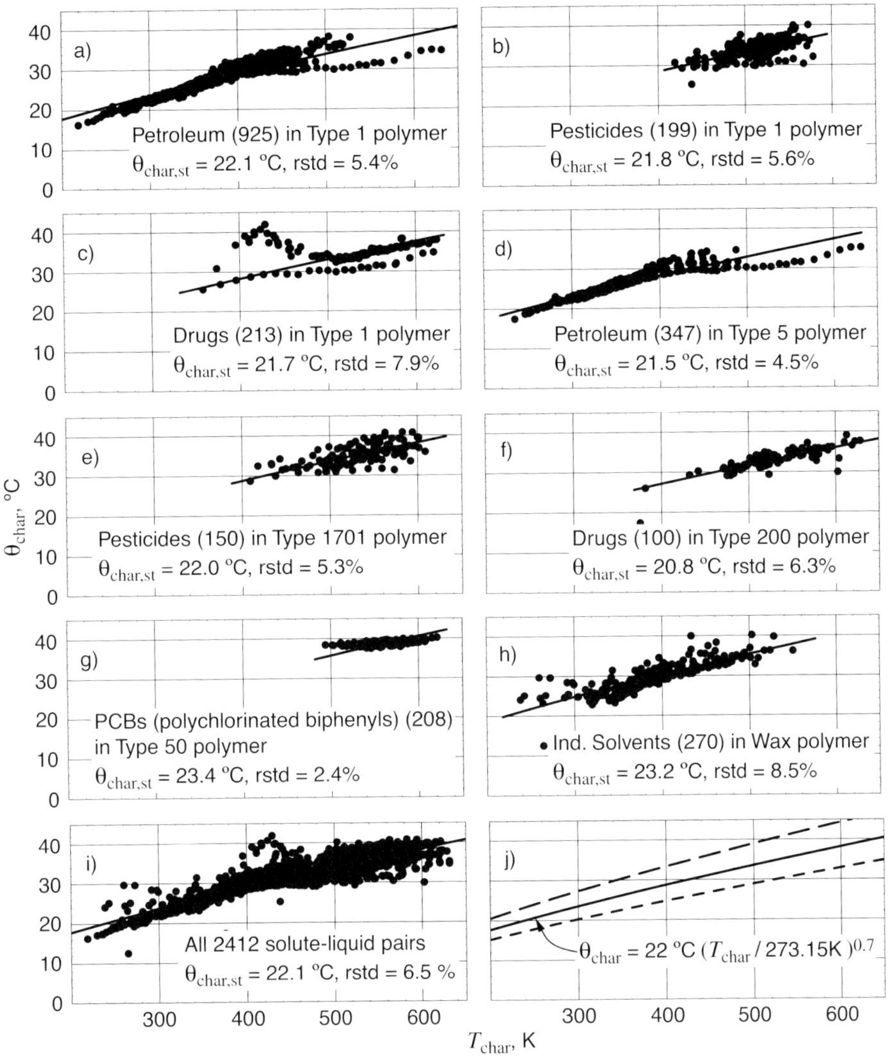

Figure 5.6 (T_{char}, θ_{char}) pairs – the dots in graphs (a)–(i) – for a number (in parentheses) of solutes in liquid polymers listed in Table 5.4. In all cases, $\varphi = 0.001$ ($\beta = 250$). Graph (i) is the summary of the graphs (a)–(h). A solid line in each graph represents the best-fit curve described in Eq. (5.38). Also shown in each collection is the relative standard deviation of departures of actual θ_{char} values from the best-fit curve. The dashed lines in (j) outline about $0.3\theta_{char}$ wide stripe containing overwhelming majority of all (T_{char},θ_{char}) pairs.

Table 5.4 Conventional type codes for some liquid polymers [7, 74, 76–78] listed in the order of increased polarity [7, 74, 78].

Type code	Polymer
1	Dimethyl polysiloxane
5	5% Diphenyl-dimethyl polysiloxane
1701	14% Cyanopropylphenyl-methyl polysiloxane
200	35% Trifluoropropylmethyl polysiloxane
50	50% Diphenyl-dimethyl polysiloxane
Wax	Polyethylene glycol

Table 5.5 Characteristic parameters of a subset of 11 (out of total 150) pesticides, Figure 5.6, in 1701 polymers, Table 5.4, at $\varphi = 0.001$[a].

#	Solute name	T_{char} (K)	θ_{char} (K)	#	Solute name	T_{char} (K)	θ_{char} (K)
35	Prometon	504.87	31.77	46	Atrazine	513.29	32.96
36	Atraton	505.82	32.20	47	Disulfoton	514.58	34.28
37	Diazinon	506.41	31.87	48	Dioxathion	515.16	34.98
38	Eicosane (C_{20})	506.92	30.89	49	Simazine	515.56	33.91
39	Profluralin	508.11	31.59	51	Lindane	518.07	36.75
40	Demeton S	508.67	33.81	52	Pronamide	518.17	32.9
41	BHC-alpha	509.01	36.67	53	Monocrotophos	519.54	32.59
42	Terbufos	509.22	34.26	54	Dichloran	520.74	36.84
43	Dicrotophos	510.98	33.21	55	Heptachlor	523.3	37.82
44	Propazine	511.73	32.48	56	Dichlone	524.57	37.85
45	PCNB	512.37	36.98				

a) The 150 solutes in the original set were numbered, #, in the ascending order of T_{char} starting with # = 1 for the solute with the lowest T_{char}.

5.5.2
Arbitrary Film Thickness

Equations (5.31) and (5.32) allow one to express characteristic parameters, T_{char} and θ_{char}, via thermodynamic parameters, T_H and g_S, whose dependence on dimensionless film thickness, φ, is described in Eqs. (5.9) and (5.10) (as before, lower case "c" in the subscripts of T_{char} and θ_{char} indicates arbitrary film thickness, whereas, as defined in Eq. (5.37), upper case "C" in the subscripts of T_{Char} and θ_{Char} marks the values of T_{char} and θ_{char} at $\varphi = 0.001$). Solving together Eqs. (5.9), (5.10), (5.31), and (5.32) for $\varphi = 0.001$ and for arbitrary φ allows one to express T_{char} and θ_{char} as, Figure 5.7,

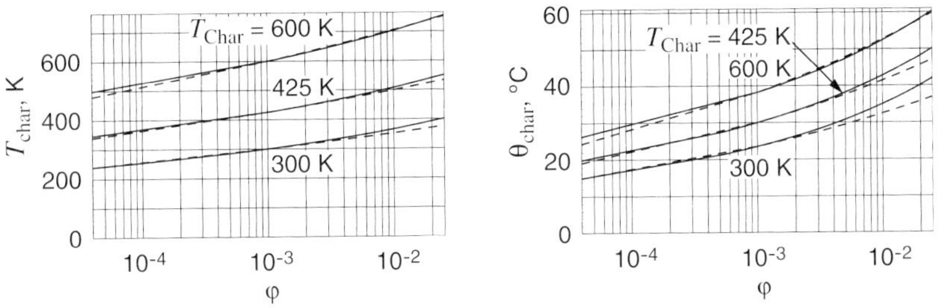

Figure 5.7 Characteristic temperature, T_{char}, and characteristic thermal constant, θ_{char}, as functions of dimensionless film thickness, φ, for the solute–liquid pairs with several values of T_{Char} – the characteristic temperature at $\varphi = 0.001$, Eq. (5.37). Solid lines represent Eqs. (5.42) and (5.43), and dashed lines represent Eqs. (5.44) and (5.45).

$$T_{char} = \frac{T_{Char}}{1-(\theta_{Char}/T_{Char})\ln(10^3\varphi)} \tag{5.42}$$

$$\theta_{char} = \frac{\theta_{Char}}{\left(1-(\theta_{Char}/T_{Char})\cdot\ln(10^3\varphi)\right)^2} \tag{5.43}$$

More suitable for the study of a column performance are simple approximations, Figure 5.7,

$$T_{char} \approx (10^3\varphi)^{0.07}\, T_{Char} \tag{5.44}$$

$$q_{char} \approx (10^3\varphi)^{0.14}\, \theta_{Char} \tag{5.45}$$

to Eqs. (5.42) and (5.43). They show that the parameters T_{char} and θ_{char} of a particular solute only weakly depend on dimensionless film thickness. Thus,

Statement 5.5

Twofold increase in dimensionless film thickness, φ, causes only about 5% increase in a solute characteristic temperature, T_{char}, and about 10% increase in its characteristic thermal constant, θ_{char}.

It can be useful to know how a change in φ affects θ_{char} for a specific T_{char} at different values of φ rather than for a specific solute–liquid pair. Combination of Eqs. (5.40), (5.44), and (5.45) and (5.45) yields, Figure 5.8, Table 5.6,

$$\theta_{char} \approx \theta_{char.st}\left(\frac{T_{char}}{T_{st}}\right)^{0.7} \tag{5.46}$$

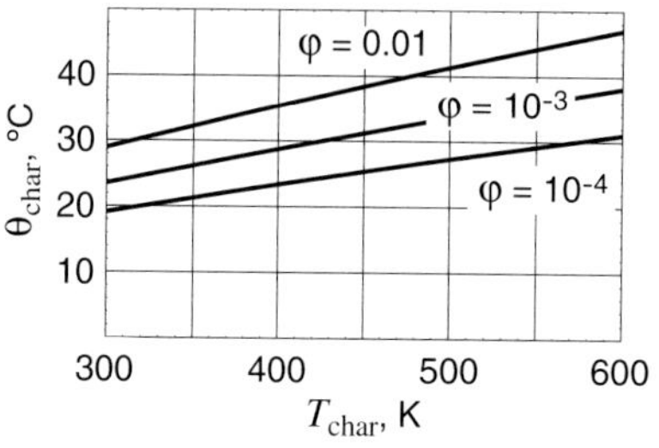

Figure 5.8 Dependence of θ_{char} on T_{char}, Eq. (5.46), for several values of dimensionless film thickness, φ.

where

$$\theta_{char,st} = \theta_{Char,st}(10^3\varphi)^{0.09} = 22\,^\circ C(10^3\varphi)^{0.09} \tag{5.47}$$

Similarly, Eq. (5.41) yields

$$|\Delta\theta_{char}| = |\theta_{char,2} - \theta_{char,1}| \leq 0.3\theta_{char} \quad \text{at fixed } T_{char} \tag{5.48}$$

In Eq. (5.45), the quantity θ_{char} is proportional to $(10^3\varphi)^{0.14}$, whereas, in Eq. (5.47), the quantity $\theta_{char,st}$ is proportional to $(10^3\varphi)^{0.09}$. This is because the quantity θ_{char} in Eq. (5.45) represents a specific solute at an arbitrary φ, whereas the quantity $\theta_{char,st}$ in Eq. (5.47) represents all solutes whose T_{char} is equal to 273.15 K at an arbitrary φ.

Example 5.1

Consider two solutes, "a" and "b", and two columns, "1" and "2." The columns have the same stationary phase film material with dimensionless thickness $\varphi_1 = 0.001$ and $\varphi_2 = 0.01$. Let the characteristic parameters of solutes "a" and "b" in column "1" be: $T_{Char,a} = T_{st} = 273.15$ K, $\theta_{Char,a} = \theta_{Char,st} = 22$ K, $T_{Char,b} = 232.49$ K, and $\theta_{Char,b} = 19.65$ K. Let us also assume that, for each solute, relationship between its θ_{Char} and T_{Char} follows the trend described in Eq. (5.40). According to Eqs. (5.44) and (5.45), characteristic parameters of the solutes in column "2" are: $T_{char,a} \approx 320.92$ K, $\theta_{char,a} \approx 30.37$ K, $T_{char,b} \approx 273.15$ K, and $\theta_{char,b} \approx 27.13$ K. Notice that, in column "2" ($\varphi_2 = 0.01$), $T_{char,a}$ and $\theta_{char,a}$ are not equal to the standard values of 273.15 and 22 K as they are in column "1." On the other hand, $T_{char,b} = T_{st}$. Therefore, $\theta_{char,b} = \theta_{char,st}$. The quantity $\theta_{char,st}$ relates to $\theta_{Char,st}$ as $\theta_{char,st}/\theta_{Char,st} = 27.13\,\text{K}/(22\,\text{K}) \approx 1.23 \approx 10^{0.09}$ as described by Eq. (5.47).

Example 5.2

The following is an estimation of a combined effect of variation of φ from 0.0001 to 0.01 and T_{char} from 300 to 600 K. According to Eqs. (5.46) and (5.47), at $\varphi = 0.0001$ and $T_{char} = 300$ K, $\theta_{char} = 19$ K and $1/\theta_{char} = 0.052$/K. At $\varphi = 0.01$ and $T_{char} = 600$ K, $\theta_{char} = 48$ K and $1/\theta_{char} = 0.021$/K. □

Table 5.6 Characteristic and binary thermal constants, θ_{char} and θ_{bin}, Eqs. (5.46) and (5.22), and the rate, $1/\theta_{char}$, of percent change in a solute retention factor, k, per 1 °C change in a column temperature at $\varphi = 0.001$.

T_{char} (°C)	25	50	75	100	125	150	175	200	225	250	275	300	Typical
θ_{char} (°C)	23.4	24.7	26.1	27.4	28.6	29.9	31.1	32.3	33.5	34.7	35.8	37	30
θ_{bin} (°C)	16.2	17.2	18.1	19	19.9	20.7	21.6	22.4	23.2	24	24.8	25.6	20
$1/\theta_{char}$ (%/°C)	4.3	4	3.8	3.7	3.5	3.3	3.2	3.1	3	2.9	2.8	2.7	3.3

As mentioned earlier, the quantity $1/\theta_{char}$ is the rate of a relative reduction in retention factor, k, per 1 °C increase in a column temperature, T (Statement 5.2). Within a wide range of dimensionless film thickness ($0.0001 \le \varphi \le 0.01$) and a wide range of solutes ($25\,°C \le T_{char} \le 325\,°C$), relative reduction in k per 1 °C increase in T ranges from 2 to 5%. The thinner is the film and the lower is the solute characteristic temperature, the more sensitive is the solute retention factor, k, to a change in T.

Solving some problems of a column performance might require inversion of a thermodynamic model in order to find T for a given k. Such inversion of ideal thermodynamic model in Eq. (5.33) is show in Eq. (5.34). Similar inversion of linearized model in Eq. (5.36) is much simpler. However, in both cases, T is a function of not only a solute retention factor, k, but also of its characteristic temperature, T_{char}, and characteristic thermal constant, θ_{char}. Strong correlation of θ_{char} with T_{char} allows one to remove parameter θ_{char} from a simple approximation of T as a function of T_{char} and k.

Let us express Eq. (5.46) as

$$\theta_{char} = \frac{T_{st}^{0.3} T_{char}^{0.7}}{\Theta_{st}} \approx 0.08(10^3\varphi)^{0.09} T_{char}^{0.7} T_{st}^{0.3} \tag{5.49}$$

where

$$\Theta_{st} = \frac{T_{st}}{\theta_{char.st}} = \frac{\Theta_{St}}{(10^3\varphi)^{0.09}}, \qquad \Theta_{St} = \frac{T_{st}}{\theta_{Char.st}} = \frac{273.15\ \text{K}}{22\ \text{K}} \approx 12.4 \tag{5.50}$$

The parameter Θ_{St} is a fixed dimensionless quantity. Substitution of Eq. (5.49) in Eq. (5.36) yields

$$\ln k = \frac{(1 - T/T_{char})(T_{char}/T_{st})^{0.3}\Theta_{St}}{(\varphi/\varphi_{ord})^{0.09}} \tag{5.51}$$

which is a week function of T_{char}/T_{st} and φ. If k is not a very small quantity, then the ratio T/T_{char} can be approximated as, Figure 5.9,

$$\frac{T}{T_{char}} \approx k^{-0.08} \qquad (k \ge 0.1) \tag{5.52}$$

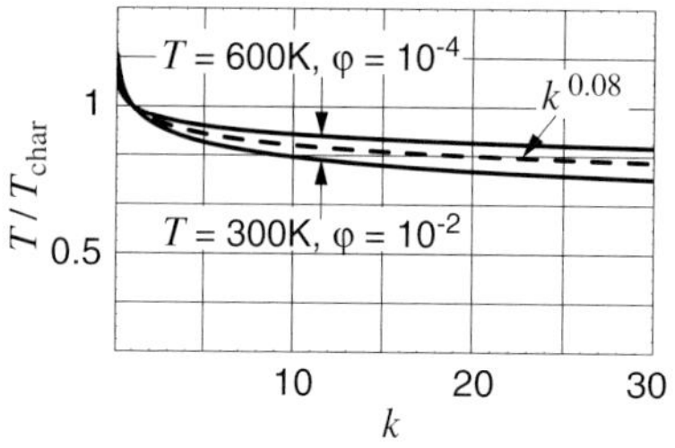

Figure 5.9 The ratio T/T_{char} as a function of solute retention factor, *k*. Solid lines represent lower (at $T=300\,°C$ and $\varphi=0.01$) and upper (at $T=600\,°C$ and $\varphi=0.0001$) bounds of T/T_{char} in Eq. (5.51). Dashed line represents Eq. (5.52).

Equation (5.52) results from the linearized model of a solute–liquid interaction. In the case of the ideal thermodynamic model, $T/T_{Char} \approx k^{-0.06}$. While this approximation is different from Eq. (5.52), the difference is relatively small. Thus, at $k=10$, the difference is about 4.5%.

Additional details on the accuracy of Eq. (5.52) can be found in Appendix 5.A. Too small and too large values of *k* (compared to $k=1$) tend to increase the error of approximation by Eq. (5.52). The lower bound, $k \geq 0.1$ in Eq. (5.52) limits potential error at very small *k*. However, there is no upper bound for *k* in Eq. (5.52). This is because, at large *k*, not only the accuracy of the approximation in Eq. (5.52), but, as mentioned earlier, the accuracy of the basic ideal thermodynamic model itself is in question. At large *k*, the model and its approximations are suitable only for qualitative evaluations.

It should be also mentioned that, as follows from the error analysis in Appendix 5.A, Eq. (5.52) is not suitable for evaluation of *k* as a function of *T* and T_{char}. Indeed, according to Eq. (5.52), *k* would be proportional to $(T_{Char}/T)^{12.5}$. The large power (12.5) in this expression can be a source of unacceptably large errors.

5.5.3
Generic Solutes

As described earlier (Chapter 2), this book is concerned with general properties of GC analyses (such as the effect of the heating rate on retention times, elution order, widths, and other parameters of all peaks). This is different from specific properties concerned with prediction of retention times of particular solutes and with other similar specifics.

The fact that the interaction of a solute with a given stationary phase film can be described by two independent parameters – T_{char} and θ_{char}, or $\Delta\mathscr{G}_H$ and $\Delta\mathscr{G}_S$ – suggests that the interaction can be viewed as two-dimensional property. It is convenient to divide this property into general and specific ones.

Although parameters T_{char} and θ_{char} of a particular solute in a particular liquid stationary phase are independent from each other, there is a general trend (Figure 5.6) in their relationship described in Eq. (5.46). One can introduce a concept of generic solutes – the ones for which θ_{char} exactly follows the general trend in Eq. (5.46). Since the characteristic thermal constant (θ_{char}) of a generic solute with a given characteristic temperature (T_{char}) can be found from Eq. (5.46), its interaction with liquid stationary phase is a one-dimensional property completely described by T_{char}.

In reality, there are probably very few, if any, actual solutes that interact with stationary phases as generic ones. The concept of generic solutes is simply a conceptual approach to the study of general trends. On the other hand, parameters θ_{char} of overwhelming majority of real solutes fall within a relatively narrow range (Figure 5.6j) around θ_{char} predicted from Eq. (5.46). In that sense, almost all solutes behave more or less as generic ones.

Departure of parameter θ_{char} of any particular solute from its generic value can be viewed as specific property of the solute. This property, however, is outside of the scope of this book, although some aspects of it will be considered in the study of the factors affecting peak spacing and reversal of peak elution order.

5.5.4
Characteristic Temperatures of *n*-Alkanes

In a typical GC analysis, components of a test mixture generally elute in the order of increase in their characteristic temperatures (Chapter 8). They also generally elute in the order of increase in molecular weights. Thus, as shown in Table 5.3, *n*-alkanes with higher carbon numbers have higher characteristic temperatures and elute later than their lighter classmates. Mathematical description of these observations is necessary for evaluation of the effect of method parameters (column dimensions, gas flow rate, and so forth) on the peak broadening (Chapter 10). Unfortunately, accurate prediction of the solute elution parameters is an enormously complex task. Under these circumstances, even a very approximate description of correlation of a solute retention with its molecular weight is better than nothing as long as the description eventually allows one to establish reasonably accurate relationship between method parameters and a column performance.

There is enough experimental data to outline empirical relationship between carbon numbers of *n*-alkanes and their characteristic temperatures. This is the task for this section. Later, the relationships found here will be used for evaluation of diffusion of *n*-alkanes in carrier gases (Chapter 6) which can be used for finding the carrier gas optimal flow rates (Chapter 10). Validity of this approach has experimental confirmation. It explains experimentally found relationship between elution temperatures of *n*-alkanes and the rates of their diffusion in a gas [79] (see additional comments at the end of Chapter 6). A major manufacturer of GC instrumentation began to recommend the flow rates found from this approach [80] for the products introduced since 1994 [81–84]. Successful adoption of these recommendations in a large number of applications is the evidence of validity of the underlying approach.

Evaluation, Figure 5.10, of available experimental data [73] for retention of *n*-alkanes in apolar (type 1), moderately polar (type 5 and 1701), and polar (wax type) liquid polymers listed in Table 5.4 shows that characteristic temperature, T_{Char}, of *n*-alkane as a function of its carbon number, i, can be expressed by the following empirical formulae:

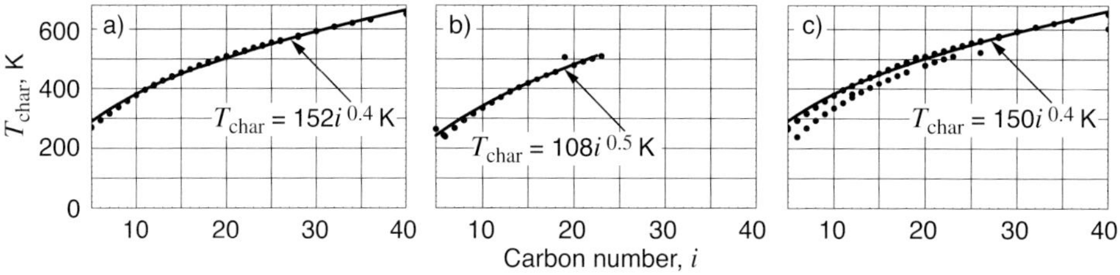

Figure 5.10 Relationship between characteristic temperature, T_{char}, and carbon number, i, of n-alkanes at $\varphi = 0.001$. (a) C_5–C_{26} and all even-numbered n-alkanes from C_{28} to C_{40} in type 1 and 5 polymers, Table 5.4, plus all even-numbered n-alkanes from C_{12} to C_{28} in type 1701 polymers (65 solute–liquid pairs); (b) C_5–C_{23} in wax-type polymers; and (c) all of the above plus C_{16}, C_{26}, and C_{40} in type 200 polymers (total of 87 solute–liquid pairs).

$$T_{Char} \approx 152i^{0.4}\ \text{K} \approx 0.556 T_{st} i^{0.4}\ (n\text{-alkanes in type 1, 5 and 1701 polymers}) \tag{5.53}$$

$$T_{Char} \approx 108i^{0.5}\ \text{K} \approx 0.395 T_{st} i^{0.5}\ (n\text{-alkanes in wax-type polymer}) \tag{5.54}$$

Figure 5.10c shows that the spread of T_{Char} values for the same n-alkane in different liquid polymers can be relatively large especially for the lightest n-alkanes. Thus, T_{Char} of C_5 in wax-type polymer is almost 50 K lower than its T_{Char} in type 1 polymer. However, the spread in majority of n-alkanes is much narrower compared with the range of T_{Char} values for all n-alkanes in Figure 5.10c. As a source of temperatures for evaluation of diffusion of n-alkanes in carrier gases (Chapter 6), Eq. (5.53) reasonably well represents the trend of dependence of T_{Char} on i for all n-alkanes in Figure 5.10c in all polymers shown in Figure 5.10. In the future, Eq. (5.53) will be treated as an empirical function $T_{Char}(i)$ for all regular n-alkanes (C_5 to C_{40}).

As an example, the T_{Char} values calculated from Eq. (5.53) are shown in Table 5.7. While there is a difference between T_{Char} values in Tables 5.3 and 5.7, the difference is relatively small. This is especially important because the data in the tables come from different sources [11, 73].

Solving Eq. (5.53) for i, followed by substitution of T_{Char} from Eq. (5.44) allows one to reconstruct carbon numbers, i, of n-alkanes from their characteristic temperatures, T_{char}, in a column with arbitrary dimensionless film thickness, φ, as

Table 5.7 Characteristic temperatures, T_{char}, of n-alkanes at $\phi = 0.001$ calculated from Eq. (5.53).

	C_5	C_6	C_8	C_{10}	C_{12}	C_{14}	C_{16}	C_{18}	C_{20}	C_{24}	C_{28}	C_{32}	C_{36}	C_{40}
T_{char} (K)	289	311	349	382	411	437	461	483	504	542	576	608	637	665
T_{char} (°C)	16	38	76	109	138	164	188	210	231	269	303	335	364	392

$$i \approx \frac{4.34}{(10^3\varphi)^{0.18}}\left(\frac{T_{\text{char}}}{T_{\text{st}}}\right)^{2.5} \quad (n\text{-alkanes}) \tag{5.55}$$

Due to Eq. (5.52), the last formula can also be transformed into a function

$$i \approx \frac{4.34k^{0.2}}{(10^3\varphi)^{0.18}}\left(\frac{T}{T_{\text{st}}}\right)^{2.5} \quad (n\text{-alkanes, } k \geq 0.1) \tag{5.56}$$

of elution temperature, T, and retention factor, k, of n-alkane.

Appendix 5.A Relative Errors in Power Functions

Approximation

$$y(x) \approx Ax^a \tag{A1}$$

of a complex function $y(x)$ by a power function Ax^a is an attractive way of simplifying complex problems whenever such approach is proper. How proper is this approach might depend on the approximation error that, in turn, depends on the sensitivity of function Ax^a to changes in its power parameter, a.

It follows directly from differentiation of function Ax^a by parameter a that a small *relative change*, $\delta(x^a)$, in Ax^a depends on a small relative change, δa, in a as

$$\delta(Ax^a) = a \ln x\, \delta a \tag{A2}$$

where

$$\delta(x^a) = \frac{d(Ax^a)}{Ax^a}, \qquad \delta a = \frac{da}{a} \tag{A3}$$

Equation (A2) shows that the smaller is the value of a, the lower is the effect of its relative change on the relative change in Ax^a. Thus, at $a = 0.1$ and x close to e (2.718), 10% change in a causes 1% change in Ax^a. Large values of x increase the sensitivity of Ax^a to changes in a. Thus, at $a = 0.1$ and $x = 39$, 10% change in a causes 3.4% change in Ax^a.

References

1 Ettre, L.S. (1993) *Pure Appl. Chem.*, **65**, 819–872.
2 Purnell, J.H. (1962) *Gas Chromatography*, John Wiley & Sons, Inc., New York.
3 Littlewood, A.B. (1970) *Gas Chromatography: Principles, Techniques, and Applications*, 2nd edn, Academic Press, New York.
4 Lee, M.L., Yang, F.J., and Bartle, K.D. (1984) *Open Tubular Gas Chromatography*, John Wiley & Sons, Inc., New York.
5 Guiochon, G. and Guillemin, C.L. (1988) *Quantitative Gas Chromatography for Laboratory Analysis and On-Line Control*, Elsevier, Amsterdam.

6 Giddings, J.C. (1991) *Unified Separation Science*, John Wiley & Sons, Inc., New York.
7 Jennings, W., Mittlefehldt, E., and Stremple, P. (1997) *Analytical Gas Chromatography*, 2nd edn, Academic Press, San Diego.
8 González, F.R. (2002) *J. Chromatogr. A*, **942**, 211–221.
9 Harris, W.E. and Habgood, H.W. (1966) *Programmed Temperature Gas Chromatography*, John Wiley & Sons, Inc., New York.
10 Moore, W.J. (1972) *Physical Chemistry*, 4th edn, Prentice-Hall, Englewood Cliffs, NJ.
11 Blumberg, L.M. and Klee, M.S. (2000) *Anal. Chem.*, **72**, 4080–4089.
12 Ambrose, D. and Purnell, J.H. (1958) in *Gas Chromatography 1958* (ed. D.H. Desty), Academic Press, New York, pp. 369–372.
13 Giddings, J.C. (1962) in *Gas Chromatography* (eds N. Brenner, J.E. Callen, and M.D. Weiss), Academic Press, New York, pp. 57–77.
14 Martire, D.E. (1989) *J. Chromatogr.*, **471**, 71–80.
15 Vezzani, S., Moretti, P., and Castello, G. (1994) *J. Chromatogr.*, **677**, 331–343.
16 Castello, G., Vezzani, S., and Moretti, P. (1996) *J. Chromatogr. A*, **742**, 151–160.
17 Tudor, E. (1997) *J. Chromatogr. A*, **779**, 287–297.
18 González, F.R. and Nardillo, A.M. (1997) *J. Chromatogr. A*, **779**, 263–274.
19 Beens, J., Tijssen, R., and Blomberg, J. (1998) *J. Chromatogr. A*, **822**, 233–251.
20 Héberger, K., Görgényi, M., and Kowalska, T. (2002) *J. Chromatogr. A*, **973**, 135–142.
21 Knox, J.H. and Saleem, M. (1969) *J. Chromatogr. Sci.*, **7**, 614–622.
22 Hyver, K.J. (1989) Chapter 1, in *High Resolution Gas Chromatography*, 3rd edn (ed. K.J. Hyver), Hewlett-Packard Co., USA.
23 Ettre, L.S. (1984) *Chromatographia*, **18**, 477–488.
24 Ravindranath, B. (1989) *Principles and Practices of Chromatography*, Ellis Horwood Limited, Chichester, UK.
25 Bautz, D.E., Dolan, J.W., Raddatz, L.R., and Snyder, L.R. (1990) *Anal. Chem.*, **62**, 1560–1567.
26 Bautz, D.E., Dolan, J.W., and Snyder, L.R. (1991) *J. Chromatogr.*, **541**, 1–19.
27 Dolan, J.W., Snyder, L.R., and Bautz, D.E. (1991) *J. Chromatogr.*, **541**, 21–34.
28 Abbay, G.N., Barry, E.F., Leepipatpiboon, S., Ramstad, T., Roman, M.C., Siergiej, R.W., Snyder, L.R., and Winniford, W.L. (1991) *LC-GC*, **9**, 100–114.
29 Fóti, G. and Kováts, E. (2000) *J. High Resolut. Chromatogr.*, **23**, 119–126.
30 Blumberg, L.M. and Klee, M.S. (2001) *Anal. Chem.*, **73**, 684–685.
31 Blumberg, L.M. and Klee, M.S. (2001) *J. Chromatogr. A*, **918/1**, 113–120.
32 Blumberg, L.M. and Klee, M.S. (2001) *J. Chromatogr. A*, **933**, 13–26.
33 Weast, R.C., Astle, M.J., and Beyer, W.H. (1988) *CRC Handbook of Chemistry and Physics*, 69th edn, CRC Press, Boca Raton, FL.
34 Giddings, J.C. (1962) *J. Chem. Educ.*, **39**, 569–573.
35 Fialkov, A.B., Gordin, A., and Amirav, A. (2003) *J. Chromatogr. A*, **991**, 217–240.
36 Grushka, E. (1970) *Anal. Chem.*, **42**, 1142–1147.
37 Pauschmann, H.H.(April 1998) Letter to L. M. Blumberg.
38 Blumberg, L.M. and Klee, M.S. (2000) *J. Micro. Sep.*, **12**, 508–514.
39 Blumberg, L.M., David, F., Klee, M.S., and Sandra, P. (2008) *J. Chromatogr. A*, **1188**, 2–16.
40 Dal Nogare, S. and Langlois, W.E. (1960) *Anal. Chem.*, **32**, 767–770.
41 Giddings, J.C. (1960) *J. Chromatogr.*, **4**, 11–20.
42 Dal Nogare, S. and Juvet, R.S. (1962) *Gas-Liquid Chromatography. Theory and Practice*, John Wiley & Sons, Inc., New York.
43 Snyder, L.R. (1964) *J. Chromatogr.*, **13**, 415–434.
44 Snyder, L.R. and Saunders, D.L. (1969) *J. Chromatogr. Sci.*, **7**, 195–208.
45 Scott, R.P.W. and Kucera, P. (1973) *Anal. Chem.*, **45**, 749–754.
46 Snyder, L.R., Dolan, J.W., and Gant, J.R. (1979) *J. Chromatogr.*, **165**, 3–30.
47 Snyder, L.R. (1992) *Chromatography, 5th Edition: Fundamentals and Applications of Chromatography and Related Differential Migration Methods. Part A: Fundamentals*

and Techniques (ed. E. Heftmann), Elsevier, Amsterdam, pp. A1–A68.

48 Kaliszan, R., Bączek, T., Buciński, A., Buszewski, B., and Sztupecka, M. (2003) *J. Sep. Sci.*, **26**, 271–282.

49 Hao, W., Zhang, X., and Hou, K. (2006) *Anal. Chem.*, **78**, 7828–7840.

50 Snyder, L.R. and Dolan, J.W. (2006) *High-Performance Gradient Elution: Practical Application of the Linear-Solvent-Strength Model*, John Wiley & Sons, Inc., Hoboken, NJ.

51 Lee, M.L., Vassilaros, D.L., White, C.M., and Novotny, M. (1979) *Anal. Chem.*, **51**, 768–774.

52 Bermejo, J., Blanco, C.G., Diez, M.A., and Guíllén, M.D. (1987) *J. High Resolut. Chromatogr.*, **10**, 461–463.

53 Vassilaros, D.L., Kong, R.C., Later, D.W., and Lee, M.L. (1982) *J. Chromatogr.*, **252**, 1–20.

54 Anderson, W.H. and Stafford, D.T. (1983) *J. High Resolut. Chromatogr.*, **6**, 247–254.

55 Lubeck, A.J. and Sutton, D.L. (1983) *J. High Resolut. Chromatogr.*, **6**, 328–332.

56 Lubeck, A.J. and Sutton, D.L. (1984) *J. High Resolut. Chromatogr.*, **7**, 542–544.

57 Hayes, P.C. Jr. and Pitzer, E.W. (1985) *J. High Resolut. Chromatogr.*, **8**, 230–242.

58 Lora-Tamayo, C., Rams, M.A., and Chacon, J.M.R. (1986) *J. Chromatogr.*, **374**, 73–85.

59 Rostad, C.E. and Pereira, W.E. (1986) *J. High Resolut. Chromatogr.*, **9**, 328–334.

60 Weber, L. (1986) *J. High Resolut. Chromatogr.*, **9**, 446–451.

61 Premecz, J.E. and Ford, M.E. (1987) *J. Chromatogr.*, **388**, 23–35.

62 Saxton, W.L. (1987) *J. Chromatogr.*, **393**, 175–194.

63 Dose, E.V. (1987) *Anal. Chem.*, **59**, 2414–2419.

64 Laub, R.J. and Purnell, J.H. (1988) *J. High Resolut. Chromatogr.*, **11**, 649–660.

65 Phillips, A.M., Logan, B.K., and Stafford, D.T. (1990) *J. High Resolut. Chromatogr.*, **13**, 754–758.

66 White, C.M., Hackett, J., Anderson, R.R., Kail, S., and Spoke, P.S. (1992) *J. High Resolut. Chromatogr.*, **15**, 105–120.

67 Bruno, T.J. and Caciari, M. (1994) *J. Chromatogr. A*, **672**, 149–158.

68 Bruno, T.J. and Caciari, M. (1994) *J. Chromatogr. A*, **679**, 123–132.

69 Bruno, T.J. and Caciari, M. (1994) *J. Chromatogr. A*, **686**, 245–251.

70 Bruno, T.J., Wertz, K.H., and Caciari, M. (1995) *J. Chromatogr. A*, **708**, 293–302.

71 Snijders, H., Janssen, H.-G., and Cramers, C.A. (1995) *J. Chromatogr.*, **718**, 339–355.

72 Bruno, T.J. and Wertz, K.H. (1996) *J. Chromatogr. A*, **736**, 175–184.

73 Analytical Innovations Inc . (2002) Pro EzGC, version 2.20 for Windows, Distributed by Restek Corp. (cat #21487), Bellefonte, PA.

74 Restek Corporation (2003) *2003 Chromatography Products*, Restek Corp., USA.

75 Rohrschneider, L. (2001) *J. Sep. Sci.*, **24**, 3–9.

76 Agilent Technologies (2002) *Chromatography and Spectroscopy Supplies: Reference Guide 2002–2003*, Agilent Technologies, Inc., USA.

77 Alltech Associates Inc . (2003) *Alltech Chromatography Sourcebook*, Alltech Associates Inc., USA.

78 Supelco (1999) *Chromatography Products for Analysis and Purification*, Supelco, USA.

79 Blumberg, L.M., Wilson, W.H., and Klee, M.S. (1999) *J. Chromatogr. A*, **842/1-2**, 15–28.

80 Blumberg, L.M. (1999) *J. High Resolut. Chromatogr.*, **22**, 403–413.

81 Hewlett-Packard Company (1995) HP 6890 Series Gas Chromatograph - Operating Manual (Second Edition, HP Part Number G1530-90310), Hewlett-Packard Company, Little Falls, Wilmington.

82 Agilent Technologies (1998) GC Method Translation Freeware, version 2.0.a.c, http://www.chem.agilent.com/en-US/Support/Downloads/Utilities/Pages/GcMethodTranslation.aspx Agilent Technologies, Inc., Wilmington, DE.

83 Agilent Technologies (2004) 6850 Series Control Module User Information (Part No. G2629-90329), Agilent Technologies.

84 Agilent Technologies (2007) Agilent 7890A Gas Chromatograph Advanced User Guide (Part No. G3430-90015), Agilent Technologies.

6
Molecular Properties of Ideal Gas

Unlike the stationary phase, *carrier gas* does not affect the potential separation performance of GC analysis. However, it significantly affects the speed of analysis and other operational parameters such as optimal flow rate of a gas, pressure, and so forth. These parameters directly depend on two molecular properties of carrier gas – its viscosity and diffusivity – also known as the gas *transport properties*.

The diffusivity [1, 2] (diffusion coefficient [1–8], coefficient of diffusion [9], and diffusion constant [10]) is measured in units or length^2/time (such as cm^2/s) and represents the rate of the transport of mass of one material into another along the material concentration gradient. The viscosity is measured in units or pressure × time (such as μPa s) and represents the resistance to transport of a mechanical momentum along its gradient. A detailed analysis of these properties can be found in textbooks [1, 5, 7, 8, 11] and in other sources [12–14]. This chapter is concerned with the evaluation of numerical values of the transport properties of a gas and with their dependence on each other and on other known molecular properties (molecular weight, average molecular speed, and so forth) of a gas [1, 5, 7, 13, 15].

Numerical data obtained from simple theories for some molecular properties of a gas can be substantially different from the experimental data. Simpler and more accurate results can be obtained from empirical formulae [13, 14, 16–25] that are typically used for the evaluation of operational parameters in GC. On the other hand, the empirical formulae have their own shortcomings. While providing sufficiently accurate numerical data, they typically treat the diffusivity and the viscosity as two independent properties, ignoring their mutual interdependence, and, occasionally, their dependence on other molecular properties of the gas. As a result, in many cases, the studies of the gas-related GC parameters do not go far enough in revealing the dependence of GC performance on the gas properties.

In this chapter, the empirical formulae for the viscosity and the diffusivity are reconciled with each other and with other molecular properties of gases. To achieve this, a rather complex dependence of a solute diffusivity in a gas on the solute and on the gas properties was reduced to a product of gas- and solute-independent parameters. Previously, this approach has been used for solving several problems in fast GC [26–29] and in comprehensive GC × GC [30–32].

Temperature-Programmed Gas Chromatography. Leonid M. Blumberg

ISBN: 978-3-527-32642-6

6.1 Theory

As mentioned earlier, the set of parameters

$$T_{st} = 273.15\ \text{K} \tag{6.1}$$

$$p_{st} = 1\ \text{atm} \tag{6.2}$$

is sometimes treated as standard temperature and pressure (STP) condition for measurement of gas volumes, densities, and so forth, and can be used as the reference temperature and pressure in many evaluations.

Under regular conditions (Chapter 2), a carrier gas – helium, hydrogen, nitrogen, and argon – behaves almost as an ideal gas [1, 5, 14, 18, 23, 33, 34] (Appendix 6.A.1); its state is governed by the ideal gas law, its viscosity is independent of pressure, and diffusivity of any solute in the gas is inversely proportional to the pressure.

Let m, M, p, T, and V be, respectively, the total mass, the molar mass, the pressure, the (absolute) temperature, and the volume of a gas in a closed container. The state of ideal gas can be described by the ideal gas law

$$pV = \frac{m\mathscr{R}T}{M} \tag{6.3}$$

where $\mathscr{R} = 8.31447\ \text{J K}^{-1}\ \text{mol}^{-1}$ is the molar gas constant. For a pure gas (only the helium, only the hydrogen, and so forth), the molar mass, M, expressed in units of g/mol is numerically equal to the gas molecular weight.

The number, $\mathscr{A} = 6.02214 \times 10^{23}/\text{mol}$, of molecules in one mole of a gas is the Avogadro's number. At $m = M$ mol, $p = p_{st}$, and $T = T_{st}$, Eq. (6.3) yields for the molar volume, $\mathscr{V}_{st}$, of ideal gas at standard pressure and temperature:

$$\mathscr{V}_{st} = 22.414 \times 10^{-3}\ \text{m}^3/\text{mol} = 22.41410\ \text{L/mol} \tag{6.4}$$

It follows directly from Eq. (6.3) that the density, $\varrho = m/V$, of ideal gas at an arbitrary pressure and temperature can be found as

$$\varrho = \frac{Mp}{\mathscr{R}T} \tag{6.5}$$

Properties of random motions of a gas molecules can be expressed via their average molecular speed (an average of directionless velocities of the molecules), υ; the mean free path, λ; or the mean time between collisions, t_{mol}. For a pure gas, these quantities can be found as [1, 5, 15] (see Table 6.1)

$$\upsilon = \sqrt{\frac{8\mathscr{R}T}{\pi M}}, \qquad \lambda = \frac{\mathscr{R}T}{\sqrt{2}\pi\mathscr{A}p\phi^2}, \qquad t_{mol} = \frac{5\sqrt{\pi M\mathscr{R}T}}{64\mathscr{A}p\phi^2} \tag{6.6}$$

Table 6.1 Molecular properties of gases at 1 atm.

Gas, molecular weight	**He, $M = 4.003$**				**H_2, $M = 2.016$**				**N_2, $M = 28.01$**				**Ar, $M = 39.95$**			
Temperature (°C)	0	25	150	300	0	25	150	300	0	25	150	300	0	25	150	300
Viscosity, $\eta \sim T^{\xi}$ (μPa s), Eq. (6.20), Table 6.3	18.7	19.8	25.2	31.1	8.4	8.9	11.3	14.0	16.8	17.9	23.0	28.5	21.4	22.8	29.6	37.2
Collision diameter, $\phi \sim T^{1/4-\xi/2}$ (pm), Eq. (6.22)	217	216	209	203	274	271	262	254	372	369	356	344	361	357	342	329
Average molecular speed, $\upsilon \sim T^{1/2}$ (km/s), Eq. (6.6)	1.20	1.26	1.50	1.74	1.69	1.77	2.11	2.45	0.45	0.47	0.57	0.66	0.38	0.40	0.47	0.55
Mean free path, $\lambda \sim T^{1/2+\xi}/p$ (nm), Eq. (6.12)	177	197	298	427	112	124	189	272	60	67	103	148	64	72	111	162
Mean time between collisions, $t_{mol} \sim T^{\xi}/p$ (ps), Eq. (6.10)	145	154	196	241	65	69	88	109	131	139	178	221	165	177	230	289
Self-diffusivity, $D_g \sim T^{1+\xi}/p$ (cm^2/s), Eq. (6.16)	1.26	1.46	2.63	4.38	1.12	1.29	2.35	3.93	0.16	0.19	0.34	0.57	0.14	0.17	0.31	0.53

where ϕ is collision diameter of a gas molecule. It follows from these formulae and from Eq. (6.5) that

$$5\pi\lambda = 16\upsilon t_{mol} \tag{6.7}$$

$$\pi\upsilon^2\varrho = 8p \tag{6.8}$$

Except for the collision diameter, all parameters on the right-hand sides of Eq. (6.6) are either *a priori* known quantities ($\mathscr{A}$, $\mathscr{R}$, and M) or are experimental conditions (p and T). If viscosity, η, of the gas is known then collision diameter can be obtained from expression [1] (see Table 6.1)

$$\phi = \left(\frac{5}{16\mathscr{A}\eta}\right)^{1/2}\left(\frac{M\mathscr{R}T}{\pi}\right)^{1/4} \tag{6.9}$$

As shown below, η for a regular carrier gas (Chapter 2) is roughly proportional to T^{ξ} where $\xi \approx 0.7$. This, according to Eq. (6.9), means that collision diameter, ϕ, of a gas molecule is roughly proportional to $T^{-0.1}$. This parameter is difficult to theoretically predict with sufficient accuracy. With that, comes difficulty of the theoretical prediction of gas viscosity, η, and other parameters related to ϕ.

Equation (6.9) leads to the following relation [15] between t_{mol} in Eq. (6.6) and η

$$4pt_{mol} = \pi\eta \tag{6.10}$$

Relations between η and other molecular properties of a gas are

$$5p\lambda = 4\upsilon\eta \tag{6.11}$$

$$\lambda = \frac{\eta}{p}\sqrt{\frac{128\mathscr{R}T}{25\pi M}} \tag{6.12}$$

$$\eta = \frac{5\sqrt{M\mathscr{R}T}}{16\sqrt{\pi}\mathscr{A}\phi^2} \tag{6.13}$$

Another transport property of a gas is its self-diffusivity and the diffusivity of other solutes in it.

If two different compounds, say "1" and "2" – both soluble in each other – are placed in an empty tube from opposite ends, then random motions of the molecules of the compounds would cause their mixing known as molecular diffusion. Let z and t be the distance along the tube and time, respectively. Then, according to Fick's second law of diffusion [7, 8, 35], the molar fraction, c_1, of compound "1" in vicinity of z at time t can be found as a solution of partial differential equation:

$$\frac{\partial c_1}{\partial t} = D\frac{\partial^2 c_1}{\partial z^2} \tag{6.14}$$

where D is diffusivity [1, 2] (diffusion coefficient [1–8], coefficient of diffusion [9], and diffusion constant [10]) of compound "1" in compound "2." The quantity D depends on the properties of both compounds. In GC, the molecular diffusion of solutes in a carrier gas plays a positive role of transporting the solutes to the stationary phase and back into the main flow stream of the carrier gas. On the other hand, the diffusion is also the root cause of undesirable broadening of the solute zones migrating along the column (Chapter 10).

To describe the self-diffusivity of a gas, one can consider mixing one group of "marked" molecules of the gas with another group of "differently marked" molecules of the same gas. In this case, the diffusion coefficient is the self-diffusivity, D_g, (coefficient of self-diffusion) of the gas. It can be expressed via other molecular properties of the gas as [1]

$$D_g = \frac{3}{8\sqrt{\pi}\mathcal{A}\phi^2 p}\sqrt{\frac{(\mathcal{R}T)^3}{M}} \tag{6.15}$$

or, due to Eqs. (6.6), (6.11), and (6.13), as (see Table 6.1)

$$D_g = \frac{6\mathcal{R}T\eta}{5Mp} \tag{6.16}$$

$$D_g = \frac{3\pi}{20}\frac{\eta\upsilon^2}{p} \approx \frac{\eta}{p}(0.686\upsilon)^2 \tag{6.17}$$

$$D_g = \frac{15\pi}{64}\frac{p\lambda^2}{\eta} \approx \frac{p}{\eta}(0.858\lambda)^2 \tag{6.18}$$

$$D_g = \frac{3\pi}{16}\lambda\upsilon \approx 0.589\lambda\upsilon \tag{6.19}$$

No simple and adequately accurate formulae for the calculation of gas viscosity, η, or the diffusivity, D, of a solute in a gas are known from theory. As a result, both η and D are typically predicted in GC from empirical formulae based on experimental data.

6.2 Gas Viscosity and Related Parameters – Empirical Formulae

The viscosity, η, of a regular carrier gas is practically independent of pressure [14, 18], but substantially depends on temperature, T, Figure 6.1.

This chapter is based on the evaluation of several sources of experimental data [13, 14, 20, 23] and empirical formulae for η [13, 20, 22, 24, 25]. The data source [13] compiled by Touloukian *et al.* is treated here as the primary source. Not only does this source provide detailed compilations of viscosity data for each carrier gas at different T, but it also supports the data in several ways. It evaluates the original data sources, describes the statistical treatment of the original data, and specifies the boundaries of the expected errors (see Table 6.2) in the recommended data. Other compilations

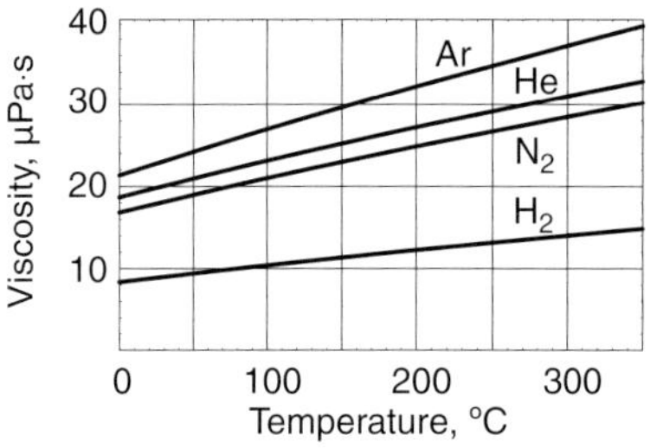

Figure 6.1 Temperature dependence of gas viscosity.

Table 6.2 The largest %-errors in viscosity data in the primary source [13] (primary errors) and in Eq. (6.20) (the latter represent the largest %-difference between the data in the source and the data obtained from Eq. (6.20)).

Gas	He	H_2	N_2	Ar
Primary errors [13]	1	2	2	No data
Errors in Eq. (6.20) at $300\text{ K} \leq T \leq 600\text{ K}$	0.2	0.2	0.8	0.8

either do not mention their original sources [20, 23], do not specify the errors in the original data [14, 23], or provide only a limited number of data points for some [14] or for all [23] gases.

The gas viscosity, η, as a function of its absolute temperature, T, can be expressed as [22, 25] (see also Appendix 6.A.2)

$$\eta = \left(\frac{T}{T_{st}}\right)^{\xi} \eta_{st} \tag{6.20}$$

where η_{st} is η at $T = T_{st} = 273.15$ K. Numerical values for gas-specific constants η_{st} and ξ, Table 6.3, were obtained from the least-squares fit [36] of $\ln\eta$ in Eq. (6.20) to the logarithms of viscosity data in the primary source [13]. The difference between the data for η in Eq. (6.20) and in the source [13] is shown in Figure 6.2. The bounds of the difference are listed in Table 6.2. Numerical values of η for several temperatures are listed in Table 6.4.

Table 6.3 Gas viscosity parameters in Eq. (6.20).

Gas	He	H_2	N_2	Ar
η_{st} (μPa s)	18.69	8.362	16.84	21.35
ξ	0.685	0.698	0.710	0.750

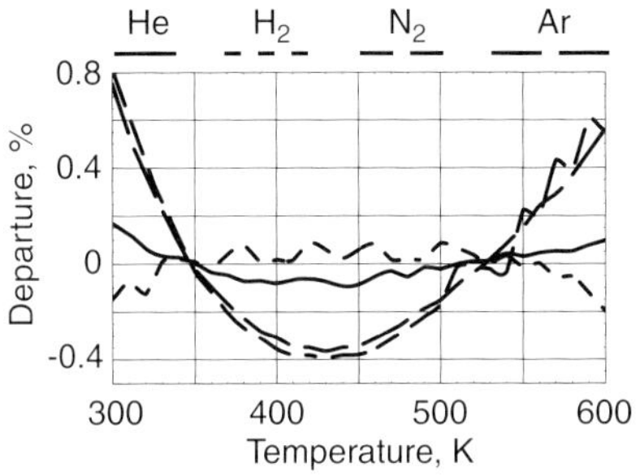

Figure 6.2 Departure of viscosity, η, in Eq. (6.20) from the data in the source [13].

Note 6.1

The parameters η_{st} and ξ for He, H_2, and N_2 in Table 6.3 are slightly different from widely known recommendations by Hinshaw and Ettre [25]. For helium at high temperature, the difference, Figure 6.3, between the Hinshaw–Ettre data and the date in the primary source [13] for this book is statistically significant. For that reason, and to provide a mutually consistent set of data for all regular gases, all parameters in Table 6.3 were regenerated anew. □

It might be useful to know the rate of a relative increase, $d\eta/\eta$, in η per unit of increase in T. As follows from the transformations $d\eta/\eta = ((d\eta/d\,T)/\eta)\,dT = (\xi/T)dT$, this rate is equal to ξ/T.

Table 6.4 Gas viscosity, η, μPa s, calculated from Eq. (6.20).

T (°C)	He	H_2	N_2	Ar	T (°C)	He	H_2	N_2	Ar
0	18.68	8.366	16.83	21.35	160	25.62	11.54	23.35	30.16
10	19.15	8.578	17.27	21.93	170	26.02	11.72	23.74	30.68
20	19.61	8.788	17.70	22.51	180	26.42	11.91	24.11	31.20
30	20.06	8.996	18.13	23.08	190	26.82	12.09	24.49	31.72
40	20.51	9.202	18.55	23.65	200	27.21	12.27	24.87	32.23
50	20.96	9.406	18.97	24.21	210	27.61	12.45	25.24	32.74
60	21.40	9.607	19.38	24.77	220	28.00	12.63	25.61	33.25
70	21.84	9.808	19.79	25.33	230	28.38	12.81	25.97	33.75
80	22.27	10.01	20.20	25.88	240	28.77	12.98	26.34	34.25
90	22.70	10.20	20.61	26.43	250	29.15	13.16	26.70	34.75
100	23.13	10.40	21.01	26.97	260	29.53	13.33	27.07	35.25
110	23.55	10.59	21.41	27.51	270	29.91	13.51	27.42	35.74
120	23.97	10.78	21.80	28.05	280	30.29	13.68	27.78	36.24
130	24.39	10.97	22.19	28.58	290	30.66	13.85	28.14	36.73
140	24.80	11.16	22.58	29.11	300	31.03	14.02	28.49	37.21
150	25.21	11.35	22.97	29.64	310	31.40	14.19	28.84	37.70

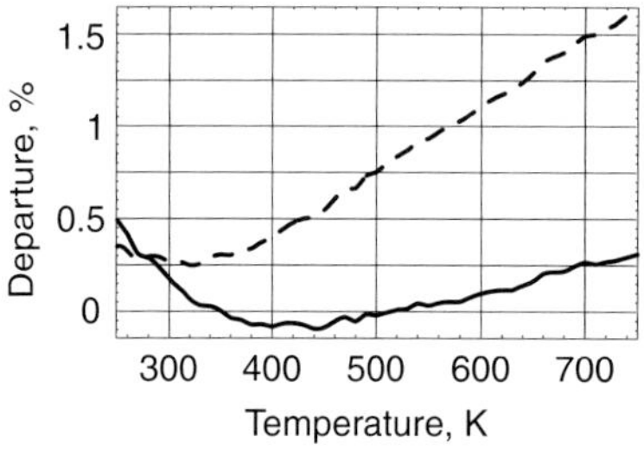

Figure 6.3 The departure of η in Eq. (6.20) from the data in the primary source [13]. Parameters η_{st} and ξ in Eq. (6.20) were taken from Table 6.3 (solid lines) and from the literature [25] (dashed lines).

Example 6.1

For the gases listed in Table 6.3, the rate, ξ/T, of a relative increase, $d\eta/\eta$, in η per 1 °C increase in temperature, T, can be estimated as $0.7/(300\ \mathrm{K}) \approx 0.0023/\mathrm{K}$ at $T = 300$ K and as $0.7/(600\ \mathrm{K}) \approx 0.0012/\mathrm{K}$ at $T = 600$ K. This means that, for $25\ °\mathrm{C} \leq T \leq 325\ °\mathrm{C}$, η increases by approximately 0.12–0.23% per each 1 °C increase in T. □

As mentioned earlier, there is no sufficiently simple and accurate theoretical formula for the prediction of molecular collision diameter, ϕ, and related properties of the gas. If viscosity of a gas is known, its self-diffusivity, D_g, and collision diameter, ϕ, can be derived from Eqs. (6.20), (6.16), and (6.9) as (see Table 6.1)

$$D_g = D_{g,st} \frac{p_{st}}{p} \left(\frac{T}{T_{st}}\right)^{1+\xi} \tag{6.21}$$

$$\phi = \phi_{st} \left(\frac{T}{T_{st}}\right)^{0.25-\xi/2} \tag{6.22}$$

where

$$D_{g,st} = \frac{6\mathcal{R} T_{st} \eta_{st}}{5 M p_{st}} \tag{6.23}$$

is D_g at standard temperature and pressure, and

$$\phi_{st} = \left(\frac{5}{16\mathcal{A}\eta_{st}}\right)^{1/2} \left(\frac{M\mathcal{R}T_{st}}{\pi}\right)^{1/4} \tag{6.24}$$

is ϕ at standard temperature. After exclusion of quantity η_{st} from the last two formulae, $D_{g,st}$ can be expressed via ϕ_{st} as

$$D_{g,st} = \frac{3}{8\sqrt{\pi}\mathcal{A} p_{st} \phi_{st}^2} \sqrt{\frac{(\mathcal{R}T_{st})^3}{M}} \tag{6.25}$$

which is in agreement with Eq. (6.15).

The power, $1 + \xi$, of temperature in Eq. (6.21) consists of two components. If ϕ was independent of T then, according to Eq. (6.15), D_g would change in proportion with $T^{1.5}$. However, according to Eq. (6.22), ϕ changes in proportion with $T^{0.25-\xi/2}$ where, for helium, hydrogen, nitrogen, and argon, quantity $0.25 - \xi/2$ is close to -0.1. This, according to Eq. (6.15), raises the power of the temperature dependence of D_g to approximately 1.7 or, more specifically, to $1 + \xi$ as described in Eq. (6.21).

6.3
Empirical Formulae for Solute Diffusivity in a Gas

The diffusivity, D, of a solute in a gas can be found from Fuller–Giddings empirical formula [16, 17, 19]

$$D = \frac{10^{-3}\sqrt{1/M_g + 1/M_{sol}}}{\left(V_g^{1/3} + (\Sigma V_{sol})^{1/3}\right)^2}\frac{\text{atm}}{p}\left(\frac{T}{\text{K}}\right)^{1.75}\frac{\text{cm}^2}{\text{s}} \tag{6.26}$$

where T is the absolute temperature (in kelvin, K); M_g and M_{sol} are the molecular weights of a gas and a solute, respectively; p is pressure; and quantities V_g and V_{sol} are dimensionless empirical quantities [19] known as the molecular diffusion volume of a gas and the atomic diffusion volume increments of a solute, respectively. The parameters V_g for several gases and V_{sol} for the atoms of n-alkanes (chemical formula $C_iH_{2(1+i)}$, $i = 1, 2, \ldots$) are listed in Table 6.5. Unlike the molecular parameters in theoretical formulae, the values of the quantities V_g and V_{sol} in Eq. (6.26) have no absolute physical meaning. Rather, these are relative empirical quantities normalized in the way that leads to a simple quotient, 10^{-3}, in the numerator of Eq. (6.26). The standard deviation of the differences between the values calculated from Eq. (6.26) and more than 500 experimental data points is estimated as [17] 6.7%. The measurement accuracy of the experimental data is unknown.

Table 6.5 Dimensionless quantities V_g and V_{sol} [19] in Eq. (6.26)[a].

Molecular diffusion volumes, V_g, of a gas		Atomic diffusion volume increments, V_{sol}, for some atoms	
He	2.67	C	15.9
H_2	6.12	H	2.31
N_2	18.5		
Ar	16.2		

a) Note: More data is available in the source [19].

6.3.1 Simplified Formulae

Diffusivity, D, of a solute in a carrier gas is an important parameter of column performance. Thus, optimal flow rate of a carrier gas is proportional [37] to D, optimal gas velocity is a strong function [11, 26, 38–40] of D, and so forth.

To use Eq. (6.26) in the evaluations of column operation in a temperature-programmed analysis, one needs to know gas parameters M_g and V_g, and solute parameters M_{sol} and $\sum V_{sol}$. However, this is not all. Equation (6.26) shows that temperature, T, strongly affects solute diffusivity, D. In a temperature-programmed analysis, diffusivity of a particular solute migrating through a column changes with the temperature. Giddings has shown [41, 42] that the net effect of the temperature change can be expressed by replacing the changing temperature, T, in Eq. (6.26) with a single temperature that is close to the solute elution temperature, T_R. This means that, in order to utilize Eq. (6.26) for the calculation of diffusivity of a given solute in a temperature-programmed analysis, one needs to know three solute parameters – M_{sol}, $\sum V_{sol}$, and T_R. Of these three parameters, $\sum V_{sol}$ is unknown for the majority of the solutes, and parameter T_R, being a complex function of several factors, is difficult to predict. This makes Eq. (6.26) unsuitable for the evaluation of solute diffusivity in practical GC analyses and, therefore, unsuitable for the evaluation of column performance. To adapt it to the needs of evaluating column performance in GC, Eq. (6.26) should be modified into the form that does not rely on parameters M_{sol}, $\sum V_{sol}$, and T_R of particular solutes.

In the following analysis, Eq. (6.26) is simplified through several approximation steps. The goal of these modifications is to arrive at a form where D is expressed as a function of a carrier gas and column temperature. The combined effect of all approximations leading to the final formulae for D is evaluated at the end of this section.

In GC, and especially when carrier gas is helium or hydrogen, the molecular weight, M_{sol}, of a solute is typically much larger than that of a gas. Similarly, quantity ΣV_{sol} representing the molecular diffusion volume of a solute in Eq. (6.26) is larger than quantity V_g representing the molecular diffusion volume of a gas. In other words,

$$M_{sol} \gg M_g, \ \Sigma V_{sol} \gg V_g \tag{6.27}$$

As a result, Eq. (6.26) can be approximated as

$$D = \frac{10^{-3}}{\sqrt{M_g}(\Sigma V_{sol})^{2/3}} \frac{\text{atm}}{p} \left(\frac{T}{\text{K}}\right)^{1.75} \frac{\text{cm}^2}{\text{s}} \tag{6.28}$$

Equation (6.28) has several interesting implications.

Consider the ratio, D_b/D_a, of diffusivities of an arbitrary solute in two different gases "a" and "b." It follows from Eq. (6.28) that (see Table 6.6)

$$\frac{D_b}{D_a} \approx \left(\frac{M_{g,a}}{M_{g,b}}\right)^{1/2} \tag{6.29}$$

Table 6.6 Molecular weights, M_g, of several gases[a].

Gas	He	H_2	N_2	Ar
M_g	4.003	2.016	28.01	39.95
$(M_{hydrogen}/M_g)^{1/2}$	0.71	1	0.27	0.22

a) Note: $M_{hydrogen}$ is molecular weight of hydrogen.

Statement 6.1

The ratio of diffusivities of an arbitrary solute in two different gases is essentially independent of the solute and depends only on the molecular weights of the gases.

Equation (6.28) also shows that D can be expressed as a product

$$D = X_{sol} X_g T^{1.75} p^{-1} \tag{6.30}$$

where X_{sol} and X_g are mutually independent terms representing only the properties of a solute and a gas, respectively. Comparing this formula with Eqs. (6.25) and (6.28) suggests that D can also be expressed as

$$D = X_{sol} \phi_{st}^2 D_{g,st} \frac{p_{st}}{p} \left(\frac{T}{T_{st}} \right)^{1.75} \tag{6.31}$$

where $D_{g,st}$ is gas self-diffusivity at standard temperature (T_{st}) and pressure (p_{st}), and ϕ_{st} is the molecular collision diameter of the gas at T_{st}. Furthermore, for the gases listed in Table 6.3, the power, $1 + \xi$, of the temperature term in Eq. (6.21) is confined within a narrow range $1.685 \leq 1 + \xi \leq 1.75$. This means that the difference between the powers of the temperature terms in Eqs. (6.31) and (6.21) is minor and can be ignored. As a result, Eq. (6.31) can be expressed as

$$D = X_{sol} \phi_{st}^2 D_g \tag{6.32}$$

where X_{sol} and ϕ_{st} are the fixed (independent of pressure and temperature) parameters of a solute and a gas, respectively, and D_g, Eq. (6.21), is the gas self-diffusivity that depends on the gas pressure and temperature.

Now that the dependence of solute diffusivity on gas pressure and temperature is represented by a single gas-dependent term – the gas self-diffusivity, D_g, – the two remaining fixed terms in Eq. (6.32) could be combined into one term. Equation (6.32) can be expressed as [43]

$$D = D_f D_g \tag{6.33}$$

where the empirical diffusivity factor, D_f, depends on a solute and a gas (Figure 6.4), but does not depend on pressure or temperature. Substitution of Eq. (6.21) in

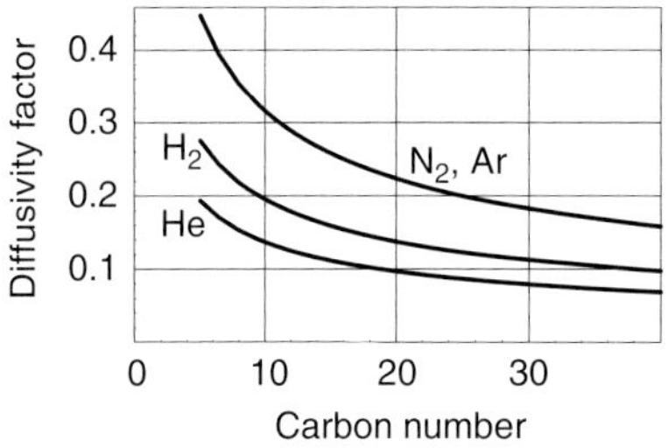

Figure 6.4 Diffusivity factor, D_f, – the ratio, $D_f = D/D_g$, of diffusivity, D, of n-alkanes to self-diffusivity, D_g, of the gas – for several gases.

Eq. (6.33) yields

$$D = \frac{p_{st}}{p} \cdot \left(\frac{T}{T_{st}}\right)^{1+\xi} D_f D_{g,st} \tag{6.34}$$

This formula can also be arranged as

$$D = \frac{p_{st}}{p} D_{pst} = \frac{p_{st}}{p}\left(\frac{T}{T_{st}}\right)^{1+\xi} D_{st}, \quad D_{pst} = D_{st}\left(\frac{T}{T_{st}}\right)^{1+\xi}, \quad D_{st} = D_f D_{g,st} \tag{6.35}$$

where D_{pst} is D at $p = p_{st}$ and at an arbitrary T, whereas D_{st} is D_{pst} at $T = T_{st}$ (which also means that D_{st} is D at 1 atm and 273.15 K).

6.3.2 Diffusivity of n-Alkanes

There is only one unknown parameter, D_f, in Eqs. (6.33)–(6.35). To reconcile D_f with experimental data for D, one of these formulae should be reconciled with Eq. (6.26). Let us reconcile Eqs. (6.34) and (6.26). The reconciliation involves additional errors in a final formula for D. The acceptable errors for D are defined by GC parameters that are direct functions of D. Among these parameters are carrier gas, optimal velocity [11, 26, 38–40], optimal flow rate [37], and peak widths [4, 43].

As mentioned earlier, Eq. (6.26) is based on two sets of parameters – (M_g, V_g) and (M_{sol}, ΣV_{sol}, T_R) – representing carrier gas and the solute, respectively. For four of the most widely used carrier gases, parameters M_g and V_g are listed in Tables 6.1 and 6.5. Since the number of practically useful carrier gases is small, it is acceptable for a study of a column performance to rely on a specific pair of parameters M_g and V_g in Eq. (6.26) for each gas. In effect, this means using a special version of Eq. (6.26) and, as a result, a special value of D_f in Eq. (6.34) for each gas.

For solute parameters M_{sol}, ΣV_{sol}, and T_R, the situation is more complex. Not only is the number of possible combinations of parameters M_{sol}, ΣV_{sol}, and T_R enormously large, but, for a majority of the solutes that can be analyzed by GC, the parameters ΣV_{sol} are unknown and parameters T_R are difficult to predict.

On the other hand, experimentally based recommendations [40, 44–47] for optimal gas velocity in capillary columns recognize the dependence of velocity on carrier gas, column dimensions, and temperature, but are not typically concerned with the composition of the test mixture or with the type of stationary phase. The same is true for optimal flow rates [29, 37]. Experimental data also shows that closely spaced peaks have more or less equal widths. Since the optimal flow parameters as well as the peak widths are direct functions of solute diffusivity, one can conclude that the difference in the diffusivities of closely spaced solutes is not significant. This suggests that it should be possible to choose one class of solutes that can represent all other solutes so that the diffusivity of a member of the class will represent the diffusivities of all solutes eluting closely with it.

Parameters of *n*-alkanes are among the best-known, and, if there is an *n*-alkane in a test mixture, its peak width is typically indistinguishable from the widths of closely spaced peaks. This justifies using the diffusivities of *n*-alkanes to represent the diffusivities of all closely eluting solutes. Using the data from Table 6.5, one can express Eq. (6.26) for *n*-alkanes (chemical formula $C_iH_{2(1+i)}$) as (Tables 6.7–6.10)

$$D = \frac{10^{-3}\sqrt{1/M_g + 1/(2.02 + 14.03i)}}{(V_g^{1/3} + (4.62 + 20.52i)^{1/3})^2}\frac{\text{atm}}{p}\left(\frac{T}{\text{K}}\right)^{1.75}\frac{\text{cm}^2}{\text{s}} \tag{6.36}$$

where i is the carbon number of *n*-alkane.

The diffusivity factor, D_f, in Eq. (6.34) can be found from reconciling Eq. (6.34) with Eq. (6.36). After several trials and errors, it has been found that, for *n*-alkanes most frequently analyzed in GC [38] (C_5–C_{40}, Eq. (2.12)), D_f can be approximated by a simple formula, Figure 6.4,

$$D_f = \gamma_{D1}^2/\sqrt{i}, \qquad 5 \le i \le 40 \tag{6.37}$$

where γ_{D1} is an empirical gas-dependent constant listed in Table 6.11. The lower limit, $i \ge 5$, to the carbon number is imposed to comply with Eq. (6.27).

Table 6.7 Diffusivity, D, cm^2/s, of *n*-alkanes in helium at 1 atm (calculated from Eq. (6.36)).

T (°C)	C_5	C_{10}	C_{15}	C_{20}	C_{25}	C_{30}	C_{35}	C_{40}
0	0.25	0.173	0.139	0.118	0.104	0.094	0.086	0.079
50	0.336	0.232	0.186	0.158	0.14	0.126	0.115	0.107
100	0.432	0.299	0.239	0.204	0.18	0.162	0.148	0.137
150	0.538	0.372	0.298	0.254	0.224	0.202	0.185	0.171
200	0.654	0.453	0.363	0.309	0.272	0.245	0.224	0.208
250	0.78	0.54	0.432	0.368	0.324	0.292	0.268	0.248
300	0.915	0.633	0.507	0.432	0.381	0.343	0.314	0.29
350	1.059	0.733	0.587	0.5	0.441	0.397	0.363	0.336

Table 6.8 Diffusivity, D, cm^2/s, of n-alkanes in hydrogen at 1 atm (calculated from Eq. (6.36)).

T (°C)	C_5	C_{10}	C_{15}	C_{20}	C_{25}	C_{30}	C_{35}	C_{40}
0	0.303	0.216	0.175	0.15	0.133	0.121	0.111	0.103
50	0.406	0.289	0.235	0.202	0.179	0.162	0.149	0.138
100	0.523	0.372	0.302	0.26	0.23	0.209	0.192	0.178
150	0.651	0.464	0.376	0.323	0.287	0.26	0.239	0.222
200	0.792	0.564	0.458	0.393	0.349	0.316	0.29	0.27
250	0.944	0.672	0.546	0.469	0.416	0.377	0.346	0.321
300	1.107	0.789	0.64	0.55	0.488	0.442	0.406	0.377
350	1.282	0.913	0.741	0.637	0.565	0.512	0.47	0.436

Table 6.9 Diffusivity, D, cm^2/s, of n-alkanes in nitrogen at 1 atm (calculated from Eq. (6.36)).

T (°C)	C_5	C_{10}	C_{15}	C_{20}	C_{25}	C_{30}	C_{35}	C_{40}
0	0.075	0.051	0.041	0.036	0.032	0.029	0.026	0.025
50	0.1	0.069	0.056	0.048	0.042	0.038	0.035	0.033
100	0.129	0.089	0.072	0.061	0.055	0.05	0.046	0.042
150	0.161	0.111	0.089	0.077	0.068	0.062	0.057	0.053
200	0.195	0.135	0.108	0.093	0.083	0.075	0.069	0.064
250	0.233	0.16	0.129	0.111	0.099	0.089	0.082	0.077
300	0.273	0.188	0.152	0.13	0.116	0.105	0.097	0.09
350	0.316	0.218	0.176	0.151	0.134	0.121	0.112	0.104

Table 6.10 Diffusivity, D, cm^2/s, of n-alkanes in argon at 1 atm (calculated from Eq. (6.36)).

T (°C)	C_5	C_{10}	C_{15}	C_{20}	C_{25}	C_{30}	C_{35}	C_{40}
0	0.068	0.046	0.036	0.031	0.027	0.025	0.023	0.021
50	0.092	0.061	0.049	0.042	0.037	0.033	0.031	0.028
100	0.118	0.079	0.063	0.054	0.047	0.043	0.039	0.037
150	0.147	0.098	0.078	0.067	0.059	0.053	0.049	0.046
200	0.179	0.12	0.095	0.081	0.072	0.065	0.06	0.055
250	0.213	0.143	0.114	0.097	0.086	0.077	0.071	0.066
300	0.25	0.167	0.133	0.114	0.101	0.091	0.083	0.077
350	0.289	0.194	0.154	0.132	0.116	0.105	0.097	0.09

Note 6.2

The arrangement of Eq. (6.37) so that D_f is proportional to the square, γ_{D1}^2, of an empirical quantity γ_{D1} was made with the intention of simplifying the forthcoming formulae. For example, when a column outlet is at vacuum, optimal average gas velocity and minimal width of a peak are proportional [28] to $D_f^{1/2}$ and $1/D_f^{1/2}$, respectively, and, therefore, to γ_{D1} and to $1/\gamma_{D1}$ [43]. □

Table 6.11 Diffusion properties of *n*-alkanes in several gases[a].

Gas	He	H_2	N_2	Ar
ξ (empirical quantity in Eq. (6.20))	0.685	0.698	0.710	0.750
γ_{D1} (empirical quantity in Eq. (6.37))	0.658	0.785	1.00	0.99
$D_{g,st}$ (gas self-diffusivity at $p = p_{st}$, $T = T_{st}$) (cm^2/s)	1.26	1.12	0.162	0.144
$D/D_{hydrogen}$	0.79	1	0.24	0.21

a) Note: Parameters $D_{g,st}$ and ξ are reproduced from Table 6.1 and quantity D is obtained from Eq. (6.38).

Substitution of Eq. (6.37) in Eqs. (6.33) and (6.34) yields

$$D = D_f D_g = \frac{\gamma_{D1}^2}{\sqrt{i}} D_g, \qquad 5 \le i \le 40 \tag{6.38}$$

$$D = \frac{p_{st}}{p}\left(\frac{T}{T_{st}}\right)^{1+\xi} \frac{\gamma_{D1}^2 D_{g,st}}{\sqrt{i}} \qquad 5 \le i \le 40 \tag{6.39}$$

Equation (6.38), being more compact than Eq. (6.39), shows the key factors affecting D. It highlights the fact that the difference in the diffusivity of *n*-alkanes in different gases comes mostly from the difference in the gas self-diffusivity, D_g, in the carbon number, i, and in the empirical gas-independent parameter, γ_{D1}. The ratios $D/D_{hydrogen}$ calculated from Eq. (6.38) for several gases are listed in Table 6.11. In accordance with Statement 6.1, these ratios are close to the ratios $(M_{hydrogen}/M)^{1/2}$ in Table 6.6.

The errors of the approximation of Eq. (6.36) by Eq. (6.39) are illustrated in Figure 6.5. When evaluating the significance of the errors, the following facts should be taken into account. Equation (6.36) is a special case of Eq. (6.26) developed in ref. [17] where the standard deviation of the differences between the results calculated from Eq. (6.26) and the experimental data (over 500 experimental data points) was estimated as 6.7%. Some actual differences were spread all the way up to and beyond 40%. There were also unreported measurement errors. In view of these facts, it would be reasonable to conclude that the additional differences, Figure 6.5, between the data from Eqs. (6.39) and (6.26) are relatively minor.

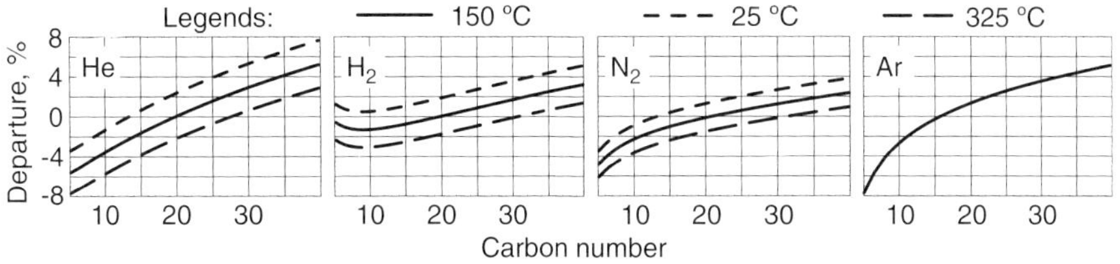

Figure 6.5 Departure of diffusivity, D, of an *n*-alkane in Eq. (6.39) from D in Eq. (6.36).

Equation (6.39) is much simpler than Eq. (6.26): all combinations of solute parameters, M_{sol} and ΣV_{sol}, in Eq. (6.26) are represented in Eq. (6.26) by a simple numerical parameter i – the carbon number of a representative n-alkane. Still, Eq. (6.39) is not the end of the road. It expresses the diffusivity, D, of n-alkane as a function of its carbon number, i, and temperature, T. This means that, in the case of a temperature-programmed analysis, D is a function of a pair i and T_R where T_R is elution temperature of the i-carbon n-alkane. Since T_R and i are mutually dependent parameters, Eq. (6.39) cannot be used without knowing how one of these parameters depends on the other. This dependence is described in Eqs. (5.55) and (5.56) (Chapter 5).

Substitution of Eqs. (5.55) and (5.56) in Eq. (6.39) yields

$$D = \gamma_D^2 D_{g,st}(10^3\varphi)^{0.09}\frac{p_{st}}{p}\left(\frac{T_{char}}{T_{st}}\right)^{-1.25}\left(\frac{T}{T_{st}}\right)^{1+\xi}, \quad (n\text{-alkane}) \tag{6.40}$$

$$D = \frac{\gamma_D^2 D_{g,st}(10^3\varphi)^{0.09}}{k^{0.1}}\frac{p_{st}}{p}\left(\frac{T}{T_{st}}\right)^{\xi-0.25}, \quad (n\text{-alkane},\ k \geq 0.1) \tag{6.41}$$

where (see Table 6.12)

Table 6.12 Molecular and other properties of ideal gas compiled from Table 6.1 and Table 6.11[a].

Gas	He	H_2	N_2	Ar
Molecular properties:				
M_g (molar mass) (g/mol)	4.003	2.016	28.01	39.95
λ_{st} (mean free path at $p = p_{st}$, $T = T_{st}$) (μm)	0.177	0.112	0.0604	0.0641
υ_{st} (average molecular speed at $T = T_{st}$) (km/s)	1.20	1.69	0.454	0.38
η_{st} (viscosity at $T = T_{st}$), (μPa s)	18.69	8.362	16.84	21.35
$D_{g,st}$ (self-diffusivity at $p = p_{st}$, $T = T_{st}$) (cm^2/s)	1.26	1.12	0.162	0.144
Empirical quantities:				
ξ (empirical quantity in Eqs. (6.20), (6.40), and (6.41))	0.685	0.698	0.710	0.750
γ_D (empirical quantity in Eqs. (6.40) and (6.41))	0.456	0.544	0.693	0.686
Combinations:				
$\gamma_D \lambda_{st}$ (μm)	0.081	0.061	0.042	0.044
$(\gamma_D \lambda_{st})/(\gamma_D \lambda_{st})_{hydrogen}$	1.33	1	0.689	0.721
$\gamma_D \upsilon_{st}$ (km/s)	0.548	0.92	0.315	0.261
$(\gamma_D \upsilon_{st})/(\gamma_D \upsilon_{st})_{hydrogen}$	0.596	1	0.342	0.284
$\gamma_D^2 D_{g,st}$ (cm^2/s)	0.262	0.331	0.078	0.068
$D/D_{hydrogen} = \gamma_D^2 D_{g,st}/(\gamma_D^2 D_{g,st})_{hydrogen}$ (Eqs. (6.40) and (6.41))	0.79	1	0.24	0.21
$D/D_{hydrogen} = (M_g/M_{hydrogen})^{1/2}$ (Eqs. (6.28) and (6.29))	0.71	1	0.27	0.22

a) Note: Quantities $D_{g,st}$, M_g, γ_D, λ_{st}, υ_{st}, η_{st}, and ξ are parameters of a gas. Quantities D and $D_{hydrogen}$ are diffusivities of *the same* n-alkane in an arbitrary gas and in hydrogen, respectively.

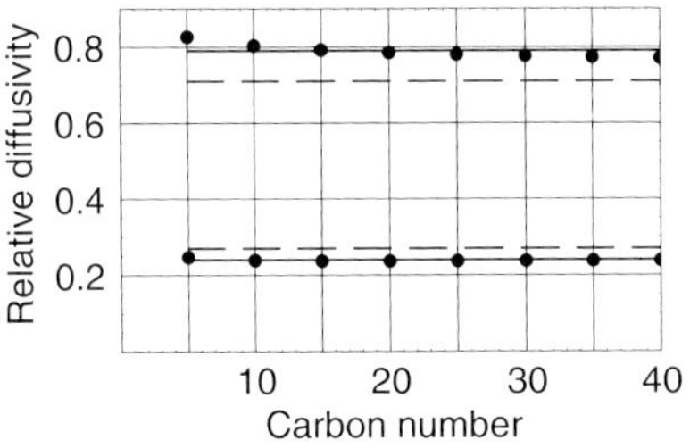

Figure 6.6 Relative diffusivity, D/D_{hydrogen}, of an *n*-alkane in helium (upper lines and dots) and in nitrogen (lower lines and dots). D_{hydrogen} is diffusivity of the same *n*-alkane in hydrogen. Dots were calculated from original formula [16, 17, 19], Eq. (6.36). Solid lines represent quantities $\gamma_D^2 D_{\text{g.st}}/(\gamma_D^2 D_{\text{g.st}})_{\text{hydrogen}}$, while dashed lines represent quantities $(M/M_{\text{hydrogen}})^{1/2}$.

$$\gamma_D = \frac{\gamma_{D1}}{4.34^{1/4}} = 0.693\gamma_{D1} \tag{6.42}$$

and φ is dimensionless thickness of stationary phase film, Eq. (2.6). Other parameters in Eqs. (6.40) and (6.41) are described below. The parameters of several carrier gases (listed in previous Tables 6.1 and 6.11) that are important for the evaluation of column operation are recompiled in Table 6.12. The ratios D/D_{hydrogen} in Table 6.12 were calculated as $D/D_{\text{hydrogen}} = (M_g/M_{\text{hydrogen}})^{1/2}$ and as $D/D_{\text{hydrogen}} = \gamma_D^2 D_{\text{g.st}}/(\gamma_D^2 D_{\text{g.st}})_{\text{hydrogen}}$. The first formula is simpler and relies on the most basic molecular properties of the gases – their molecular weights. However, this formula is less accurate than the second one as shown in Figure 6.6.

The quantity T_{char} in Eq. (6.40) is a solute characteristic temperature (Chapter 5). As the temperature at which a solute elutes with retention factor (k) equal to one, T_{char} is a key elution parameter of a given solute in a given column. Equation (6.40) describes diffusivity (D) of a solute as a function of its characteristic temperature T_{char} and a variable T (a column temperature in GC).

Equation (6.41) requires no solute parameters. Instead, it expresses the solute diffusivity as a function of two variables, T and k. Unlike it is in Eq. (6.40), the diffusivity in Eq. (6.41) is not the diffusivity of a particular solute, but of any solute that, in a specific method of GC analysis (a specific column, a specific temperature program, and so forth), elutes at temperature T with retention factor k. In a single-ramp temperature program, k is a known function of a heating rate, R_T (Chapter 8, Ref. [48]). In this case, therefore, D in Eq. (6.41) can be expressed as a function of two *a priori* known variables, R_T and T. Furthermore, all solutes eluting during a long heating ramp elute with roughly equal retention factors, k (Chapter 8, Refs. [48, 49]). As k in Eq. (6.41) becomes a fixed quantity, diffusivity becomes a function of a single variable, T. This leads to another interesting fact regarding the diffusivity in a temperature-programmed analysis.

To be more specific in the forthcoming observations, let us approximate ξ for the gases listed in Table 6.12, as $\xi \approx 0.7$. At a fixed k, Eq. (6.41) yields $D \approx \sim T^{0.45}$ (D is approximately proportional to $T^{0.45}$). This might appear as a contradiction with approximate proportionality $D \approx \sim T^{1.7}$ in other formulae for D including Eq. (6.26) Actually, this is not a contradiction. Equation (6.26) describes diffusivity of a particular solute as a function of its temperature, T. Equation (6.41), on the other hand, describes diffusivity of a solute eluting at temperature T. During a heating ramp, different solutes elute at different temperatures. As a general trend, solutes with larger molecules elute at higher temperatures. It follows from Eq. (6.26) that solutes with larger molecules generally have lower diffusivity at a given T. Dependence $D \approx \sim T^{0.45}$ results from two conflicting factors. Higher T increases D of any particular solute approximately in proportion with $T^{1.7}$. On the other hand, the fact that an increase in T raises molecular weights of eluting solutes tends to reduce D at a given T. The net result of the two conflicting factors appears as the dependence $D \approx \sim T^{0.45}$. Similar results have been experimentally obtained elsewhere [50].

One of the steps in transformation of Eq. (6.26) into Eqs. (6.40) and (6.41) was replacement of the term $(T/T_{st})^{1.75}$ in Eq. (6.31) with the term $(T/T_{st})^{1+\xi}$ in Eq. (6.34). The replacement slightly changes empirical formulae for diffusivities (D) of all solutes in a carrier gas. On the other hand, the replacement made it possible to relate a column performance to self-diffusivity (D_g) of a carrier gas and to other molecular parameters of the gas. Similar approach can be applied to description of the general trend in relationship between two parameters – characteristic temperature (T_{char}) and characteristic thermal constant (θ_{char}) – describing the interaction of the solutes with stationary phase liquid film in a column. The empirical formula describing the trend (Figure 5.6) is expressed in Eq. (5.46). Replacing in this formula the term $(T_{char}/T_{st})^{0.7}$ with slightly different term $(T_{char}/T_{st})^{\xi}$ makes a negligible change in the description of the trend. On the other hand, by utilizing the similarity of the trend with the temperature-dependence (Eq. (6.20)) in gas viscosity, it becomes possible to significantly simplify analysis of a column operation. Equation (5.46) becomes

$$\theta_{char} \approx \theta_{char.st} \cdot \left(\frac{T_{char}}{T_{st}}\right)^{\xi} \qquad (6.43)$$

Appendix 6.A

6.A.1 Ideal Gas

In this book, a gas is considered to be ideal if (a) relations between its pressure, p, temperature, T, and volume, V, are in agreement with the ideal gas law, Eq. (6.3); (b) its viscosity, η, is independent of p; and (c) diffusivity, D, of any solute in the gas is inversely proportional to p.

The state of a gas – the degree of its compliance with the ideal gas law – can be found from van der Waals equation [1, 5], Figure 6.A.1,

$$\left(p + \frac{a}{V_1^2}\right)(V_1 - b) = \mathcal{R}T \qquad (A1)$$

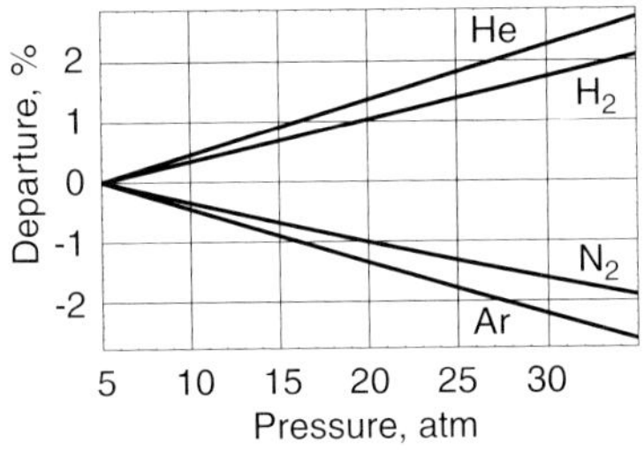

Figure 6.A.1 Departure of a volume, V_1, of one mole of real gas, Eq. (A1), from the volume, V, of the same mass of ideal gas, Eq. (6.3), at 300 K. (For helium, hydrogen, nitrogen, and argon at temperature between 300 and 600 K, the highest departure takes place at 300 K.)

Table 6.A.1 Quantities a and b in Eq. (A1).

	He	H_2	N_2	Ar
a, (liter/mole)^2 atm	0.03 412	0.2444	1.390	1.345
b, liter/mole	0.02 370	0.02 661	0.03 913	0.03 219

where a and b are gas-specific constants [23], Table 6.A.1, and V_1 is the volume of 1 mole of the gas.

It appears (see Figure 6.A.1) that, when pressure is not higher than several tens of atmospheres, a departure of the state of helium, hydrogen, nitrogen, and argon from the ideal gas law, Eq. (6.3), is insignificant. Although the evaluations of the departure of gas viscosity and diffusivity from ideal behavior are more complex [33], they lead to similar conclusions. Experimental data [14] also confirm nearly ideal behavior of the gas viscosity at up to 30 atm pressure.

6.A.2 Gas Viscosity in an Extended Temperature Range

Self-diffusivity, D_g, of a carrier gas and its viscosity, η, govern gas-dependent factors of column performance in GC. Among these factors is speed of analysis, optimal flow rate, minimal peak width, pressure requirement, and so forth. Viscosity of a particular gas depends on its temperature, T. It is important for a study of column performance to have a simple expression of the dependence of η on T even if that simplicity comes at the expense of some degradation in the accuracy. The empirical formula in Eq. (6.20) of the main text has been designed to satisfy that goal.

Unlike its self-diffusivity, gas viscosity also affects operational parameters of each GC analysis. In contemporary GC instrumentation, the settings of some pneumatic parameters of a carrier gas such as the gas velocity or the flow rate is typically implemented not via direct measurement of these parameters, but via their calculation from measured column pressure. Gas viscosity is a key gas parameter affecting its calculated flow parameters at a given pressure. The accuracy and the reproducibility of the settings of the calculated flow parameters across a wide temperature range might

Table 6.A.2 Gas viscosity parameters in Eq. (A2)[a].

Gas	He	H_2	N_2	Ar
η_{st0} (μPa s)	18.63	8.382	16.62	21.04
ξ_0	0.6958	0.6892	0.7665	0.8131
ξ_1	−0.0071	0.005	−0.0378	−0.0426

a) Note: η_{st0} is viscosity at $T_{st} = 273.15\,K = 0\,°C$.

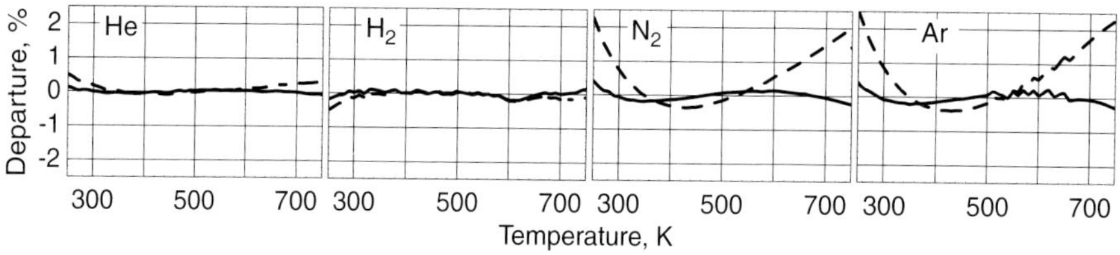

Figure 6.A.2 Departure of viscosity, η, in Eq. (A2) (—) and Eq. (6.20) (– – –) from the data in the source [13].

directly affect the highly important accuracy and the reproducibility of the retention times of all peaks in a chromatogram. This demands the highest possible accuracy of formula for η. Because of computerized calculations of set points in contemporary GC instruments, the simplicity of the formula is not of the highest importance.

Empirical formula,

$$\eta = \eta_{st0}\left(\frac{T}{T_{st}}\right)^{\xi(T)}, \qquad \xi(T) = \xi_0 + \xi_1\frac{T - T_{st}}{T_{st}} \tag{A2}$$

covers an

$$\text{extended temperature range}: 250\ \text{K} \le T \le 750\ \text{K} \tag{A3}$$

The gas-dependent parameters η_{st0}, ξ_0, and ξ_1 in Eq. (A2) are listed in Table 6.A.2. They were found from the least-squares fit of lnη in Eq. (A2) to the logarithms of viscosity data in the source [13]. Equation (A2) has higher accuracy within the

Table 6.A.3 The largest errors, %, in viscosity data in the primary source [13] (primary errors) and in Eqs. (6.20) and (A2) [a].

Gas	He	H_2	N_2	Ar
Primary errors [13]	1	2	2	no data
Errors in Eq. (6.20), 300 K ≤ T ≤ 600 K	0.2	0.2	0.8	0.8
Errors in Eq. (6.20), 250 K ≤ T ≤ 750 K	0.5	0.5	2	2.5
Errors in Eq. (6.2), 250 K ≤ T ≤ 750 K	0.1	0.2	0.4	0.4

a) Note: The latter represent the largest %-difference between the data in the source and the data obtained from Eqs. (6.20) and (A,2), respectively.

Table 6.A.4 Gas viscosity, η, μPa s, calculated from Eq. (A2).

T (°C)	He	H_2	N_2	Ar	*T* (°C)	He	H_2	N_2	Ar
−20	17.67	7.96	15.67	19.76	230	28.40	12.80	26.03	33.82
−10	18.16	8.17	16.15	20.40	240	28.78	12.98	26.39	34.31
0	18.63	8.38	16.62	21.03	250	29.16	13.16	26.74	34.79
10	19.11	8.59	17.08	21.66	260	29.54	13.33	27.09	35.27
20	19.57	8.80	17.54	22.28	270	29.92	13.51	27.43	35.74
30	20.03	9.01	18.00	22.89	280	30.29	13.68	27.77	36.20
40	20.49	9.21	18.44	23.50	290	30.66	13.86	28.11	36.66
50	20.94	9.41	18.89	24.09	300	31.03	14.03	28.44	37.12
60	21.39	9.61	19.33	24.69	310	31.39	14.20	28.77	37.57
70	21.83	9.81	19.76	25.27	320	31.76	14.37	29.09	38.01
80	22.27	10.01	20.19	25.85	330	32.12	14.54	29.41	38.45
90	22.70	10.20	20.61	26.42	340	32.47	14.71	29.73	38.89
100	23.13	10.40	21.02	26.99	350	32.83	14.88	30.05	39.32
110	23.56	10.59	21.44	27.55	360	33.18	15.05	30.36	39.75
120	23.98	10.78	21.84	28.10	370	33.54	15.21	30.66	40.17
130	24.40	10.97	22.25	28.65	380	33.89	15.38	30.97	40.59
140	24.82	11.16	22.65	29.20	390	34.23	15.55	31.27	41.00
150	25.23	11.35	23.04	29.73	400	34.58	15.71	31.56	41.40
160	25.63	11.53	23.43	30.26	410	34.92	15.88	31.86	41.81
170	26.04	11.72	23.81	30.79	420	35.26	16.04	32.15	42.21
180	26.44	11.90	24.19	31.31	430	35.60	16.20	32.43	42.60
190	26.84	12.08	24.57	31.82	440	35.94	16.37	32.72	42.99
200	27.23	12.27	24.94	32.33	450	36.27	16.53	33.00	43.37
210	27.63	12.45	25.31	32.83	460	36.61	16.69	33.28	43.75
220	28.01	12.63	25.67	33.33	470	36.94	16.85	33.55	44.13

extended temperature range, Figure 6.A.2 and Table 6.A.3, than Eq. (6.20) has within a narrower range. Numerical data for η calculated from Eq. (A2) are listed in Table 6.A.4.

References

1 Kauzmann, W. (1966) *Kinetic Theory of Gases*, W. A. Benjamin, New York.
2 Zauderer, E. (1989) *Partial Differential Equations of Applied Mathematics*, 2nd edn, John Wiley & Sons, Inc., New York.
3 Lapidus, L. and Amundson, N.R. (1952) *J. Phys. Chem.*, **56**, 984–988.
4 Golay, M.J.E. (1958) *Gas Chromatography 1958* (ed. D.H. Desty), Academic Press, New York, pp. 36–55.
5 Moore, W.J. (1972) *Physical Chemistry*, 4th edn, Prentice-Hall, Englewood Cliffs, NJ.
6 Cussler, E.L. (1984) *Diffusion, Mass Transfer in Fluid Systems*, Cambridge University Press, London.
7 Crank, J. (1989) *The Mathematics of Diffusion*, 2nd edn, Clarendon Press, Oxford.
8 Giddings, J.C. (1991) *Unified Separation Science*, John Wiley & Sons, Inc., New York.

9 Einstein, A. (1956) *Investigations on the Theory of the Brownian Movement*, Dover Publications, New York.
10 Golay, M.J.E. (1980) *J. Chromatogr.*, **196**, 349–354.
11 Guiochon, G. and Guillemin, C.L. (1988) *Quantitative Gas Chromatography for Laboratory Analysis and On-Line Control*, Elsevier, Amsterdam.
12 Guiochon, G. (1966) *Chromatographic Review*, Vol. 8 (ed. M. Lederer), Elsevier, Amsterdam, pp. 1–47.
13 Touloukian, Y.S., Saxena, S.C., and Hestermans, P. (1975) *Viscosity*, IFI/ Plenum, New York.
14 L'AIR LIQUIDE (1976) *Gas Encyclopedia*, Elsevier, Amsterdam.
15 Touloukian, Y.S., Saxena, S.C., and Hestermans, P. (1970) *Thermal Conductivity. Nonmetallic Liquids and Gases*, IFI/Plenum, New York.
16 Fuller, E.N. and Giddings, J.C. (1965) *Journal of Gas Chromatography*, **3**, 222–227.
17 Fuller, E.N., Schettler, P.D., and Giddings, J.C. (1966) *Ind. Eng. Chem.*, **58**, 19–27.
18 Schettler, P.D., Eikelberger, M., and Giddings, J.C. (1967) *Anal. Chem.*, **39**, 146–157.
19 Fuller, E.N., Ensley, K., and Giddings, J.C. (1969) *J. Phys. Chem.*, **73**, 3679–3685.
20 Watson, J.T.R. (1972) *Viscosity of Gases in Metric Units*, Her Majesty's Stationery Office, Edinburgh.
21 Maynard, V.R. and Grushka, E. (1975) *Advances in Gas Chromatography*, vol. **12** (eds J.C. Giddings, E. Grushka, R.A. Keller, and J. Cazes), Marcel Dekker, New York, pp. 99–140.
22 Ettre, L.S. (1984) *Chromatographia*, **18**, 243–248.
23 Weast, R.C., Astle, M.J., and Beyer, W.H. (1988) *CRC Handbook of Chemistry and Physics*, 69th edn, CRC Press, Boca Raton, FL.
24 Hawkes, S.J. (1993) *Chromatographia*, **37**, 399–401.
25 Hinshaw, J.V. and Ettre, L.S. (1997) *J. High Resolut. Chromatogr.*, **20**, 471–481.
26 Blumberg, L.M. (1997) *J. High Resolut. Chromatogr.*, **20**, 597–604.
27 Blumberg, L.M. (1997) *J. High Resolut. Chromatogr.*, **20**, 704.
28 Blumberg, L.M. (1997) *J. High Resolut. Chromatogr.*, **20**, 679–687.
29 Klee, M.S. and Blumberg, L.M. (2002) *J. Chromatogr. Sci.*, **40**, 234–247.
30 Blumberg, L.M. (2002) Proceedings of the 25th International Symposium on Capillary Chromatography (CD ROM), Palazzo dei Congressi, Riva del Garda, Italy, I.O.P.M.S., Kortrijk, Belgium (ed. P. Sandra).
31 Blumberg, L.M. (2003) *J. Chromatogr. A*, **985**, 29–38.
32 Blumberg, L.M., David, F., Klee, M.S., and Sandra, P. (2008) *J. Chromatogr. A*, **1188**, 2–16.
33 Perry, R.H., Green, d.W., and Maloney, J.O. (1984) *Perry's Chemical Engineer's Handbook*, 6th edn, McGraw-Hill Book Company, New York.
34 Fóti, G. and Kováts, E. (2000) *J. High Resolut. Chromatogr.*, **23**, 119–126.
35 Fick, A. (1855) *Ann D. Phys. U. Chem.*, **94**, 59–86.
36 Daniel, C., Wood, F.S., and Gorman, J.W. (1971) *Fitting Equations to Data*, John Wiley & Sons, Inc., New York.
37 Blumberg, L.M. (1999) *J. High Resolut. Chromatogr.*, **22**, 403–413.
38 Lee, M.L., Yang, F.J., and Bartle, K.D. (1984) *Open Tubular Gas Chromatography*, John Wiley & Sons, Inc., New York.
39 Ettre, L.S. and Hinshaw, J.V. (1993) *Basic Relations of Gas Chromatography*, Advanstar, Cleveland, OH.
40 Jennings, W., Mittlefehldt, E., and Stremple, P. (1997) *Analytical Gas Chromatography*, 2nd edn, Academic Press, San Diego.
41 Giddings, J.C. (1962) *Gas Chromatography* (eds N. Brenner, J.E. Callen, and M.D. Weiss), Academic Press, New York, pp. 57–77.
42 Giddings, J.C. (1962) *J. Chem. Educ.*, **39**, 569–573.
43 Blumberg, L.M. and Berger, T.A. (1993) *Anal. Chem.*, **65**, 2686–2689.
44 Littlewood, A.B. (1970) *Gas Chromatography: Principles, Techniques, and Applications*, 2nd edn, Academic Press, New York.

45 Grob, K. and Tschour, R. (1990) *J. High Resolut. Chromatogr.*, **13**, 193–194.

46 Hinshaw, J.V. (2001) *LC-GC*, **19**, 1056–1064.

47 Restek Corporation (2003) *2003 Chromatography Products*, Restek Corp., USA.

48 Blumberg, L.M. and Klee, M.S. (2001) *J. Chromatogr. A*, **918/1**, 113–120.

49 Blumberg, L.M. and Klee, M.S. (2000) *Anal. Chem.*, **72**, 4080–4089.

50 Blumberg, L.M., Wilson, W.H., and Klee, M.S. (1999) *J. Chromatogr. A*, **842/1-2**, 15–28.

7
Flow of Ideal Gas

There is no need to mention in this chapter that a chromatographic column has stationary phase. To emphasize this fact, the term *tube* is used instead of the term *column*. Like a GC column, a tube can have either an open inner space (an open tube) or it can be packed with some material (a packed tube), Figure 2.1. The pneumatic parameters – flow rate, velocity, pressure, and so forth – of a gas in a tube describe various aspects of the gas flow in the tube [1–14]. When a tube is a GC column, these properties directly affect operation of GC analysis.

It is assumed throughout this book that the flow in a tube is laminar [2, 4, 6, 7, 15] (smooth) as is typically the case for the column flow in GC (Appendix 7.A.1).

7.1
Flow of Gas in a Tube

7.1.1
One-Dimensional Model of a Tube

A tube is a three-dimensional structure. Longitudinal velocity of a gas flow through a tube can be different at different locations within the tube. For example, the local longitudinal velocity, u_r, in a given cross section of an open round tube has a parabolic profile [5, 6, 8], Figure 7.1,

$$u_r = \left(1 - \left(\frac{2r}{d}\right)^2\right) u_{r,\max} \tag{7.1}$$

as a function of the tube radius, *r*. It changes from zero at the inner walls of the tube to the maximum at the center of the cross section.

The fact that the gas velocity in a tube is not the same across the tube's cross section has important consequences in GC. However, it might be of no interest for the study of the flow phenomena as such. Basic concepts of the gas flow could be expressed via its cross-sectional average velocity [3, 5, 6, 8], *u*, (briefly, velocity). According to a general concept of an average, Eq. (3.15), *u* at some cross section along the tube can be

Temperature-Programmed Gas Chromatography. Leonid M. Blumberg

ISBN: 978-3-527-32642-6

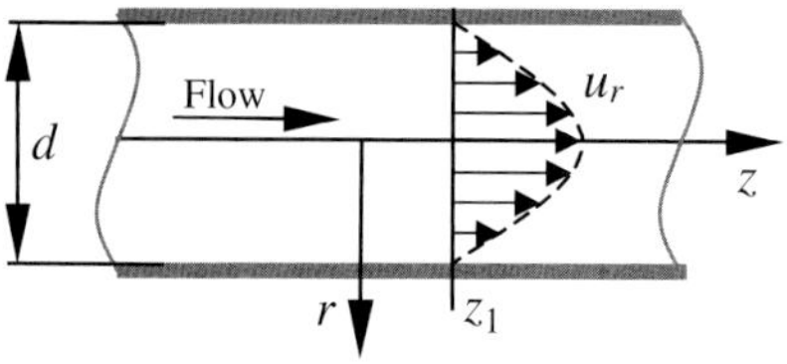

Figure 7.1 Parabolic profile, Eq. (7.1), of longitudinal velocities, u_r, of a (laminar) flow through a cross section at z_1 in an open round tube. The length of each u_r arrow is proportional to u_r at the radial distance, r, from the tube axis ($0 \leq r \leq d/2$). The dashed line is a parabola.

found as

$$u = \frac{1}{A_{\text{gas}}} \int_{A_{\text{gas}}} u_r \mathrm{d}A \tag{7.2}$$

where A_{gas} is the interparticle [16] (interstitial [16, 17], available for the flow) portion of the cross-sectional area of the tube.

In an open round tube, $A_{\text{gas}} = \pi d^2/4$, where d is the tube internal diameter, Figure 2.1. Because of the presence of packing material in any cross section of a packed tube, A_{gas} in it is smaller than that in an open tube with the same diameter, d. It is difficult to find an exact value of A_{gas} for each cross section of a packed tube. In an average tube, A_{gas} can be estimated as $A_{\text{gas}} = \varepsilon\pi d^2/4$, where ε is the porosity [5, 6, 8] of the packing which, for a compact random packing can be estimated as [8]

$$\varepsilon \approx 0.4 \tag{7.3}$$

These observations suggest that A_{gas} in an open and in a packed tubes can be expressed as

$$A_{\text{gas}} = \frac{\pi d^2}{4} \cdot \begin{cases} 1, & \text{open tube} \\ \varepsilon, & \text{packed tube} \end{cases} \tag{7.4}$$

To find gas velocity, u, in a round tube, an area increment, dA, in Eq. (7.2) can be arranged as the area, $\mathrm{d}A = 2\pi r\mathrm{d}r$, of a d$r$-wide ring. This together with Eq. (7.4) transforms Eq. (7.2) as

$$u = \frac{4}{\pi d^2} \int_0^{d/2} 2\pi r u_r \mathrm{d}r \tag{7.5}$$

The latter, after the substitution of Eq. (7.1), yields

$$u = \frac{u_{r,\max}}{2} \tag{7.6}$$

Dealing with a single (cross-sectional average) velocity, u, instead of the local velocities, u_r, at each point of a cross section of an open or a packed tube is equivalent to representing a three-dimensional tube by a one-dimensional model reducing the tube to its longitudinal (axial) dimension corresponding to the axis of the distances, z,

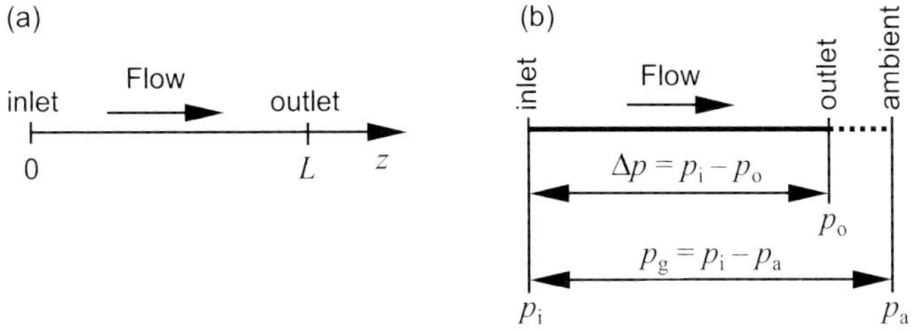

Figure 7.2 One-dimensional model of a *L*-long tube (a) and several pressure metrics (b).

from the inlet of the tube, Figure 7.2a. Without specifically mentioning this fact, the one-dimensional model of a tube was already used in previous chapters where longitudinal specific amount, a, (mass per length) of a gas or of a solute – rather than, say, a solute density (mass per volume) or other three-dimensional equivalents of the density – was considered.

One-dimensional model of a tube allows for a simple illustration of various pressure metrics along a tube, Figure 7.2b. Some of them are [16] local pressure, p, at an arbitrary point, z; the ambient pressure, p_a; the inlet pressure, p_i; the outlet pressure, p_o; the gauge pressure (head pressure), p_g; pressure drop, Δp; relative pressure (compression ratio [8]), P; and relative pressure drop, ΔP. The quantities p, p_a, p_i, and p_o are the absolute pressures. The quantities p_g and Δp are the differential pressures. The differential and the relative pressures are defined as, Figure 7.2b,

$$p_g = p_i - p_a \tag{7.7}$$

$$\Delta p = p_i - p_o \tag{7.8}$$

$$P = \frac{p_i}{p_o} \tag{7.9}$$

$$\Delta P = \frac{\Delta p}{p_o} \tag{7.10}$$

The latter two relate as

$$\Delta P = P - 1 \tag{7.11}$$

Frequently, both p_a and p_o are almost the same as the standard pressure, $p_{st} = 1$ atm, Eq. (6.2). However, p_a can be different from p_{st} (in high mountains, e.g.). Also, p_o can be lower than p_a (e.g., in GC-MS where column outlet is at vacuum, that is $p_o = 0$) or higher than p_a (e.g., due to pneumatic resistance in a GC detector).

7.1.2
Gas Velocity

Generally, local pressure, p, and gas velocity (u) can be functions of distance (z) from the tub inlet (Figure 7.2). Relationship between u and pressure gradient, $\partial p/\partial z$, along the tube is governed by Darcy's law [6, 8, 17–19]:

$$u = -\frac{B_o}{\eta}\frac{\partial p}{\partial z} \tag{7.12}$$

where η is the gas viscosity, and

$$B_o = \begin{cases} d^2/32, & \text{open round tube} \\ \dfrac{\varepsilon^2 d_p^2}{270(1-\varepsilon)^2}, & \text{packed tube} \end{cases} \tag{7.13}$$

is the specific permeability [5, 6, 8] of the tube. The negative sign in Eq. (7.12) indicates that the direction of u is opposite to the direction of the pressure gradient. For a packed tube, the substitution of B_o from Eq. (7.13) in Eq. (7.12) leads to the semiempirical Kozeny–Carman–Giddings formula [5, 6, 8, 20, 21],

$$u = \frac{\varepsilon^2 d_p^2}{270\,(1-\varepsilon)^2\eta}\cdot\frac{\partial p}{\partial z} \tag{7.14}$$

Note 7.1

Typically, quotient 180 rather than 270 is used in citations of Kozeny–Carman formula [5, 6, 8]. Giddings [8] suggested, however, that the "equation appears to work better for chromatographic materials if 180 is replaced by 270." □

7.1.3 Flow Rate

There are several ways [14] to express the flow rate of a gas at an arbitrary coordinate, z, along a tube. The flow rate metric most widely accepted in GC is a hybrid of the mass flow rate (mass per time),

$$F_m = \frac{\mathrm{d}m}{\mathrm{d}t} \tag{7.15}$$

and the volumetric flow rate (volume per time)

$$F_V = \frac{\mathrm{d}V}{\mathrm{d}t} \tag{7.16}$$

where $\mathrm{d}m$ and $\mathrm{d}V$ are, respectively, the mass and the volume of the gas transferred through the cross section at z during the time $\mathrm{d}t$.

An important advantage of using the volumetric flow rate comes from its simple relation to the gas velocity, u. Indeed, quantity $\mathrm{d}V$ in Eq. (7.16) can be expressed as $\mathrm{d}V = A_{gas}\mathrm{d}z = A_{gas}u\mathrm{d}t$. This allows one to write Eq. (7.16) as $F_V = A_{gas}u$ which, due to Eq. (7.4), becomes

$$F_V = A_{gas}u = \frac{\pi d^2 u}{4}\cdot\begin{cases} 1, & \text{open tube} \\ \varepsilon, & \text{packed tube} \end{cases} \tag{7.17}$$

For a round open tube, this, due to Eqs. (7.12) and (7.13), becomes *Poiseuille equation* [5] (*Hagen–Poiseuille equation* [8])

$$F_V = -\frac{\pi d^4}{128\eta} \cdot \frac{\partial p}{\partial z} \quad \text{(open tube)} \tag{7.18}$$

The volume of the same amount (mass, number of moles, and so forth) of gas depends on its pressure, p, and temperature, T. As a result, depending on p and T, the same flow rate F_V might correspond to the flow of a different amount of gas per the same time. This makes quantity F_V unsuitable as a flow rate metric in GC.

The mass flow rate, F_m, does not have this shortcoming. Unfortunately, in order to find F_m from u, it is necessary to know the gas density, ϱ, or its equivalent – a quantity that is not necessary in the case of volumetric flow rate. The quantities $\mathrm{d}m$ and $\mathrm{d}V$ in Eqs. (7.15) and (7.16) relate as $\mathrm{d}m = \varrho \mathrm{d}V$. Therefore,

$$F_m = \varrho F_V \tag{7.19}$$

To identify the flow rate metric that has the advantages of both F_m and F_V, one can express ϱ in Eq. (7.19) via Eq. (6.5). Equation (7.19) becomes

$$F_m = \frac{MpF_V}{\mathscr{R}T} \tag{7.20}$$

where M is the molar mass of a gas and $\mathscr{R}$ is the molar gas constant. Because M is fixed for a given gas and $\mathscr{R}$ is the same for all ideal gases, quantity $F_V p/T$ in Eq. (7.20) is proportional to F_m, and, therefore, is independent of p and T. For two mass flow rates, $F_{m,1}$ and $F_{m,2}$,

$$F_{m,2} = F_{m,1} \quad \text{implies} \quad \frac{p_2 F_{V,2}}{T_2} = \frac{p_1 F_{V,1}}{T_1} \tag{7.21}$$

This statement has several important implications.

Let V_n be the volume of the gas that flows through the tube's cross section at coordinate z and that is measured at standard pressure, $p_{st} = 1$ atm, and at some reference temperature, T_{ref} (typically 0 or 25 °C), regardless of actual pressure and temperature at z. It follows from Eq. (7.21) that the normalized (pressure and temperature adjusted) volumetric flow rate, F, defined as

$$F = \frac{\mathrm{d}V_n}{\mathrm{d}t} \tag{7.22}$$

can be found as

$$F = \frac{p}{p_{st}} \frac{T_{ref}}{T} F_V \tag{7.23}$$

where F_V is the (actual) volumetric flow rate at z measured at actual p and T existing at z. Substitution of F_V from Eq. (7.23) in Eq. (7.20) leads to expression

$$F_m = \frac{Mp_{st}F}{\mathscr{R}T_{ref}} \tag{7.24}$$

On the right-hand site of it, all but F are the fixed quantities. This means that F is proportional to F_m, and, hence, can be used as a metric of the rate of the flow of the amount of gas. One can also interpret F as the mass flow rate expressed in units of the normalized volume per time. From now on, unless otherwise is explicitly stated, the term volumetric flow rate (briefly, flow rate, flow) will always be understood as the normalized volumetric flow rate, F, defined in Eq. (7.22). Substitution of Eq. (7.17) in Eq. (7.23) and accounting for Eq. (7.4) yields

$$F = \frac{A_{\mathrm{gas}} p u T_{\mathrm{ref}}}{p_{\mathrm{st}} T} = \gamma_{FA} p u \tag{7.25}$$

where

$$\gamma_{FA} = \frac{A_{\mathrm{gas}} T_{\mathrm{ref}}}{p_{\mathrm{st}} T} = \frac{\pi d^2 T_{\mathrm{ref}}}{4 p_{\mathrm{st}} T} \cdot \begin{cases} 1, & \text{open tube} \\ \varepsilon, & \text{packed tube} \end{cases} \tag{7.26}$$

Equation (7.25) allows one to find F directly from gas velocity, u, at any coordinate, z, with known pressure, p, and temperature, T. Conversely, if F is known, Eq. (7.25) can be used for the calculation of u at any z with known p and T.

Two reference temperatures, T_{ref}, in Eq. (7.23) are most frequently used for the measurement of the flow rate (F) in GC. This is either 0 °C (273.15 K) or 25 °C (298.15 K). As mentioned earlier, the flow measurement conditions of 0 °C and 1 atm are known as the conditions of standard temperature and pressure. The flow measurement conditions of 25 °C and 1 atm are sometimes called as the conditions of normal temperature and pressure [22]. In view of that, 25 °C (298.15 K) temperature can be called as the normal temperature,

$$T_{\mathrm{norm}} = 298.15\ \mathrm{K}\ (25\ ^\circ\mathrm{C}) \tag{7.27}$$

Flow measurement conditions are not always explicitly specified in GC. As a result, the meaning of the flow rate units, say mL/min, becomes ambiguous, leading to confusions and inconsistent interpretations of flow rate parameters. Some of these problems were discussed in the literature [23]. In this book, milliliter of gas measured at standard conditions is called the standard milliliter and denoted as $\mathrm{mL_{st}}$. Milliliter of gas measured at normal conditions is called the normal milliliters and denoted as $\mathrm{mL_{norm}}$. It follows from Eq. (7.25) that the units $\mathrm{mL_{norm}}$/min (also known as sccm –standard cubic centimeters per minute) and $\mathrm{mL_{st}}$/min relate as

$$1\ \mathrm{mL_{st}/min} = \frac{273.15\ \mathrm{K}}{298.15\ \mathrm{K}}\ \mathrm{mL_{norm}/min} \approx 0.916\ \mathrm{mL_{norm}/min} \tag{7.28}$$

For example, $F = 0.8\ \mathrm{mL_{norm}/min}$ is the same as $F \approx 0.733\ \mathrm{mL_{st}/min}$.

In this book, the flow rate is measured at normal conditions and expressed in the units of $\mathrm{mL_{norm}}$/min. In view of that, unless otherwise is explicitly stated, it is assumed that

$$1\ \mathrm{mL/min} = 1\ \mathrm{mL_{norm}/min} \tag{7.29}$$

7.1.4
Mass-Conserving Flow

Generally, not only the parameters such as temperature, pressure, and so forth can vary in time, t, and along a tube coordinate, z, but even the tube dimensions (its cross section, for example) can vary with t (in a tube with soft walls, for example) and with z (in a tube consisting of several sections having different diameters, for example). A parameter (tube cross section, gas flow, pressure, temperature, and so forth) can be static (fixed in time) or dynamic (changing with time), and it can also be uniform (the same along a tube) or nonuniform (coordinate-dependent). Additional discussion of these properties can be found in Chapter 4.

A mass-conserving flow of a gas in a tube is the one where there is no delivery or leakage of the gas through the walls of the tube, and where the only places of the gas inflow and outflow are the tube's inlet and outlet.

Let V_1 and V_2 be the volumes of a gas in a portion of a tube between its inlet and an arbitrary coordinate z at the time instants t_1 and t_2, respectively. If V_1 and V_2 are measured at the same normal conditions as those in the measurement of the flow rate (F) in the tube then, in a mass-conserving flow,

$$V_2 - V_1 = \int_{t_1}^{t_2} F_i dt - \int_{t_1}^{t_2} F dt \tag{7.30}$$

where F_i and F are, respectively, the inlet flow rate and flow rate through the cross section at z. In a static flow, F at any z does not change with time. As a result,

$$\int_{t_1}^{t_2} F dt = (t_2 - t_1)F \tag{7.31}$$

at any z including inlet an outlet. The right-hand side of Eq. (7.30) becomes

$$\int_{t_1}^{t_2} F_i dt - \int_{t_1}^{t_2} F dt = (t_2 - t_1)(F_i - F) \tag{7.32}$$

Also in a static flow, $V_1 = V_2$. As a result, $F = F_i$. In other words, if the flow is static and mass-conserving, then it is the same through any cross section of the tube, that is

$$F = F_i = F_o \tag{7.33}$$

where F_o is the outlet flow rate. If a tube itself is also uniform, then A_{gas} is the same in any cross section. Equations (7.25) and (7.33) allow one to conclude that

Statement 7.1

In a static mass-conserving flow in a uniform tube, quantity pu/T is uniform.

Particularly,

$$\frac{p\,u}{T} = \frac{p_i u_i}{T_i} = \frac{p_o u_o}{T_o} \tag{7.34}$$

where (p, T, u), (p_i, u_i, T_i), and (p_o, u_o, T_o) are, respectively, the gas parameters at an arbitrary coordinate, z, along the tube, Figure 7.2, at the *inlet* of the tube ($z = 0$), and at the *outlet* of the tube ($z = L$). For uniformly heated gas, Eqs. (7.25) and (7.34) yields

$$pu = p_i u_i = p_o u_o = \frac{F}{\gamma_{FA}} \tag{7.35}$$

or, in a differential form,

$$\frac{\partial}{\partial z}(pu) = \frac{\partial F}{\partial z} = 0 \tag{7.36}$$

It is assumed through the rest of this chapter that a tube has (a) a uniform cross-sectional area, A_{gas}, and (b) is uniformly heated. In addition to that, the flow of gas through the tube is (c) static and (d) mass-conserving. When all the conditions (a)–(d) exist,

Statement 7.2

The flow of a gas in a tube is governed by the Darcy's law, Eq. (7.12), constrained by Eq. (7.35).

Darcy's law, Eq. (7.12), can be also expressed as [9]

$$u = -\frac{L}{\Omega}\frac{\partial p}{\partial z} \tag{7.37}$$

where L is the length of a tube, and $\Omega = L\eta/B_o$ is the pneumatic resistance of the tube that, due to Eq. (7.13), can be expressed as

$$\Omega = \frac{L\eta}{B_o} = \frac{\gamma_\Omega L\eta}{d_{cx}^2} = L\eta \cdot \begin{cases} \dfrac{32}{d^2}, & \text{open round tube} \\[2ex] \dfrac{270(1-\varepsilon)^2}{\varepsilon^2 d_p^2}, & \text{packed tube} \end{cases} \tag{7.38}$$

where d is the internal diameter of a tube, and d_p is the particle size of a packing material in a packed tube, Figure 2.1,

$$d_{cx} = \begin{cases} d, & \text{open round tube} \\ d_p, & \text{packed tube} \end{cases} \tag{7.39}$$

is characteristic cross-sectional dimension of a tube, and

$$\gamma_\Omega = \begin{cases} 32, & \text{open round tube} \\ 270(1-\varepsilon)^2/\varepsilon^2, & \text{packed tube} \end{cases} \tag{7.40}$$

Because of the estimate $\varepsilon \approx 0.4$, Eq. (7.3), the ratio, $\gamma_{\Omega,\text{pack}}/\gamma_{\Omega,\text{open}}$, of quantities γ_Ω in a packed and in an open round tube can be estimated as

$$\frac{\gamma_{\Omega,\text{pack}}}{\gamma_{\Omega,\text{open}}} \approx 19 \tag{7.41}$$

This means that, assuming that the same gas at the same temperature is used in both cases, pneumatic resistance, Ω, of a packed tube is 19 times higher than that of equally long open tube having the same characteristic cross-sectional dimension. Due to Eq. (7.37), this implies that if, under the above-mentioned conditions, equal pressures were used in both cases, then the gas velocity, u, in the packed tube would be 19 time lower than it was in the open tube.

7.2 Pneumatic Parameters

7.2.1 Energy Flux

The gas velocity (u) is an important pneumatic parameter in chromatography because it most directly affects the peak broadening (Chapter 10). Unfortunately, quantity u as a metric of the gas flow has several substantial shortcomings. One of them is that, due to the gas decompression along the column (see below), u can be different at different locations along the column. This creates difficulties in evaluation of a column operation as a function of parameters of GC analysis (column dimensions, pressures, and so forth). To avoid these difficulties, several workers [14, 24–27] used the product $p\,u$ as a single pneumatic variable in evaluation of a column performance. Let us denote the product as

$$J = p\,u \tag{7.42}$$

Measured in units of energy/area/time, quantity J can be interpreted as the energy flux in a tube [14]. Equations (7.35), (7.37) and (7.36) can be expressed as

$$J = p\,u = p_\text{i} u_\text{i} = p_\text{o} u_\text{o} = \frac{F}{\gamma_{FA}} \tag{7.43}$$

$$J = -\frac{L}{\Omega} p \frac{\partial p}{\partial z} \tag{7.44}$$

$$\frac{\partial J}{\partial z} = \frac{\partial F}{\partial z} = 0 \tag{7.45}$$

Equation (7.44) can be viewed as an alternative form of the Darcy's law constrained under static conditions by Eq. (7.45).

The gas velocity (u), flow rate (F), and energy flux (J) are metrics of flow of a gas in a tube. Equation (7.45) suggests that, similarly to the flow rate, the energy flux at any time instant is the same at all locations along the tube. However, the relation of the flow metric, J, to the metric, u, is simpler than that of the flow rate, F, to u. For one thing, unlike the relationship of F to u, the relationship of J to u does not depend on tube dimensions and does not involve an arbitrary reference temperature, T_{ref}.

To find how J relates to parameters of a gas itself and to its pressure, let us notice that Darcy's law in Eq. (7.44) can be written as

$$J\mathrm{d}z = -\frac{Lp}{\Omega}\mathrm{d}p \tag{7.46}$$

Because J is independent of z as Eq. (7.45) suggests, integration of Eq. (7.46) from $z = 0$ and $p = p_i$ to an arbitrary z along the tube and to a corresponding p, Figure 7.2, yields

$$J = \frac{p_i^2 - p^2}{2\Omega} \cdot \frac{L}{z} \tag{7.47}$$

At $z = L$ and $p = p_o$, this together with Eq. (7.43) allows one to rewrite Eq. (7.47) as

$$J = \frac{p_i^2 - p_o^2}{2\Omega} \tag{7.48}$$

$$u = \frac{p_i^2 - p_o^2}{2\Omega p} \tag{7.49}$$

$$u_i = \frac{p_i^2 - p_o^2}{2\Omega p_i} \tag{7.50}$$

$$u_o = \frac{p_i^2 - p_o^2}{2\Omega p_o} \tag{7.51}$$

Equation (7.48) can be also expressed as

$$p_i = \sqrt{p_o^2 + 2\Omega J} \tag{7.52}$$

7.2.2
Specific Flow Rate

As mentioned earlier, energy flux, J, in Eq. (7.42) has been used in several studies of performance of a GC column. It provides the simplest way for avoiding the use of the distance-dependent gas velocity, u, as a variable in equations for performance of GC analyses. However, further study suggests that J is not the best metric of gas flow in

GC. For one thing, J is an abstract rather than a practical flow metric. It is not practical to measure a flow of a fluid in a tube in units of energy/area/time.

The quantity J addresses special needs of GC caused by the gas decompression along a column, and, if implemented in GC, would probably be useful for GC alone. On the other hand, the gas flow rate, F, – another distance-independent flow metric – is too complex for theoretical studies. Especially disturbing is the need to deal with the reference temperature, T_{ref}, – a somewhat arbitrary parameter from theoretical point of view.

On top of these shortcomings, there is something else that is very important from theoretical point of view. Chromatographically optimal flow rate is proportional to the internal diameter, d, of a capillary column [28] (Chapter 10). This together with Eqs. (7.43) and (7.26) implies that optimal energy flux is inversely proportional to d. In both cases, a change in a column diameter changes the optimal value of the flow metric. It is desirable to construct the flow metric in such a way that its optimal value is independent of column dimensions. This makes it possible to independently explore the effect of the column dimensions and the value of the flow metric on a column operation. Since optimal flow rate in a capillary column is proportional to its diameter, d, the desirable flow metric for an open tube should be proportional to F/d. Similar consideration applies to packed tubes.

Let us call the flow rate per unit of a tube characteristic cross-sectional dimension (d_{cx}) Eq. (7.39) as specific flow rate, f. In an open tube, f can be defined as

$$f = \frac{T}{T_{\mathrm{ref}}} \cdot \frac{F}{d} \tag{7.53}$$

As will be seen shortly, the factor T/T_{ref} in this definition removes from f the temperature normalization in F. With this, the need to deal with specifics of practical measurement of F is removed from theoretical studies where flow is measured by f. Equation (7.53) can be extended to specific flow rate in a packed tube as

$$f = \frac{T}{T_{\mathrm{ref}}} \left(\frac{d_{\mathrm{p}}}{d}\right)^2 \frac{F}{\varepsilon d_{\mathrm{p}}} = \frac{T}{T_{\mathrm{ref}}} \frac{F}{\varepsilon \gamma_{\mathrm{p}}^2 d_{\mathrm{p}}} \tag{7.54}$$

where

$$\gamma_{\mathrm{p}} = \frac{d}{d_{\mathrm{p}}} \tag{7.55}$$

Relationship, Eq. (7.54), between F and f in a packed tube depends on the tube internal diameter, d, and on the particles size, d_{p}, of the packing material. As a result, the relationship is not as straightforward as Eq. (7.53) for an open tube. Typically, internal diameter of a packed column tracks the particle size of its packing material, so that the smaller is d_{p}, the smaller is d. In all packed tubes with the same ratio d/d_{p}, F is proportional to f as it is in an open tube. Equations (7.53) and (7.54) can be combined as

$$F = \gamma_F f \tag{7.56}$$

where

$$\gamma_F = d_{cx}\frac{T_{ref}}{T}\begin{cases}1, & \text{open tube}\\ \varepsilon\gamma_p^2, & \text{packed tube}\end{cases} \tag{7.57}$$

Combining Eq. (7.56) with Eqs. (7.25) and (7.26), one can also express f via u and J:

$$f = \frac{\pi d_{cx} p u}{4 p_{st}} = \frac{\pi d_{cx} J}{4 p_{st}} \tag{7.58}$$

As a result, Eq. (7.43) can be extended as

$$J = pu = p_i u_i = p_o u_o = \frac{F}{\gamma_{FA}} = \frac{4 p_{st}}{\pi d_{cx}} f \tag{7.59}$$

Due to Eqs. (7.48), (7.58) and (7.38), f can be also found as

$$f = \frac{\pi d_{cx}(p_i^2 - p_o^2)}{8 p_{st} \Omega} \tag{7.60}$$

$$f = \frac{\pi d_{cx}^3 (p_i^2 - p_o^2)}{8 \gamma_\Omega L \eta p_{st}} \tag{7.61}$$

Equation (7.58) shows that, in a given tube, specific flow rate, f, is proportional to the gas energy flux, J. Like the energy flux and the gas velocity, u, quantity f is independent of such secondary factors as the reference temperature, T_{ref}, of measurement of the flow rate and packing porosity, ε, in a tube. This simplifies the theoretical utilization of quantity f as a chromatographically convenient metric of a column flow. On the other hand, gas flow rate, F, – a practically convenient metric of the gas flow in a GC column – relates to f in a simple way described in Eq. (7.56).

7.2.3
Spatial Profiles of Pressure and Velocity

Darcy's law indicates that, in order for the flow of a gas in a tube to exist, there has to be a negative pressure gradient along the tube, Figure 7.2. The negative gradient causes reduction in the pressure, p, along the z-axis. That, in turn, causes reduction in the gas density in the same direction. To maintain a mass-conserving flow, the density reduction with the increase in z must be compensated by the increase in the gas velocity, u, with the increase in z. As a result, both p and u are not uniform, but are subjects to spatial variations.

Due to Eqs. (7.43) and (7.47), inlet velocity, u_i, of a carrier gas can be expressed as

$$u_i = \frac{p_i^2 - p^2}{2 p_i \Omega}\cdot\frac{L}{z} \tag{7.62}$$

where p is pressure at an arbitrary location along the tube. It follows from the last formula that

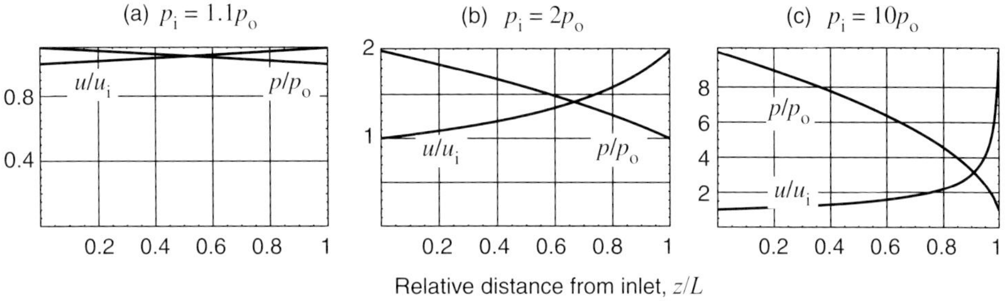

Figure 7.3 Pressure, p, Eq. (7.66), and velocity, u, Eq. (7.67), profiles in mass-conserving flow of a gas: (a) weak decompression ($\Delta p \ll p_o$), (b) intermediate decompression ($\Delta p \approx p_o$), and (c) strong decompression of the gas ($\Delta p \gg p_o$).

$$p = \sqrt{p_i^2 - \frac{2p_i u_i z\Omega}{L}} \tag{7.63}$$

To identify the key factors in these formulae, let us denote

$$L_{ext} = \frac{p_i^2}{p_i^2 - p_o^2} L = \frac{P^2}{P^2 - 1} L = \frac{L}{j_2} \tag{7.64}$$

where

$$j_2 = \frac{L}{L_{ext}} = \frac{p_i^2 - p_o^2}{p_i^2} = \frac{P^2 - 1}{P^2} \tag{7.65}$$

Interpretation of quantity L_{ext} is coming shortly. Substitution of Eq. (7.50) in Eq. (7.63) and substitution of the result in the formula $u = u_i p_i / p$ that follows from Eq. (7.43) yield the following two formulae, Figure 7.3:

$$p = p_i \sqrt{1 - \frac{z}{L_{ext}}} \tag{7.66}$$

$$u = \frac{u_i}{\sqrt{1 - z/L_{ext}}} \tag{7.67}$$

Figure 7.3 exposes two distinct profiles of p and u corresponding to two conditions of the gas decompression along the tube:

$$\text{Decompression is} \begin{cases} \text{weak} & \text{when} \quad \Delta p \ll p_o \\ \text{strong} & \text{when} \quad \Delta p \gg p_o \end{cases} \tag{7.68}$$

When the decompression is weak, both p and u remain nearly uniform, Figure 7.3a, (nearly constant along the entire length of the tube). Equation (7.37) yields

$$u = \frac{\Delta p}{\Omega} \ (\Delta p \approx p_o) \tag{7.69}$$

When the decompression is strong, large pressure drop along the tube causes proportionally large increase in the gas velocity, Figure 7.3c, so that the product pu in Eq. (7.59) remains the same at any z. Starting with a relatively low acceleration near the inlet, Figure 7.3c, the gas velocity rapidly accelerates near the outlet. As a result, the outlet velocity, u_o, of a gas is significantly higher than its inlet velocity, u_i.

A case where $\Delta p \approx p_o$, Figure 7.3b, can be treated as an intermediate between the large and the weak decompression.

Note 7.2

Conditions that are designated here as the weak/strong decompression of a gas along a tube are also known as the conditions of low-/high-pressure drop along a tube [29–32], low-/high-pressure gradients along a tube [24, 25], and noncompressible/compressible flow. These terms could be confusing because what eventually counts is the degree of change in gas velocity, u, resulted from the change in a gas density along a tube.

The same pressure drop of, say, 10 kPa would cause only a weak decompression when a tube outlet is at atmospheric pressure while causing a very strong decompression when a tube outlet is at vacuum. Therefore, the same pressure drop of, say, 10 kPa could be considered as low in some conditions, and as high in others. Similarly, the same pressure gradient could be treated as low or high depending on other conditions.

The term *compressible flow* is also not very relevant to the existence or the absence of a gas decompression along a tube. Indeed, although an ideal gas is always compressible, its actual decompression might be insignificant or very significant depending on other conditions. □

Conditions of the weak and the strong decompression in Eq. (7.68) can be expressed via specific flow rate, f, rather than pressure. It follows from Eq. (7.60) that

$$f \ll \frac{\pi d_{cx} p_o^2}{8 p_{st} \Omega} \quad \text{at} \quad \Delta p \ll p_o \text{ (weak decompression)} \tag{7.70}$$

$$f \gg \frac{\pi d_{cx} p_o^2}{8 p_{st} \Omega} \quad \text{at} \quad \Delta p \gg p_o \text{ (strong decompression)} \tag{7.71}$$

7.2.4
Critical Length of a Tube

Pressure and flow conditions that identify the boundaries between the strong and the weak decompression of a gas in a tube are described in Eqs. (7.68), (7.70), and (7.71). When a tube operates under *a priori* known flow conditions, the degree of the gas decompression in it can be deduced directly for the tube dimensions. It is known, for example, that gas decompression in a short capillary column with internal diameter of 0.53 mm is typically weak. On the other hand, the decompression in a long column with 0.1 mm diameter is typically strong. Apparently, for each column diameter and carrier gas type, the critical length, L_{crit}, of a column can be identified. If, under normal operational conditions, actual tube is substantially longer than L_{crit}, then the gas decompression in the tube is strong, and vice versa.

As described in Eq. (7.68), the gas decompression in a tube is strong if the pressure drop (Δp) across the tube is significantly larger that the outlet pressure, p_o, ($\Delta p \gg p_o$). Conversely, the decompression is weak if $\Delta p \ll p_o$. This suggests that transition from strong to weak decompression takes place somewhere in the vicinity of $\Delta p = p_o$.

Let us explicitly define the borderline inlet pressure, $p_{i,B}$, and the borderline pressure drop, Δp_B as

$$p_{i,B} = P_B p_o, \quad \Delta p_B = p_{i,B} - p_o = (P_B - 1) p_o \tag{7.72}$$

where P_B is the relative borderline inlet pressure. The previous observation suggests that Δp_B should be close to one and, therefore, P_B should be close to two. To be specific, we assume that

$$P_B = 2 \tag{7.73}$$

The last formula becomes

$$\Delta p_B = p_o, \quad p_{i,B} = p_o + \Delta p_B = 2p_o \tag{7.74}$$

This particular choice of the borderline condition is somewhat arbitrary. However, it is simple and has chromatographic justification (Chapter 10).

Note 7.3

Different treatment of the borderline conditions with slightly different outcome of $\Delta p_B \approx 0.8\ p_o$ and $p_{i,B} \approx 1.8\ p_o$ has been evaluated elsewhere [14]. □

The critical length, L_{crit}, of a column with a given specific flow rate, f, is the length that requires the borderline pressure for that value of f. It follows from Eq. (7.61) that the length, L_{crit}, of a tube that yields a predetermined f at a borderline pressure can be found as

$$L_{crit} = \frac{\pi d_{cx}^3 p_o^2 (P_B^2 - 1)}{8\gamma_\Omega\ \eta p_{st} f} \tag{7.75}$$

which, at $P_B = 2$ (Eq. (7.73)), becomes

$$L_{crit} = \frac{3\pi d_{cx}^3 p_o^2}{8\gamma_\Omega\ \eta p_{st} f} \tag{7.76}$$

The critical length, L_{crit}, of a tube can be associated not only with a given specific flow rate in it but also with a predetermined values of other flow parameters such as flow rate, F, and outlet velocity, u_o. Thus, due to Eq. (7.59), Eq. (7.76) can be expressed as

$$L_{\text{crit}} = \frac{3d_{\text{cx}}^2 p_o}{2\gamma_\Omega \eta u_o} \tag{7.77}$$

Depending on the following conditions for a given f, an L-long tube can be viewed as being pneumatically short or pneumatically long:

$$\text{a tube is} \begin{cases} \text{pneumatically short} & \text{if} \quad L \ll L_{\text{crit}} \\ \text{pneumatically long} & \text{if} \quad L \gg L_{\text{crit}} \end{cases} \tag{7.78}$$

In a pneumatically short tube, the decompression is weak, whereas it is strong in a pneumatically long tube. In other words, conditions in Eqs. (7.78) and (7.68) for a tube to be pneumatically short and to have the weak gas decompression in it are equivalent. Similarly equivalent are the conditions of pneumatically long tube and of the strong decompression in it.

Example 7.1

Suppose that a desirable specific flow rate of hydrogen is 10 mL/min/mm (this is optimal specific flow of hydrogen in a capillary column [28, 33]). Consider an open tube that has internal diameter $d = 0.1$ mm and operates at $p_o = 1$ atm and 100 °C ($\eta = 10.40$ µPa s, Table 6.4). Equation (7.76) yields $L_{\text{crit}} = 2.15$ m. A tube that is substantially shorter than 2.15 m is pneumatically short. The gas decompression in it is weak. A tube that is substantially longer than 2.15 m is pneumatically long. The gas decompression in it is strong. □

It follows from Eq. (7.76) that, at $p_o = 0$ (vacuum at the outlet), $L_{\text{crit}} = 0$. Therefore,

Statement 7.3

Any tube with its outlet at vacuum is pneumatically long (has strong gas decompression) at any nonzero flow rate.

A required column pressure can be found directly from the column actual length, L, and critical length, L_{crit}. Indeed, it follows from Eqs. (7.75), (7.61), and (7.73) that

$$P = \sqrt{1 + \frac{L}{L_{\text{crit}}}(P_B^2 - 1)} = \sqrt{1 + 3L/L_{\text{crit}}} \tag{7.79}$$

Example 7.2

In Example 7.1, $L_{\text{crit}} = 2.15$ m. It follows from Eq. (7.79) that, for a 10-m-long tube of Example 7.1, $P = 3.87$ or, equivalently $p_i = 3.87$ atm, $\Delta p = 2.87$ atm $= 280.8$ kPa. □

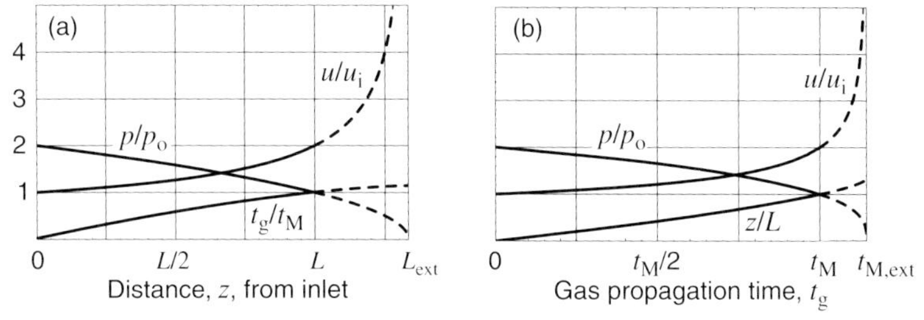

Figure 7.4 Spatial (a) and temporal (b) profiles of parameters of a static flow with $p_i = 2p_o$ (the same as in Figure 7.3b). Quantity t_g in (a) as well as quantities p, t_g, z, and u in (b) are parameters of a narrow gas packet that enters the tube at time $t_g = 0$ and migrates toward the outlet with the gas stream. The spatial profiles of gas pressure, p, and velocity, u, in (a) are described in Eqs. (7.66) and (7.67). Their temporal profiles in (b) are described in Eqs. (7.99) and (7.111). Relations between the migration time, t_g, of a narrow gas packet and its distance, z, from the inlet are described in Eqs. (7.106), (7.107) and (7.108). According to Eqs. (7.64) and (7.101), $L_{ext} = 4L/3$ and $t_{M,ext} = 8t_M/7$. Dashed lines represent the profiles in the vacuum extension of the tube (outside its actual length, L).

7.2.5
Vacuum-Extended Length and Related Parameters

So far, interpretation of quantity L_{ext} in Eqs. (7.66) and (7.67) has not been discussed. The quantity L_{ext} is defined in Eq. (7.64) so that $L_{ext} \geq L$, Figure 7.4a. The coordinate z in Eqs. (7.66) and (7.67) represents the distance from the inlet in an L-long tube only when $0 \leq z \leq L$. At $z = L_{ext}$, Eqs. (7.66) and (7.67) yield $p = 0$ and $u = \infty$ suggesting that L_{ext} is the length of a tube that, when its outlet is at vacuum, has for $0 \leq z \leq L$ exactly the same spatial profiles of p and u as those in the actual L-long tube with the actual outlet pressure, p_o. So described L_{ext}-long tube can be treated as a vacuum extension of an actual L-long prototype, and L_{ext} – as its vacuum-extended length. When both the L-long prototype tube having outlet at p_o and its L_{ext}-long vacuum extension having outlet at vacuum operate at the same inlet pressure, p_i, both have the same spatial profiles of pressure, p, and gas velocity, u, for all coordinates, z, between 0 and L, that is for $0 \leq z \leq L$. In the extreme cases of the weak and the strong gas decompression along a tube, L_{ext} can be estimated as

$$L_{ext} = \frac{p_o}{2\Delta p} L \gg L \quad \text{at} \quad \Delta p \ll p_o \tag{7.80}$$

$$L_{ext} \approx L \quad \text{at} \quad \Delta p \gg p_o \tag{7.81}$$

$$L_{ext} = L, \quad \text{at} \quad p_o = 0 \quad \text{(outlet at vacuum)} \tag{7.82}$$

Utilization of a concept of a vacuum extension of a tube simplifies descriptions of many pneumatic relations and makes them conceptually more transparent (compare, for example, Eqs. (7.63) and (7.66)). Vacuum-extended versions of other length-related parameters of a tube can be also introduced with similar advantages.

Based on Eq. (7.38), a tube's vacuum-extended pneumatic resistance, Ω_{ext}, can be defined as

$$\Omega_{ext} = \frac{\gamma_\Omega L_{ext} \eta}{d_{cx}^2} = L_{ext}\eta \cdot \begin{cases} \dfrac{32}{d^2}, & \text{open round tube} \\ \dfrac{270(1-\varepsilon)^2}{\varepsilon^2 d_p^2}, & \text{packed tube} \end{cases} \tag{7.83}$$

This, according to Eq. (7.64), implies

$$\Omega_{ext} = \frac{p_i^2}{p_i^2 - p_o^2}\Omega = \frac{P^2}{P^2-1}\Omega = \frac{\Omega}{j_2} \tag{7.84}$$

The latter allows one to rewrite Eq. (7.50) for inlet velocity, u_i, (which is the same in vacuum extension of a tube as in its L-long prototype) as

$$u_i = \frac{p_i}{2\Omega_{ext}} \tag{7.85}$$

This shows that the inlet velocity, u_i, of a gas in a tube with an arbitrary length, L, is equal to one-half of the inlet pressure divided by the vacuum-extended pneumatic resistance of the tube – a relation that is simpler and more transparent than Eq. (7.50).

Other vacuum-extended parameters of a tube will be introduced later.

7.2.6 Hold-up Time

The hold-up time [9, 16, 17, 34–40] (air time, dead time [41–50], void time [13], methane time, t_M) is the time,

$$t_M = \int_0^L \frac{dz}{u} \tag{7.86}$$

that, in a static flow, takes for a narrow gas packet, or, in an average, for any gas molecule to travel from the inlet to the outlet of the tube.

There are several ways to find t_M from other pneumatic parameters. Substitution of Eqs. (7.67) and (7.50) in Eq. (7.86) yields

$$t_M = \frac{4\Omega L(p_i^3 - p_o^3)}{3(p_i^2 - p_o^2)^2} \tag{7.87}$$

This can be also expressed as

$$t_M = \frac{\Omega L}{j_H \Delta p} \tag{7.88}$$

where, Figure 7.5a,

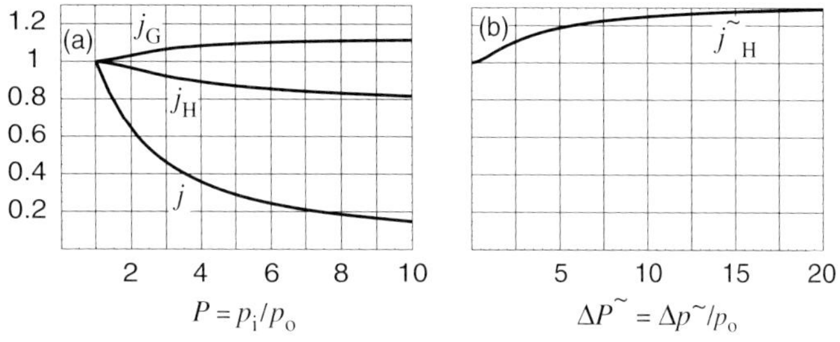

Figure 7.5 Gas compressibility factors j, j_H, $\tilde{j}_H$, and j_G, Eqs. (7.114), (7.89), and (10.87).

$$j_H = \frac{3(p_i^2 - p_o^2)(p_i + p_o)}{4(p_i^3 - p_o^3)} = \frac{3(p_i + p_o)^2}{4(p_i^2 + p_i p_o + p_o^2)} \tag{7.89}$$

$$j_H = \frac{3(P^2 - 1)(1 + P)}{4(P^3 - 1)} = \frac{3(1 + P)^2}{4(1 + P + P^2)} \tag{7.90}$$

$$j_H = \frac{3}{4}\frac{\Delta p^2 + 4p_o\Delta p + 4p_o^2}{\Delta p^2 + 3p_o\Delta p + 3p_o^2} = \frac{3}{4}\frac{4 + 4(\Delta p/p_o) + (\Delta p/p_o)^2}{3 + 3(\Delta p/p_o) + (\Delta p/p_o)^2} \tag{7.91}$$

$$j_H = \frac{3}{4}\frac{(2 + \Delta P)^2}{3 + 3\Delta P + \Delta P^2} \tag{7.92}$$

The notations P, Δp, and ΔP are described in Eqs. (7.9), (7.8), and (7.10), respectively.

In the form that appears in the right-hand side of Eq. (7.91), the quantity j_H is known from Halász *et al.* [51] since 1964. It will be called below as the *Halász compressibility* factor. The quantity j_H is confined, Figure 7.5a, within a narrow range of

$$\frac{3}{4} \le j_H \le 1 \tag{7.93}$$

approaching the highest value of one in the case of a weak gas decompression where p_i and p_o are nearly the same, and dropping to the lowest value of 0.75 in the case of a vacuum-outlet operations where $p_o = 0$.

Sometimes, it can be convenient (Appendix 7.A.3) to consider quantity

$$\tilde{j}_H = \frac{1}{j_H} \tag{7.94}$$

As the inversion of quantity j_H, quantity $\tilde{j}_H$ is confined, Figure 7.5b, within the bounds of

$$1 \le \tilde{j}_H \le \frac{4}{3} \tag{7.95}$$

Back to the hold-up time, t_M. Substitution of Eq. (7.38) in Eq. (7.88) yields

$$t_M = \frac{\gamma_\Omega \ell^2 \eta}{j_H \Delta p} = \frac{\ell^2 \eta}{j_H \Delta p} \cdot \begin{cases} 32, & \text{open round tube} \\ 270(1-\varepsilon)^2/\varepsilon^2, & \text{packed tube} \end{cases} \tag{7.96}$$

where

$$\ell = \frac{L}{d_{cx}} = \begin{cases} L/d, & \text{open round tube} \\ L/d_p, & \text{packed tube} \end{cases} \tag{7.97}$$

is the dimensionless length of the tube. Equation (7.96) shows that, when a column pressure is fixed, only the ratio, ℓ, of the tube length and it characteristic cross-sectional dimension (but not each of these parameters independently) affect t_M. This means, among other things, that the measurement of the hold-up time alone does not allow one to find the length or the inner diameter of an open tube (particle diameter of packing material in a packed tube). It only allows one to find the ratio of the two parameters.

7.2.7 Temporal Profiles

The concepts described here represent a special case of the concepts described in Chapter 4.

Consider again a static gas flow in a tube (p, u, and so forth do not change with time, t). It is important to emphasize that the static nature of the flow – the absence of change in a gas parameter with time – does not preclude its spatial variation [52–54], Eqs. (7.66) and (7.67) – being different at different coordinates z.

There can be another perspective, however. Instead of considering velocity, u, of a gas flow at any location z along a tube, one can imagine a narrow packet of "marked" gas molecules and observe its migration from inlet to outlet of the tube. Let us call velocity of such packet as *gas propagation velocity*. The gas propagation velocity changes in time from the inlet velocity, u_i, at the time zero of the packet entry into the tube to the outlet velocity, u_o, at the time t_M when the packet leaves the tube. This means that, in a decompressing flow, the gas propagation velocity can change in time even if the flow is static.

Let $t_g = t_g(z)$ be the gas propagation time in a tube – the time that it takes for a gas packet to travel from the inlet to an arbitrary coordinate, z, along a tube. At $z = L$, one has

$$t_g(L) = t_M \tag{7.98}$$

The gas propagation velocity, u, can be found as (Appendix 7.A.2, see also Ref. [10]), Figure 7.4b,

$$u = \frac{u_i}{\left(1-(t_g/t_{M,ext})\right)^{1/3}} \tag{7.99}$$

where

$$t_{M,ext} = t_g\big|_{z=L_{ext}} = \frac{2L_{ext}}{3u_i} \tag{7.100}$$

is the vacuum-extended hold-up time – the hold-up time in (L_{ext}-long) vacuum extension of a prototype (L-long) tube.

In view of Eqs. (7.50), (7.64), (7.88), (7.89), and (7.9), Eq. (7.100) can be arranged as

$$t_{\mathrm{M,ext}} = \frac{p_{\mathrm{i}}^3}{p_{\mathrm{i}}^3 - p_{\mathrm{o}}^3} t_{\mathrm{M}} = \frac{P^3}{P^3-1} t_{\mathrm{M}} = \frac{t_M}{j_3} \tag{7.101}$$

where

$$j_3 = \frac{t_M}{t_{\mathrm{M,ext}}} = \frac{p_{\mathrm{i}}^3 - p_{\mathrm{o}}^3}{p_{\mathrm{i}}^3} = \frac{P^3-1}{P^3} \tag{7.102}$$

or, after the substitution of Eqs. (7.96), (7.97), (7.8) (7.89), and (7.64),

$$t_{\mathrm{M,ext}} = \frac{4\gamma_\Omega \ell_{\mathrm{ext}}^2 \eta}{3p_i} = \frac{4\ell_{\mathrm{ext}}^2 \eta}{3p_i} \cdot \begin{cases} 32 & \text{open round tube} \\ 270(1-\varepsilon)^2/\varepsilon^2 & \text{packed tube} \end{cases} \tag{7.103}$$

where

$$\ell_{\mathrm{ext}} = \frac{L_{\mathrm{ext}}}{d_{\mathrm{cx}}} = \begin{cases} L_{\mathrm{ext}}/d, & \text{open round tube} \\ L_{\mathrm{ext}}/d_p, & \text{packed tube} \end{cases} \tag{7.104}$$

is vacuum-extended dimensionless length of the tube.

It follows from Eq. (7.101) that, Figure 7.4b, $t_{\mathrm{M,ext}} \geq t_{\mathrm{M}}$. In the case of a weak decompression along a prototype tube, $t_{\mathrm{M,ext}} \gg t_{\mathrm{M}}$. In this case, $|\Delta p| \ll p_{\mathrm{o}}$ and $t_{\mathrm{M,ext}}$ can be estimated as $t_{\mathrm{M,ext}} \approx p_{\mathrm{o}} t_{\mathrm{M}}/(3\Delta p)$. In the opposite case of a strong gas decompression along a prototype tube where $\Delta p \gg p_{\mathrm{o}}$, $t_{\mathrm{M,ext}}$ can only be slightly greater than t_{M}, and, in the extreme case of a vacuum outlet in a prototype tube ($p_{\mathrm{o}} = 0$), $t_{\mathrm{M,ext}} = t_{\mathrm{M}}$.

How the distance, z, traveled by a gas packet relates to its propagation time, t_{g}? By the definition, the velocity, u, of a gas packet located at z is the derivative, $\mathrm{d}z/\mathrm{d}t_{\mathrm{g}}$, that is

$$u = \frac{\mathrm{d}z}{\mathrm{d}t_{\mathrm{g}}} \tag{7.105}$$

Equation (7.99) becomes $\mathrm{d}z/\mathrm{d}t_{\mathrm{g}} = u_{\mathrm{i}}/(1 - t_{\mathrm{g}}/t_{\mathrm{M,ext}})^{1/3}$ or $\mathrm{d}z = (u_{\mathrm{i}}/(1 - t_{\mathrm{g}}/t_{\mathrm{M,ext}})^{1/3})\mathrm{d}t_{\mathrm{g}}$. After the integration of the last formula and accounting for Eq. (7.100), one has

$$\left(1 - \frac{z}{L_{\mathrm{ext}}}\right)^3 = \left(1 - \frac{t_{\mathrm{g}}}{t_{\mathrm{M,ext}}}\right)^2 \tag{7.106}$$

Solutions of this equation for t_{g} and for z are, Figure 7.4:

$$t_{\mathrm{g}} = \left(1 - \left(1 - \frac{z}{L_{\mathrm{ext}}}\right)^{3/2}\right) \cdot t_{\mathrm{M,ext}} \tag{7.107}$$

$$z = \left(1 - \left(1 - \frac{t_{\mathrm{g}}}{t_{\mathrm{M,ext}}}\right)^{2/3}\right) \cdot L_{\mathrm{ext}} \tag{7.108}$$

At $z = L$, $t_{\mathrm{g}} = t_{\mathrm{M}}$, and vice versa. In view of that, the last two formulae can be rearranged as

$$t_g = \frac{1-(1-j_2(z/L))^{3/2}}{1-(1-j_2)^{3/2}} t_M \tag{7.109}$$

$$z = \frac{1-(1-j_3(t_g/t_M))^{2/3}}{1-(1-j_3)^{2/3}} L \tag{7.110}$$

The quantities j_2 and j_3 are defined in Eqs. (7.65) and (7.102).

Not only the velocity, u, and the location, z, of a gas packet change with propagation time, t_g, required for the packet to reach the coordinate z, but also the pressure, p, within the packet changes with t_g. Due to Eqs. (7.43) and (7.99), one has, Figure 7.4b,

$$p = \left(1 - \frac{t_g}{t_{M,ext}}\right)^{1/3} p_i \tag{7.111}$$

7.2.8
Averages

Using general definition, Eq. (3.15), of an average of an arbitrary function, one can find the following averages for gas pressure and velocity.

The spatial average pressure (briefly, average pressure) is given as [10, 55, 56]

$$\bar{p} = \langle p \rangle_z = \frac{1}{L}\int_0^L p dz \tag{7.112}$$

After the substitution of p from Eq. (7.66), one has

$$\bar{p} = \frac{p_o}{j} \tag{7.113}$$

where, Figure 7.5a,

$$j = \frac{3(p_i^2 - p_o^2)p_o}{2(p_i^3 - p_o^3)} = \frac{3(p_i + p_o)p_o}{2(p_i^2 + p_i p_o + p_o^2)} \tag{7.114}$$

or, in view of notation $P = p_i/p_o$, Eq. (7.9),

$$j = \frac{3(P^2 - 1)}{2(P^3 - 1)} = \frac{3(1 + P)}{2(1 + P + P^2)} \tag{7.115}$$

The quantity j in this formula is the *James–Martin compressibility factor* [1, 8, 9, 11, 16, 17, 57–64] originally described by James and Martin [1, 57] in 1952.

Note 7.4

The quantity j in Eqs. (7.114) and (7.115) is known under several names such as the *compressibility correction factor* [16, 59, 61], the *James–Martin pressure gradient correction factor* [8], and so forth [9, 16, 17, 58]. The term compressibility correction factor (recommended by IUPAC [16]) for j appears to be ambiguous because j is not the only quantity that solely depends on the compressibility of an ideal gas and/or describes the effects of the compressibility on pneumatic parameters. Earlier introduced, Eqs. (7.89), (7.90), (7.91), and (7.92), *Halász compressibility factor*, j_{H}, (also known [59] as the *pressure correction factor* of Halász, or the *reduced pressure correction factor*) is also a sole function of a gas compressibility. Both j and j_{H} reflect (different aspects of) the compressibility of ideal gas, and neither seems to be more suitable than the other to be treated as the compressibility factor or as the compressibility correction factor. There are other compressibility factors, Figure 7.5, that are useful in GC [59]. To adopt a uniform approach to the naming of various compressibility factors, it seems to be appropriate to recognize all of them for what they are – the compressibility factors – and to distinguishing them by the names of their original authors. Thus, j is James–Martin compressibility factor, j_{H} is Halász compressibility factor, j_{G} is *Giddings compressibility factor* (Chapter 10), and so forth. □

The temporal average velocity (briefly, average velocity),

$$\bar{u} = \langle u \rangle_t = \frac{1}{t_{\mathrm{M}}} \int_0^{t_{\mathrm{M}}} u \mathrm{d}t_{\mathrm{g}} \tag{7.116}$$

of a gas in a tube is the average of the gas propagation velocity over the time required for a packet of the gas molecules to travel from the inlet to the outlet of the tube. It should be noted that what is briefly called here and in the literature as gas velocity, u, is actually a cross-sectional average gas velocity (see the text preceding the definition of quantity u in Eq. (7.2)). Therefore, both names for quantity $\bar{u}$ in Eq. (7.116) – the temporal average velocity and its shorter version average velocity – are themselves the short names for what can be more accurately called as the *temporal average of cross-sectional average gas velocity*. Golay [3, 65–69] called quantity u in Eq. (7.2), that is, the cross-sectional average gas velocity, as the average gas velocity. In this book, the term gas velocity always means quantity u defined in Eq. (7.2). Accordingly, the term average velocity always means quantity $\bar{u}$ defined in Eq. (7.116). Both terms are in line with IUPAC recommendations [16] and with broadly accepted use of these terms in the literature.

Because, by the definition, $u = \mathrm{d}z/\mathrm{d}t_{\mathrm{g}}$, Eq. (7.116) yields

$$\bar{u} = \frac{L}{t_{\mathrm{M}}} \tag{7.117}$$

This formula can be viewed as an alternative definition of quantity $\bar{u}$ [6, 16, 17, 59]. However, it is Eq. (7.116) that shows that $\bar{u}$ is the temporal average. Inversion of Eq. 7.1Eq (7.117) yields

$$t_{\mathrm{M}} = \frac{L}{\bar{u}} \tag{7.118}$$

The quantity $\bar{u}$ can be expressed via other parameters. Substitution of Eq. (7.88) in Eq. (7.117) yields [6, 9, 51]

$$\bar{u} = \frac{j_H \Delta p}{\Omega} \tag{7.119}$$

which after the substitution of Eq. (7.89) becomes

$$\bar{u} = \frac{3(p_i^2 - p_o^2)^2}{4\Omega(p_i^3 - p_o^3)} \tag{7.120}$$

Combination of this formula with Eqs. (7.51) and (7.114) yields a well-known formula [6, 16, 17, 59]

$$\bar{u} = j u_o \tag{7.121}$$

This formula together with Eq. (7.113) can be viewed as alternative definitions of James–Martin compressibility factor (j).

The quantity j can be also used to describe relationship between inlet and average velocities (u_i and $\bar{u}$). Due to Eqs. (7.59) and (7.9), Eq. (7.121) can be arranged as

$$\bar{u} = j\, P\, u_i \tag{7.122}$$

This together with Eq. (7.115) implies

$$u_i \leq \bar{u} \leq 1.5\, u_i \tag{7.123}$$

meaning that $\bar{u}$ is not far different from u_i. The difference increases with the increase in the degree of the gas decompression along the tube, and is the highest when $p_o = 0$.

It has already been established, Eqs. (7.36) or (7.59), that the product $p\,u$ is the same for any coordinate, z, along a tube. What about the product $\bar{p} \cdot \bar{u}$? From Eqs. (7.113) and (7.119), one has

$$\bar{p}\, \bar{u} = \frac{p_o \Delta p j_H}{\Omega j} \tag{7.124}$$

It follows from Eqs. (7.89) and (7.114) that the ratio j_H/j in the above formula can be found as

$$\frac{j_H}{j} = \frac{p_i + p_o}{2p_o} \tag{7.125}$$

Substitution of this in Eq. (7.124) yields $\bar{p}\bar{u} = \Delta p(p_i + p_o)/(2\Omega) = (p_i^2 - p_o^2)/(2\Omega)$ which due to Eq. (7.51) indicates that $\bar{p}\bar{u} = p_o\, u_o$. The latter allows one to extend Eq. (7.59) as

$$J = pu = \bar{p}\, \bar{u} = p_i u_i = p_o u_o = \frac{F}{\gamma_{FA}} = \frac{4p_{st}}{\pi d_{cx}} f \tag{7.126}$$

Let us notice that the adopted here conventional notations [12, 16, 59, 60, 62, 70] $\bar{p}$ and $\bar{u}$, though similar in form, reflect different concepts. While $\bar{p}$ is the spatial average, Eq. (7.112), $\bar{u}$ is the temporal one, Eq. (7.116).

Another average that is useful [8] in GC is the spatial average velocity [10, 55, 56]

$$\langle u \rangle_s = \langle u \rangle_z = \frac{1}{L}\int_0^L u \mathrm{d}\, z \tag{7.127}$$

of a gas. Substitution of Eq. (7.67) in Eq. (7.127) and accounting for Eqs. (7.50) and (7.64) yields

$$\langle u \rangle_s = \frac{\Delta p}{\Omega} \tag{7.128}$$

Due to Eqs. (7.119) and (7.128), one also has [10]

$$\frac{\bar{u}}{\langle u \rangle_s} = j_H \tag{7.129}$$

This, due to Eq. (7.93), implies

$$0.75\langle u \rangle_s \le \bar{u} \le \langle u \rangle_s \tag{7.130}$$

Therefore, the temporal, $\bar{u}$, and the spatial, $\langle u \rangle_s$, velocities of a gas are not very much different from each other. Accounting for Eq. (7.128), one can write

$$\bar{u} \approx \langle u \rangle_s = \frac{\Delta p}{\Omega} \tag{7.131}$$

7.2.9 Virtual Pressure

It seems natural for gas velocity to be proportional to pressure drop, Δp, along the tube and inversely proportional to the pneumatic resistance, Ω, of the tube. That is exactly how the velocity, u, in the case of a weak gas decompression depends, Eq. (7.69), on Δp and Ω, and, more generally, how the spatial average velocity, $\langle u \rangle_s$, depends, Eq. (7.128), on Δp and Ω.

Far more important than $\langle u \rangle_s$ in GC is the temporal average velocity, $\bar{u}$. It is tempting to find a suitable concept of pressure drop, $\Delta\tilde{p}$, so that, similarly to Eq. (7.128), $\bar{u}$ can be expressed as

$$\bar{u} = \frac{\Delta\tilde{p}}{\Omega} \tag{7.132}$$

This approach has several important implications [13] for GC. For now, let us only identify $\Delta\tilde{p}$ in Eq. (7.132) and postpone the discussion of its GC implications.

The quantity $\Delta\tilde{p}$ satisfying Eq. (7.132) can be defined as

$$\Delta\tilde{p} = \Omega\bar{u} \tag{7.133}$$

and, due to Eq. (7.119), found as [13]

$$\Delta\tilde{p} = j_H \Delta p \tag{7.134}$$

This due to Eq. (7.93), implies

$$0.75\Delta p \leq \Delta\tilde{p} \leq \Delta p \tag{7.135}$$

suggesting that $\Delta\tilde{p}$ is not very much different from Δp. Substitution of Eqs. (7.91) and (7.92) in Eq. (7.134) yields exact formula for $\Delta\tilde{p}$:

$$\Delta\tilde{p} = j_{\mathrm{H}}\Delta p = \frac{3}{4}\frac{\Delta p^2 + 4p_{\mathrm{o}}\Delta p + 4p_{\mathrm{o}}^2}{\Delta p^2 + 3p_{\mathrm{o}}\Delta p + 3p_{\mathrm{o}}^2}\Delta p = \frac{3}{4}\frac{(2+\Delta P)^2}{3 + 3\Delta P + \Delta P^2}\Delta p \tag{7.136}$$

Note 7.5

Contemporary GC instruments typically measure ambient pressure, p_{a}, and gauge pressure, p_{g}. Also frequently known from the experimental conditions is outlet pressure, p_{o} (equal to zero for vacuum-outlet operations, almost equal to p_{a} in many other cases, and so forth). The other pressure parameters, Figure 7.2, such as p_{i} and Δp, are internally calculated in a GC system from Eqs. (7.7) and (7.8) if necessary. Also calculated in many cases are the gas flow rate (F), the average velocity ($\bar{u}$), and other pneumatic parameters.

While some pneumatic parameters mentioned here are measured and other are calculated in a GC system, all are typically presented to an operator as independent variables that can be set to arbitrary values within the instrument-specific range. For example, a GC instrument control panel can allow an operator to set a column gauge pressure (a measured parameter) or a flow rate (a calculated parameter). From that perspective, quantity $\Delta\tilde{p}$, Eq. (7.134), is not different from other calculated pneumatic parameters. As other parameters, it can be treated as an independent parameter that can be set to an arbitrary value within a predetermined range. □

Because it is approximately equal to the pressure drop (Δp) and because it affects $\bar{u}$ in Eq. (7.132) in the way similar to the way Δp affects $\langle u\rangle_{\mathrm{s}}$ in Eq. (7.128), quantity $\Delta\tilde{p}$ can be designated as the *virtual pressure drop* (briefly, *virtual pressure*) and can be treated just as any other pressure parameter such as p_{o}, p_{i}, Δp, and so forth.

The virtual pressure, $\Delta\tilde{p}$, allows not only to express $\bar{u}$ in Eq. (7.132) as a quantity proportional to a single pressure parameter but it also makes it possible to simplify expressions for t_{M}. Replacement of $j_{\mathrm{H}}\Delta p$ with $\Delta\tilde{p}$ in Eqs. (7.88) and (7.96) transforms these formulae into

$$t_{\mathrm{M}} = \frac{\Omega L}{\Delta\tilde{p}} \tag{7.137}$$

$$t_{\mathrm{M}} = \frac{\Upsilon_{\Omega}\ell^2\eta}{\Delta\tilde{p}} = \frac{\ell^2\eta}{\Delta\tilde{p}}\cdot\begin{cases} 32, & \text{open round tube} \\ 270(1-\varepsilon)^2/\varepsilon^2, & \text{packed tube} \end{cases} \tag{7.138}$$

Note 7.6

Equation (7.137) points to an important advantage of allowing an operator of a GC instrument to directly control the value of $\Delta\tilde{p}$.

Reproducibility of retention times can be significantly improved by adjusting gauge pressure, p_g, in order to maintain a predetermined value of hold-up time [13], t_M. The adjustment can be made before each critical analysis or a group of analyses. This treatment recommended by (American Society for Testing and Materials (ASTM) [71] is the essence of the retention time locking [13] that can be implemented either manually or with various degrees of automation [72, 73]. Because relation between p_g and t_M is complex, Eqs. (7.7), (7.88), and (7.89), manual adjustment of p_g typically involves a series of time-consuming trials and errors in changing p_g and measurement of t_M. The same result can be achieved in one step by setting of $\Delta\tilde{p}$ calculated from Eq. (7.137) for a required adjustment in t_M. □

Example 7.3

Suppose that, due to the column trimming, t_M dropped from nominal 60 s to actual 55 s. It follows from Eq. (7.137) that, in order to restore the nominal value of t_M, the relative reduction in $\Delta\tilde{p}$ should be equal to the relative reduction in t_M. Thus, if $\Delta\tilde{p} = 100$ kPa yields $t_M = 55$ s then, in order to obtain $t_M = 60$ s, $\Delta\tilde{p}$ should be reduced to $(55/60) \times 100$ kPa $= 91.7$ kPa. □

For $\Delta\tilde{p}$ to be treated as an independent parameter in a GC instrument, it is useful to express other pneumatic parameters – gas flow rate, inlet and/or outlet pressure and/or velocity, and so forth, – as functions of $\Delta\tilde{p}$. One way of doing so is through expressing inlet pressure, p_i, as a function of $\Delta\tilde{p}$. This, due to Eq. (7.132), would also allow one to express p_i as a function of $\bar{u}$. Once p_i is expressed via $\Delta\tilde{p}$ or $\bar{u}$, other parameters that can be expressed via p_i could be also expressed [9] via $\Delta\tilde{p}$ and $\bar{u}$.

To express p_i as a function of $\Delta\tilde{p}$, it is more convenient to deal with the dimensionless parameters, $P = p_i/p_o$, Eq. (7.9), and the relative virtual pressure,

$$\Delta\tilde{P} = \frac{\Delta\tilde{p}}{p_o} \tag{7.139}$$

Substitution of Eq. (7.89) in Eq. (7.134) followed by substitution of Eq. (7.8) yields

$$\Delta\tilde{p} = \frac{3(p_i^2 - p_o^2)^2}{4(p_i^3 - p_o^3)} = \frac{3(1+P)^2(P-1)p_o}{4(1+P+P^2)} \tag{7.140}$$

$$\Delta\tilde{P} = \frac{3(1+P)^2(P-1)}{4(1+P+P^2)} \tag{7.141}$$

Equation (7.141) is a cubic equation for P. Let us look for its solution in the form

$$P = \Phi(\Delta\tilde{P}) \tag{7.142}$$

Function $\Phi(\cdot)$ – a solution of equation

$$\Delta\tilde{P} = \frac{3(1+\Phi(\Delta\tilde{P}))^2(\Phi(\Delta\tilde{P})-1)}{4(1+\Phi(\Delta\tilde{P})+\Phi^2(\Delta\tilde{P}))} \tag{7.143}$$

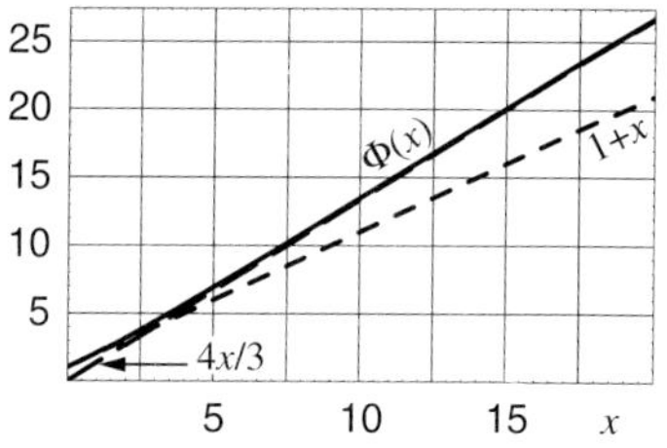

Figure 7.6 Function $\Phi(x)$, Eq. (7.144), (solid line —) and its tangent lines: $1 + x$ (short dashes - - -) – the tangent to $\Phi(x)$ at $x = 0$ and $4x/3$ (long dashes – – –) – the tangent to $\Phi(x)$ at $x = \infty$.

can be expressed as [9]

$$\Phi(x) = \frac{1}{9}(-3 + 4x + X_+^{1/3} + X_-^{1/3}) \tag{7.144}$$

where

$$X_+ = 2(X_1 + 27\sqrt{X_2}), \quad X_- = 2(X_1 - 27\sqrt{X_2})$$
$$X_1 = (3 + 8x)(36 + 3x + 4x2), \quad X_2 = x(3 + 4x)(24 + 3x + 4x2)$$

The seemingly complex formula in Eq. (7.144) describes a smooth, monotonic, nearly linear function, Figure 7.6, whose slope gradually changes from 1.0 at $x = 0$ to $4/3$ at $x = \infty$.

Expressed via the absolute parameters, Eq. (7.142) becomes

$$p_i = p_o \Phi\left(\frac{\Delta\tilde{p}}{p_o}\right) \tag{7.145}$$

Due to Eq. (7.133), the latter can be also expressed as a function of $\bar{u}$:

$$p_i = p_o \Phi\left(\frac{\Omega\bar{u}}{p_o}\right) \tag{7.146}$$

Sometimes, instead Eq. (7.142), it might be more convenient to deal with the formulae

$$\Delta P = \Delta\Phi(\Delta\tilde{P}) \tag{7.147}$$

$$\Delta p = p_o \Delta\Phi\left(\frac{\Delta\tilde{p}}{p_o}\right) \tag{7.148}$$

where

$$\Delta\Phi(x) = \Phi(x) - 1 \tag{7.149}$$

Approximations to functions $\Phi(x)$ and $\Delta\Phi(x)$ are described in Appendix 7.A.3.

7.3
Relations Between Pneumatic Parameters

So far, the flow of the text in this chapter has been subordinated to the logic of introduction of pneumatic concepts and respective parameters. Now, the focus is on relations between already introduced parameters.

It might be useful to have the formulae for the calculation of any previously introduced parameter as functions of any combination of other mutually independent parameters. This, however, would be an enormous task. One way to reduce the task could be to identify a core group of parameters, and to derive the expressions for each core parameter as a function of all mutually independent combinations of other core parameters [9]. Once the relations between the core parameters are known, the relations between the core parameters and parameters outside of the core group can be found in a few straightforward steps. The relationships between the core parameters could be also used for finding the relationships between the parameters that are outside of the core group.

The five parameters, (p_i, p_o, u_i, u_o, and $\bar{u}$), are treated here as the core group.

7.3.1
General Formulae for the Core Group

It takes two pneumatic parameters to completely describe pneumatic condition of a tube with known dimensions. Any pair from the group (p_i, p_o, u_i, u_o, and $\bar{u}$) can be a pair of mutually independent parameters [9]. There are 10 such pairs, each being a source for the calculation of three remaining parameters. This leads to the total of 30 formulae for the calculation of any core parameter as a function of any pair of core parameters. Some of these formulae and their modifications can be found in several sources [5, 6, 8]. All are compiled in a single source [9]. Several alternative modifications offered in the following listing are known from elsewhere [14].

To simplify the structure of some formulae in the following listing, the earlier introduced functions $\Phi(x)$ and $\Delta\Phi(x)$ are complimented by the function

$$\Phi_1(x) = \frac{1}{2}\left(\sqrt{\frac{x+2}{x-2/3}} - 1\right) \tag{7.150}$$

In addition to that, an abbreviation in a particular formula is described in line with respective formula.

The internal references to all 30 core formulae are compiled in Table 7.1. Several of those formulae have already been supplied in this chapter. The following are the formulae that complete the listing in Table 7.1.

$$p_i = \Omega u_i\left(\sqrt{1+\left(\frac{p_o}{\Omega u_i}\right)^2} + 1\right) \tag{7.151}$$

Table 7.1 References to the complete set of formulae describing the relations between the core group (p_i, p_o, u_i, u_o, $\bar{u}$) of pneumatic parameters of a tube[a].

	(p_i, p_o)	(p_i, u_i)	(p_i, u_o)	(p_i, $\bar{u}$)	(p_o, u_i)	(p_o, u_o)	(p_o, $\bar{u}$)	(u_i, u_o)	(u_i, $\bar{u}$)	(u_o, $\bar{u}$)
p_i					(7.151)	(7.153)	(7.146)	(7.155)	(7.158)	(7.160)
p_o		(7.154)	(7.152)	(7.157)				(7.156)	(7.161)	(7.159)
u_i	(7.50)		(7.162)	(7.164)		(7.166)	(7.168)			(7.170)
u_o	(7.51)	(7.167)		(7.169)	(7.163)		(7.165)		(7.171)	
$\bar{u}$	(7.120)	(7.172)	(7.174)		(7.175)	(7.173)		(7.176)		

a) Note: The following quantities appear in some formulae:
1) Quantity Ω, Eq. (7.38), is the tube pneumatic resistance.
2) Functions Φ and Φ_1 are defined in Eqs. (7.144) and (7.150).

$$p_o = \Omega u_o\left(\sqrt{1+\left(\frac{p_i}{\Omega u_o}\right)^2}-1\right) \tag{7.152}$$

$$p_i = p_o\sqrt{1+\frac{2\Omega u_o}{p_o}} \tag{7.153}$$

$$p_o = p_i\sqrt{1-\frac{2\Omega u_i}{p_i}} \tag{7.154}$$

$$p_i = \frac{2\Omega u_i}{1-(u_i/u_o)^2} \tag{7.155}$$

$$p_o = \frac{2\Omega u_o}{(u_o/u_i)^2-1} \tag{7.156}$$

$$p_o = p_i\Phi\left(-\frac{\Omega\bar{u}}{p_i}\right) \tag{7.157}$$

$$p_i = \frac{2\Phi_1^2(u_i/\bar{u})}{\Phi_1^2(u_i/\bar{u})-1}\Omega u_i \tag{7.158}$$

$$p_o = \frac{2\Phi_1^2(u_o/\bar{u})}{1-\Phi_1^2(u_o/\bar{u})}\Omega u_o \tag{7.159}$$

$$p_i = \frac{2\Phi_1(u_o/\bar{u})}{1-\Phi_1^2(u_o/\bar{u})}\Omega u_o \tag{7.160}$$

$$p_o = \frac{2\Phi_1(u_i/\bar{u})}{\Phi_1^2(u_i/\bar{u})-1}\Omega u_i \tag{7.161}$$

$$u_i = u_o\left(\sqrt{1+\left(\frac{\Omega u_o}{p_i}\right)^2}-\frac{\Omega u_o}{p_i}\right) \tag{7.162}$$

$$u_o = u_i\left(\sqrt{1+\left(\frac{\Omega u_i}{p_o}\right)^2}+\frac{\Omega u_i}{p_o}\right) \tag{7.163}$$

$$u_i = \frac{p_i}{2\Omega}\left(1-\Phi^2\left(-\frac{\Omega\bar{u}}{p_i}\right)\right) \tag{7.164}$$

$$u_o = \frac{p_o}{2\Omega}\left(\Phi^2\left(\frac{\Omega\bar{u}}{p_o}\right)-1\right) \tag{7.165}$$

$$u_i = \frac{u_o}{\sqrt{1+(2\Omega u_o/p_o)}} \tag{7.166}$$

$$u_o = \frac{u_i}{\sqrt{1-(2\Omega u_i/p_i)}} \tag{7.167}$$

$$u_i = \frac{p_o}{2\Omega}\left(\Phi\left(\frac{\Omega\bar{u}}{p_o}\right)-\Phi^{-1}\left(\frac{\Omega\bar{u}}{p_o}\right)\right) \tag{7.168}$$

$$u_o = \frac{p_i}{2\Omega}\left(\Phi^{-1}\left(-\frac{\Omega\bar{u}}{p_i}\right)-\Phi\left(-\frac{\Omega\bar{u}}{p_i}\right)\right) \tag{7.169}$$

$$u_i = u_o\Phi_1\left(\frac{u_o}{\bar{u}}\right) \tag{7.170}$$

$$u_o = u_i\Phi_1\left(\frac{u_i}{\bar{u}}\right) \tag{7.171}$$

$$\bar{u} = \frac{3u_i}{2}\frac{1+\sqrt{1-x}}{2-x+\sqrt{1-x}} = \frac{3u_i}{2}\frac{x}{1-(1-x)^{3/2}}, \quad x = \frac{2\Omega u_i}{p_i} \tag{7.172}$$

$$\bar{u} = \frac{3u_o}{2}\frac{1+\sqrt{1+x}}{2+x+\sqrt{1+x}} = \frac{3u_o}{2}\frac{x}{(1+x)^{3/2}-1}, \quad x = \frac{2\Omega u_o}{p_o} \tag{7.173}$$

$$\bar{u} = 3u_o\frac{x(x-\sqrt{1+x^2})^2}{(x-\sqrt{1+x^2})^3+1}, \quad x = \frac{\Omega u_o}{p_i} \tag{7.174}$$

$$\bar{u} = \frac{3u_i}{2}\frac{(x+\sqrt{1+x^2})((x+\sqrt{1+x^2})^2-1)}{(x+\sqrt{1+x^2})^3-1}, \quad x = \frac{\Omega u_i}{p_o} \qquad (7.175)$$

$$\bar{u} = \frac{3u_o}{2}\frac{1+x}{1+x+x^2} = \frac{3u_o}{2}\frac{x^2-1}{x^3-1}, \quad x = \frac{u_o}{u_i} \qquad (7.176)$$

7.3.2
General Formulae for Other Parameters

The core parameters (p_i, p_o, u_i, u_o, and $\bar{u}$) listed in Table 7.1 are not the only pneumatic parameters of interest in GC. Several useful formulae relating noncore parameters to each other and to the core parameters appeared earlier in the text. Others are provided below.

References to the formulae relating virtual pressure ($\Delta\tilde{p}$) to other parameters are compiled in Table 7.2.

By combining the formulae in Tables 7.1 and 7.2, one can obtain additional expressions describing relationship of $\Delta\tilde{p}$ to other parameters.

Not listed in Table 7.1 are specific flow rate, f, hold-up time, t_M, and other parameters. Several formulae involving these parameters were provided earlier in the text. Others can be derived with the help of formulae listed in Table 7.1. Some of them are provided below.

The component $J = \bar{p}\bar{u}$ of Eq. (7.126) combined with Eqs. (7.113), (7.114), and (7.52) allows one to express average gas velocity, $\bar{u}$, as functions,

$$\bar{u} = \bar{u}(p_o, J) = \frac{J}{\bar{p}} = \frac{3\Omega J^2}{(p_o^2 + 2\Omega J)^{3/2} - p_o^3} \qquad (7.177)$$

$$\bar{u} = \bar{u}(p_o, f) = \frac{48\Omega p_{st}^2 f^2}{\pi^2 d_{cx}^2 p_o^3}\left(\left(1 + \frac{8\Omega p_{st} f}{\pi d_{cx} p_o^2}\right)^{3/2} - 1\right)^{-1} \qquad (7.178)$$

Table 7.2 References to the formulae relating virtual pressure, $\Delta\tilde{p}$, to other parameters[a)].

	(p_o, p_i)	$(p_o, \Delta p)$	$(p_o, p_i, \Delta p)$	$(p_o, \Delta\tilde{p})$	$\Delta\tilde{p}$	$\bar{u}$
p_i				(7.145)		
Δp				(7.148)		
$\Delta\tilde{p}$	(7.140)	(7.136)	(7.134)			(7.133)
t_M					(7.137), (7.138)	
$\bar{u}$					(7.132)	

a) Note: The following quantities appear in some formulae:
1) Quantity Ω, Eq. (7.38), is the tube pneumatic resistance.
2) Halász compressibility factor, j_H, is defined in Eq. (7.88) and described in Eqs. (7.89), (7.90), (7.91) and (7.92).
3) Functions Φ and $\Delta\Phi$ are defined in Eqs. (7.144) and (7.149), respectively.

of p_o and J or p_o and f. Due to Eqs. (7.38) and (7.76), the last formula can be also expressed as a function

$$\bar{u} = \bar{u}(L_{crit}, L) = \frac{27\, d_{cx}^2 L p_o}{4\gamma_\Omega\, \eta L_{crit}^2}\left(\left(1 + \frac{3L}{L_{crit}}\right)^{3/2} - 1\right)^{-1} \quad (7.179)$$

of critical length, L_{crit}, of a tube at some predetermined flow of a gas and the tube's actual length, L. Let u_o be a predetermined outlet velocity corresponding to L_{crit} then, due to Eqs. (7.76) and (7.126), $\bar{u}$ can be also expressed as

$$\bar{u} = \bar{u}(L_{crit}, L) = \frac{9\, u_o}{2}\frac{L}{L_{crit}}\left(\left(1 + \frac{3L}{L_{crit}}\right)^{3/2} - 1\right)^{-1} \quad (7.180)$$

Comparison of Eq. (7.180) with Eq. (7.121) implies that the James–Martin compressibility factor, j, at a predetermined flow in an L-long tube can be found directly from the ratio L/L_{crit} as

$$j = \frac{9\,(L/L_{crit})}{2((1 + 3L/L_{crit})^{3/2} - 1)} \quad (7.181)$$

where L_{crit} is the critical lengths of the tube with that flow.

An important parameter of a tube is its hold-up time, t_M. Substitution of Eq. (7.178) in Eq. (7.118) for t_M and accounting for Eq. (7.38) allows one to express t_M as a function of the tube dimensions and specific flow rate, f:

$$t_M = t_M(d_{cx}, L, f) = \frac{\pi^2 d_{cx}^4 p_o^3}{48\gamma_\Omega\, \eta p_{st}^2 f^2}\left(\left(1 + \frac{8\gamma_\Omega L\, \eta p_{st} f}{\pi d_{cx}^3 p_o^2}\right)^{3/2} - 1\right) \quad (7.182)$$

Excluding L from this formula by solving it together with Eq. (7.61) replaces parameter L in this formula with inlet pressure, p_i:

$$t_M = t_M(d_{cx}, p_i, f) = \frac{\pi^2 d_{cx}^4 (p_i^3 - p_o^3)}{48\gamma_\Omega\, \eta p_{st}^2 f^2} \quad (7.183)$$

Note 7.7

As mentioned earlier, two pneumatic parameters completely describe pneumatic conditions of a tube with known dimensions. Three pneumatic parameters (f, p_i, and p_o) are given in Eq. (7.183). However, the tube length is absent from the formula. □

References to the formulae for hold-up time, t_M, are compiled in Table 7.3. Additional formulae for t_M can be obtained by combining the formulae in this table with the ones listed in Tables 7.1 and 7.2.

7.3.3 Special Case: Weak Decompression

When pressure drop across a tube is small compared with outlet pressure, gas decompression in the tube is weak. Along with that, the change in gas velocity, u,

Table 7.3 References to the formulae for hold-up time, t_M[a)].

	(p_o, p_i)	$(p_o, \Delta p)$	$\Delta\tilde{p}$	(f, p_o)	(f, p_o, p_i)	$\bar{u}$
t_M	(7.87)	(7.88), (7.96)	(7.137), (7.138)	(7.182)	(7.183)	(7.118)

a) Note: The following quantities appear in some formulae:
1) Quantity Ω, Eq. (7.38), is the tube pneumatic resistance.
2) Quantity γ_Ω is described in Eq. (7.40).
3) Halász compressibility factor, j_H, is defined in Eq. (7.88) and described in Eqs. (7.89), (7.90), (7.91) and (7.92).

and in pressure, p, along the tube is small, Figure 7.3a, and can be ignored. One can write:

$$p_i \approx p_o \approx \bar{p} \approx p \tag{7.184}$$

$$u_i \approx u_o \approx \bar{u} \approx u \tag{7.185}$$

At these conditions, it is more convenient to deal with the pressure drop, Δp, Eq. (7.8), than with inlet or outlet pressure, p_i or p_o. The entire group (p_i, $p_o u_i$, u_o, and $\bar{u}$) of parameters in Table 7.1 shrinks to only two: Δp and u. Only one of these parameters can be independent, and the whole set of formulae referenced in Table 7.1 shrinks to Eq. (7.69) and to its inversion, $\Delta p = \Omega u$, whereas Eqs. (7.60), (7.134), (7.114), (7.106), (7.118), and (7.182) converge to

$$f = \frac{\pi d_{cx} p_o \Delta p}{4 p_{st} \Omega} = \frac{\pi d_{cx}^3 p_o \Delta p}{4 \gamma_\Omega L \eta p_{st}} \tag{7.186}$$

$$\Delta\tilde{p} = \Delta p \tag{7.187}$$

$$j = 1 \tag{7.188}$$

$$\frac{z}{L} = \frac{t_g}{t_M} \tag{7.189}$$

$$t_g = \frac{z}{u} \tag{7.190}$$

$$t_M = \frac{L}{u} = \frac{\pi d_c L p_o}{4 p_{st} f} \tag{7.191}$$

When working with these formulae, one should keep in mind that they are approximations whose accuracy depends on the accuracy of the underlying conditions in Eq. (7.184) or (7.185). However, even when the gas decompression is moderate (not sufficiently weak) and, therefore, these formulae are not sufficiently accurate for the final calculations; their simplicity might be useful for obtaining estimates and guidelines.

7.3.4
Special Case: Strong Decompression

When the decompression of a carrier gas along the tube is strong ($p_i >> p_o$, Eq. (7.68) Figure 7.3c), all formulae compiled in Table 7.1 can be reduced to the following six simple relations:

$$p_i^2 = 2\Omega p_o u_o \qquad (7.192)$$

$$p_o u_o = 2\Omega u_i^2 \qquad (7.193)$$

$$9 p_o u_o = 8\Omega \bar{u}^2 \qquad (7.194)$$

$$p_i = 2\Omega u_i \qquad (7.195)$$

$$3 p_i = 4\Omega \bar{u} \qquad (7.196)$$

$$2\bar{u} = 3 u_i \qquad (7.197)$$

These relations can be divided in two groups: the first consisting of Eqs. (7.192)–(7.194), and the second consisting of Eqs. (7.195)–(7.197).

Each relation in the first group binds three parameters, and allows one to express any of those three as a function of the other two. For example, Eq. (7.192) binds p_i, p_o, and u_o, and can be used to express p_i as a function of (p_o, u_o), or p_o as a function of (p_i, u_o), or u_o as a function of (p_i, p_o).

Each relation in the second group binds two parameters, and can yield two solutions expressing one parameter as a function of the other. For example, Eq. (7.195) binds p_i and u_i, and, in addition to expressing p_i as a function of u_i, it can be arranged to express u_i as a function of p_i. Any formula in the second group follows from the formulae in the first group after exclusion of the product $p_o u_o$. For example, Eq. (7.195) follows from Eqs. (7.192) and (7.193). As a result, the formulae in the second group while providing simpler alternatives to some formulae in the first group do not add new information.

Due to Eqs. (7.126) and (7.38), one can express Eqs. (7.192)(7.194) as

$$J = \frac{p_i^2}{2\Omega} = 2\Omega u_i^2 = \frac{8\Omega \bar{u}^2}{9} \qquad (7.198)$$

or as [14, 30, 31]

$$\bar{u} = \sqrt{\frac{9J}{8\Omega}} \qquad (7.199)$$

Due to Eqs. (7.126), (7.38) and (7.76), average velocity, $\bar{u}$, in this formula can also be expressed as functions

$$\bar{u} = \sqrt{\frac{9p_{st}f}{2\pi d_{cx}\Omega}} = \sqrt{\frac{9d_{cx}p_{st}f}{2\pi\gamma_{\Omega}\eta L}} \tag{7.200}$$

$$\bar{u} = \frac{u_o}{2}\sqrt{\frac{3L_{crit}}{L}} = \frac{2p_{st}f}{\pi d_{cx}p_o}\sqrt{\frac{3L_{crit}}{L}} \tag{7.201}$$

of specific flow rate, f, in a tube or its critical length, L_{crit}, at a given u_o or f.
Other simplifications resulting from large gas decompression are

$$\Delta p = p_i \tag{7.202}$$

$$\Delta\tilde{p} = \frac{3}{4}\Delta p = \frac{3}{4}p_i \tag{7.203}$$

$$L_{ext} = L, \quad t_{M.ext} = t_M, \quad \ell_{ext} = \ell \tag{7.204}$$

$$\left(1-\frac{z}{L}\right)^3 = \left(1-\frac{t_g}{t_M}\right)^2 \tag{7.205}$$

$$j = \frac{9}{8} \tag{7.206}$$

$$t_M = \sqrt{\frac{2\pi\gamma_{\Omega}\eta L^3}{9d_{cx}p_{st}f}} \tag{7.207}$$

Equation (7.202) follows from Eq. (7.8); Eq. (7.203) – from Eq. (7.136); Eq. (7.204) – from Eqs. (7.64), (7.101), and (7.104); Eq. (7.205) – from Eqs. (7.106) and (7.204); Eq. (7.206) – from Eq. (7.114); and Eq. (7.207) – from Eq. (7.182). For an open tube, Eq. (7.207) yields due to Eqs. (7.40), (7.56), and (7.57):

$$t_M = \frac{8}{3}\sqrt{\frac{\pi\eta L^3}{p_{st}F}\cdot\frac{T_{ref}}{T}} \quad \text{(open tube)} \tag{7.208}$$

Generally, formulae listed here are approximate. Their accuracy depends on degree of the gas decompression. When the tube outlet is at vacuum ($p_o = 0$), the formulae become absolutely accurate (within the model of laminar, mass-conserving flow of uniformly heated ideal gas in a uniform tube). It should be noticed, however, that, at $p_o = 0$, Eqs. (7.192), (7.194) become indeterminate because in the product $p_o u_o$, p_o approaches zero whereas u_o approaches infinity.

Equation (7.208) reveals an interesting fact. Column diameter is not present in this formula. This means that in GC-MS ($p_o = 0$) with a capillary column, hold-up time and, therefore [13], analysis time at a given flow rate are independent of column diameter.

Appendix 7.A

7.A.1 Molecular, Laminar, and Turbulent Flow in an Open Round Tube

Gas flow in a tube can be molecular [7] when the tube's cross-sectional dimension is too small for a particular gas and its flow conditions, turbulent [2, 4, 6, 7] when the cross-sectional dimension is too large for the gas and its flow conditions, and laminar (smooth) when the conditions are between the two extremes. In capillary GC, the flow is typically laminar. Let us consider why. Material of this section relies on the material of Chapter 6.

Consider an open L-long round tube with internal diameter d. For the gas flow in the tube not to be turbulent, its dimensionless velocity, known as the *Reynolds number* [2, 4, 6, 7],

$$R_e = \frac{d\varrho u}{\eta} \tag{A1}$$

should be smaller than the critical value, $R_{e,cr}$, that can be estimated as [4, 6, 7]

$$R_{e,cr} \approx 1000 \tag{A2}$$

In Eq. (A1), u, η, and ϱ are the gas velocity, viscosity and density, respectively.

Let υ and λ be the gas average molecular speed (average of absolute values of velocities of the gas molecules) and the mean free path, respectively. The value of R_e is in the order of a product of two ratios, d/λ and u/υ [2]. Indeed, substitution of ϱ from Eq. (6.8) in Eq. (A1) yields

$$R_e = \frac{8dpu}{\pi\eta\upsilon^2} \tag{A3}$$

which, after the substitution of η from Eq. (6.11), becomes

$$R_e = \frac{32}{5\pi} \cdot \frac{d}{\lambda} \cdot \frac{u}{\upsilon} \approx 2\frac{d}{\lambda} \cdot \frac{u}{\upsilon} \tag{A4}$$

With as many variables in it as there are in Eq. (A1), Eq. (A4) is not simpler than Eq. (A1). However, Eq. (A4) provides a better insight into the factors affecting R_e. Particularly, it shows that, for a given u and υ, quantity R_e increases with the increase in d relative to λ. To estimate the value of R_e using Eq. (A4), one can take the values of υ and λ from Table 6.1. However, there is a possibility for a more general estimate.

Generally, the gas velocity, u, in GC is inversely proportional to d, and, therefore, the product $d\,u$ in Eqs. (A1) and (A4) is confined within roughly the same narrow range for all columns. Furthermore, u in Eq. (A1) is not totally independent of η, whereas u in Eq. (A4) is not totally independent of the product $\upsilon\lambda$. All these facts can be sorted out as follows.

Although different column optimization goals in GC can lead to different optimal values of gas velocity, u, it is reasonable to assume that, at optimal conditions, u can be estimated as [28, 74]

$$u \approx \frac{6D}{d} \qquad \text{(A5)}$$

where D is the diffusivity of a solute in a gas. Substitution of Eq. (A5) in Eq. (A3) yields

$$R_e \leq \frac{50Dp}{\pi\eta\upsilon^2} \qquad \text{(A6)}$$

There are no tube dimensions in this formula. This means that, for the conditions that are typical for a pneumatically optimized GC column, the ratio $R_{e,cr}/R_e$ does not depend on the tube dimensions.

To further simplify Eq. (A6), one can account for the fact that D can be expressed, Eq. (6.33), as $D = D_f D_g$, where D_f is diffusivity factor of a solute – a quantity that shows how much the diffusivity, D, of a solute in a gas is larger that the self-diffusivity, D_g, of the gas itself. After substitution of Eq. (6.33), Eq. (A6) becomes $R_e \leq 50\ D_f D_g p/(\pi v^2 \eta)$. This, after substitution of Eqs. (6.38) and (6.17), yields a simple relation

$$R_e \leq 7D_f \qquad \text{(A7)}$$

To reduce the latter relation to a numerical value, one can notice that the solute diffusivity, D, does not typically exceed the self-diffusivity, D_g, of a carrier gas (see, for example, Figure 6.4), that is $D_f \leq 1$. Equation (A7) becomes

$$R_e \leq 7 \qquad \text{(A8)}$$

This, in view of Eq. (A2), means that typical (optimal or near optimal) flow in a GC column is about two orders of magnitude lower than the level that can cause the turbulence.

Let us consider the conditions that can lead to transition from the laminar to the molecular flow.

When internal diameter, d, of an open tube is not sufficiently larger than mean free path, λ, of gas molecules, then molecular flow characterized by statistically insignificant number of collisions between the gas molecules might take place. The flow remains essentially laminar as long as [7]

$$d \geq 100\lambda \qquad \text{(A9)}$$

The transition to molecular flow in a tube appears as reduction [7] in its pneumatic resistance compared to the resistance, Ω, Eq. (7.38), to laminar flow. For an arbitrary pressure, p, the product $p\lambda$ increases with temperature, T, Eq. (6.12), and is the largest for helium, Table 6.1.

Let us consider the case of helium at high temperature of $T = 300\,^\circ\text{C}$ and the two outlet pressures: $p_o = 1$ atm and $p_o = 0$.

According to the data in Table 6.1, at $p = 1$ atm and $T = 300\,^\circ\text{C}$, $\lambda_{He} \approx 0.4\ \mu$. Equation (A9) becomes

$$d \geq 40\ \mu \qquad \text{(A10)}$$

indicating that the flow in a tube remains essentially laminar if (1) the tube has 30 μ or larger internal diameters and if (2) the tube's outlet is at atmospheric or higher pressure. It should also be taken into consideration that narrow-bore GC columns require high inlet pressure for normal operations. That, according to Eq. (6.12), proportionally reduces λ in a major portion of a column length excluding only an immediate vicinity of a column outlet and strengthening inequality in Eq. (A9). Furthermore, the fact that a significant violation of Eq. (A9) leads to a significant reduction [7] in pneumatic resistance of a tube compared to its resistance to laminar flow causes further reduction in the size of a near outlet segment where Eq. (A9) can be violated. This means that molecular flow has a self-correcting nature in the sense that the more favorable are the conditions for the molecular flow at the column outlet, the smaller is the near outlet segment where Eq. (A9) is violated. These observations lead to the following conclusion.

If a tube outlet is at atmospheric pressure and other conditions are typical for GC columns, then no transition to molecular flow exists in a tube with 30 μm or larger internal diameter.

When a tube outlet is at vacuum, the margin for the flow to remain laminar might be not as significant as it is at atmospheric outlet pressure. A more careful evaluation of conditions is required.

The inlet pressure, p_i, in an L-long tube with its outlet at vacuum can be found from Eqs. (7.195) and (7.38) as

$$p_i = \frac{64L\eta u_i}{d^2} \tag{A11}$$

For optimal or near optimal flow in a GC column, this, due to Eq. (A5), becomes

$$p_i \approx \frac{383LD_i\eta}{d^3} = \frac{383\ell D_i\eta}{d^2} \tag{A12}$$

where $\ell = L/d$ is the dimensionless length of a tube and D_i is diffusivity of a solute in a gas at inlet pressure. Due to Eqs. (6.18) and (6.38), the last formula yields

$$\frac{d}{\lambda_i} \approx 16.8\sqrt{D_f \ell} \tag{A13}$$

where λ_i is the mean free path of gas molecules at a tube inlet, and D_f, being the same as in Eq. (A7), is a quantity that depends on a solute and its temperature. It also follows from Eq. (6.38) that, for n-alkanes, the last formula can be estimated as

$$\frac{d}{\lambda_i} \approx \frac{16.8\gamma_D\sqrt{\ell}}{i^{1/4}} \tag{A14}$$

where γ_D is a constant, Table 6.12, that depends only on a gas type, and i is the carbon number of n-alkane. Equation (7.14) suggests that the higher is the carbon number, i, of n-alkane for which the gas flow in a GC column is optimized, the smaller (the least favorable for laminar flow) is the d/λ_i value.

In a temperature-programmed analysis, n-alkanes with higher carbon numbers elute at higher temperatures. Normally, each solute elutes at a temperature that is close to its characteristic temperature, T_{char} (the temperature, Chapter 5, at which $k = 1$ where k is a solute retention factor). The relationship between i and T_{char} can be estimated as shown in Eq. (5.53). Assuming that in a GC column, $T = T_{char}$, one has from Eqs. (A14) and (5.53):

$$\frac{d}{\lambda_i} \approx 11.6\gamma_D\sqrt{\ell}\left(\frac{T_{st}}{T}\right)^{5/8} \tag{A15}$$

where $T_{st} = 273.15$ K, Eq. (6.1), is standard temperature.

These conditions yield the lowest d/λ_i (the least favorable for laminar flow) for the gas that has the lowest γ_D. In the case of a GC column, the conditions are especially unfavorable when the flow is optimized for the solute that elutes at the highest temperature. For the gases listed in Table 6.12, helium has the lowest γ_D. For helium at $T = 325$ °C, Eq. (A15) yields

$$\frac{d}{\lambda_i} \approx 3.3\sqrt{\ell} \quad (\text{helium at } T = 325\,°\text{C}) \tag{A16}$$

Example 7.4

Consider a tube with $\ell = L/d = 10^3$ (for example, $L = 0.1$ m and $d = 0.1$ mm, or $L = 0.5$ m and $d = 0.5$ mm) with helium flow at $T = 325$ °C. Equation (A16) yields $d/\lambda_i \approx 100$ indicating that Eq. (A9) is satisfied at a tube inlet, but not with a wide margin. Further along the tube toward its outlet, d/λ decreases in inverse proportion to the declining pressure, p, Eq. (6.12), along the tube. Therefore, a slip flow [7] – the initial phase of transitioning from laminar to molecular flow – exists in, probably, the main portion of the tube. This means that pneumatic resistance of the tube is smaller than its resistance, Ω, Eq. (7.38), at the laminar flow. It should be emphasized, however, that the condition evaluated here represents unrealistic worst-case scenario for GC – helium at high temperature in unusually short column. □

Example 7.5

Consider a tube with $\ell = L/d = 10^4$ (for example, $L = 1$ m and $d = 0.1$ mm, or $L = 5$ m and $d = 0.5$ mm) with helium flow at $T = 325$ °C. Equation (A16) yields $d/\lambda_i \approx 330$. At a tube inlet, Eq. (A9) is satisfied with a threefold margin. The margin vanishes when pressure, p, along the tube falls below $p_i/3.3$, that is

$$p \leq 0.3p_i \tag{A17}$$

Assuming that the flow is laminar along the entire tube, the pressure profile along the tube can be found from Eq. (7.66). With $p_o = 0$ (vacuum at the outlet), Eq. (7.66) becomes $p = p_i\sqrt{1 - z/L}$, where z is the distance from the tube inlet. According to Eq. (A17), the transition to molecular flow begins when $1 - z/L$ falls below 0.1, that is when $z/L \leq 0.9$. This means that, assuming laminar flow along the entire tube, the conditions, Eq. (A9), for such flow would not be satisfied only within the last 10% section of the tube near its outlet. Due to the previously mentioned self-correcting nature of molecular flow, the section where Eq. (A9) is not satisfied should be even shorter than 10% of L. Finally, although the conditions evaluated here are realistic for practical GC (very short 5 m × 0.53 mm columns are used in some applications), they represent the worst-case realistic scenario – helium at high temperature in a very short column. □

The following observations can be made regarding an open tube if the flow in the tube is at or above chromatographic optimum.

Helium at high temperature is the most likely case for slip or molecular flow at the end of a short tube.

In a tube with its outlet at atmospheric or higher pressure, no transition to molecular flow takes place if $d \geq 30$ μ.

A noticeable transition to molecular flow can take place in a relatively short tube ($L/d \leq 10^3$) if its outlet is at vacuum.

Even if the outlet of a tube is at vacuum, no significant transition to molecular flow can take place if $L/d \geq 10^4$.

7.A.2 Gas Propagation Velocity

Time, t_g, of migration of a narrow packet of gas molecules from a tube inlet to coordinate z along the tube can be found as

$$t_g = \int_0^z \frac{dz}{u} \tag{A18}$$

where u is a coordinate-dependent gas velocity in the tube. Substitution of Eq. (7.67) for u in Eq. (7.18) yields

$$t_g = \frac{2L_{ext}}{3u_i}\left(1-\left(1-\frac{z}{L_{ext}}\right)^{3/2}\right) \tag{A19}$$

At $z = L_{ext}$, time t_g becomes vacuum-extended hold-up time, $t_{M,ext}$, that can be found as

$$t_{M,ext} = t_g\big|_{z=L_{ext}} = \frac{2L_{ext}}{3u_i} \tag{A20}$$

Solving Eq. (A19) for z and substitution of the result in Eq. (7.67) yields velocity

$$u = \frac{u_i}{\left(1-(t_g/t_{M,ext})\right)^{1/3}} \tag{A21}$$

of a narrow packet of gas molecules as a function of the packet migration time, t_g.

7.A.3 Inverse Halász Compressibility Factor and Related Formulae

The virtual pressure, $\Delta\tilde{p}$, can be found from pressure drop, Δp, along the tube as $\Delta\tilde{p} = j_H \Delta p$, Eq. (7.134), where j_H is the Halász compressibility factor. To make the concept of virtual pressure more useful for practical evaluations of pneumatic parameters of gas flow in a tube, one should be able to find other parameters from known $\Delta\tilde{p}$. A path to making this possible is through inversion of expression

Eq. (7.134) and expressing it as

$$\Delta p = \tilde{j}_H \Delta \tilde{p} \tag{A22}$$

where parameter

$$\tilde{j}_H = \frac{1}{j_H} \tag{A23}$$

can be viewed as inverse Halász compressibility factor, Figure 7.5b.

Known formulae for j_H such as Eq. (7.91) express it as a function of Δp, but not as a function of $\Delta \tilde{p}$. Therefore, Eq. (A23) does not express $\tilde{j}_H$ as a function of $\Delta \tilde{p}$, and, consequently, Eq. (A22) does not express Δp as a functions of $\Delta \tilde{p}$. To make Eq. (A22) a complete function of $\Delta \tilde{p}$, let us express it in a normalized form

$$\Delta P = \tilde{j}_H \Delta \tilde{P} \tag{A24}$$

where ΔP and $\Delta \tilde{P}$ are the relative pressure drop and relative virtual pressure defined in Eqs. (7.10) and (7.139), respectively. It follows from comparison of Eq. (A24) with Eq. (7.147) that, Figure 7.5b,

$$\tilde{j}_H = \frac{p_o}{\Delta \tilde{p}} \cdot \Delta \Phi \left(\frac{\Delta \tilde{p}}{p_o} \right) = \frac{\Delta \Phi(\Delta \tilde{P})}{\Delta \tilde{P}} \tag{A25}$$

Equations (A24) and (A22) with quantity $\tilde{j}_H$ described in Eq. (A25) express quantities Δp and ΔP as functions of $\Delta \tilde{p}$ (or of its relative version $\Delta \tilde{P}$).

Function $\Delta\Phi(x)$ in Eq. (A25) has algebraically cumbersome description, Eqs. (7.149) and (7.144), although it is a monotonic slightly curved function, Figure 7.6. The computation complexity of function $\Delta\Phi(x)$ is of almost no consequence for the computerized calculations within a GC system. However, it might be a source of inconvenience for noncomputerized evaluations. A substantial simplification is possible at the cost of a small error.

Recall that $\tilde{j}_H$ is inversion, Eq. (A23), of j_H. It is tempting to (rather than looking for the exact solution, Eq. (A25), for $\tilde{j}_H$) simply replace ΔP in Eq. (7.92) for j_H with $\Delta \tilde{P}$, and to substitute the result in Eq. (A23). The worst-case error for so found approximation,

$$\tilde{j}_{H,approx} = \frac{4}{3} \cdot \frac{3 + 3\Delta \tilde{P} + \Delta \tilde{P}^2}{(2 + \Delta \tilde{P})^2} \tag{A26}$$

to $\tilde{j}_H$ does not exceed 1.5%. Replacing $\tilde{j}_H$ in Eq. (A25) with this approximation and replacing variable ΔP with variable x in it leads to approximation

$$\Delta\Phi_{approx}(x) = \frac{4x}{3} \cdot \frac{3 + 3x + x^2}{(2 + x)^2} \tag{A27}$$

Due to Eq. (7.149), one can also obtain the approximation

$$\Phi_{approx}(x) = 1 + \Delta\Phi_{approx}(x) = 1 + \frac{4x}{3} \cdot \frac{3 + 3x + x^2}{(2 + x)^2} \tag{A28}$$

for $\Phi(x)$. The errors of these approximations do not exceed 1.5% and 1.3%, respectively.

References

1 James, A.T. and Martin, A.J.P. (1952) *Biochem. J.*, **50**, 679–690.

2 Shapiro, A.H. (1953) *The Dynamics and Thermodynamics of Compressible Fluid Flow*, Vol. I The Ronald Press Co., New York.

3 Golay, M.J.E. (1958) *Gas Chromatography 1958* (ed. D.H. Desty), Academic Press, New York, pp. 36–55.

4 Levich, V.G. (1962) *Physicochemical Hydrodynamics*, Prentice-Hall, Englewood Cliffs, NJ.

5 Purnell, J.H. (1962) *Gas Chromatography*, John Wiley & Sons, Inc., New York.

6 Guiochon, G. (1966) *Chromatographic Review*, Vol. 8 (ed. M. Lederer), Elsevier, Amsterdam, pp. 1–47.

7 Perry, R.H., Green, d.W. and Maloney, J.O. (1984) *Perry's Chemical Engineer's Handbook*, 6th edn, McGraw-Hill Book Company, New York.

8 Giddings, J.C. (1991) *Unified Separation Science*, Wiley, New York.

9 Blumberg, L.M. (1995) *Chromatographia*, **41**, 15–22.

10 Blumberg, L.M. (1996) *Chromatographia*, **43**, 73–75.

11 Blumberg, L.M. (1996) *Chromatographia*, **42**, 112–113.

12 Blumberg, L.M. (1997) *Chromatographia*, **44**, 325–329.

13 Blumberg, L.M. and Klee, M.S. (1998) *Anal. Chem.*, **70**, 3828–3839.

14 Blumberg, L.M. (1999) *J. High Resolut. Chromatogr.*, **22**, 213–216.

15 Golay, M.J.E. (1958) *Gas Chromatography 1958* (ed. D.H. Desty), Academic Press, New York, pp. 62–68.

16 Ettre, L.S. (1993) *Pure Appl. Chem.*, **65**, 819–872.

17 Guiochon, G. and Guillemin, C.L. (1988) *Quantitative Gas Chromatography for Laboratory Analysis and On-Line Control*, Elsevier, Amsterdam.

18 Darcy, H. (1856) *Les Fontaines Publiques De La Ville De Dijon*, Dalmont, Paris.

19 Darcy, H. (2004) *The Public Fountains of the City of Dijon*, Kendall Hunt Publishing Co., Dubuque, IA.

20 Kozeny, J. (1927) *S. B. Akad. Wiss. Wien.*, Abt., IIa **136**, 271–306.

21 Carman, P.C. (1937) *Trans. Inst. Chem. Eng. (London)*, **15**, 150–166.

22 Agilent Technologies (1998) GC Method Translation Freeware, version 2.0.a.c, http://www.chem.agilent.com/en-US/Support/Downloads/Utilities/Pages/GcMethodTranslation.aspx Agilent Technologies, Inc., Wilmington, DE.

23 Grob, K. (1994) *J. High Resolut. Chromatogr.*, **17**, 556.

24 Stewart, G.H., Seager, S.L., and Giddings, J.C. (1959) *Anal. Chem.*, **31**, 1738.

25 Giddings, J.C., Seager, S.L., Stucki, L.R., and Stewart, G.H. (1960) *Anal. Chem.*, **32**, 867–870.

26 Giddings, J.C. (1962) *Anal. Chem.*, **34**, 314–319.

27 DeFord, D.D., Loyd, R.J., and Ayers, B.O. (1963) *Anal. Chem.*, **35**, 426–429.

28 Blumberg, L.M. (1999) *J. High Resolut. Chromatogr.*, **22**, 403–413.

29 Blumberg, L.M. and Berger, T.A. (1993) *Anal. Chem.*, **65**, 2686–2689.

30 Blumberg, L.M. (1997) *J. High Resolut. Chromatogr.*, **20**, 597–604.

31 Blumberg, L.M. (1997) *J. High Resolut. Chromatogr.*, **20**, 704.

32 Blumberg, L.M. (1997) *J. High Resolut. Chromatogr.*, **20**, 679–687.

33 Klee, M.S. and Blumberg, L.M. (2002) *J. Chromatogr. Sci.*, **40**, 234–247.

34 Stafford, S.S. (1994) *Electronic Pressure Control in Gas Chromatography*, Hewlett-Packard Co., Wilmington, DE.

35 Le Vent, S. (1996) *J. Chromatogr. A*, **752**, 173–181.

36 Jennings, W., Mittlefehldt, E., and Stremple, P. (1997) *Analytical Gas Chromatography*, 2nd edn, Academic Press, San Diego.

37 Lebrón-Aguilar, R., Quintanilla-López, J.E., and García-Domínguez, J.A. (1997) *J. Chromatogr.*, **760**, 219–226.

38 Quintanilla-López, J.E., Lebrón-Aguilar, R., and García-Domínguez, J.A. (1997) *J. Chromatogr. A*, **767** (1–2), 127–136.

39 González, F.R. (1999) *J. Chromatogr. A*, **832**, 165–172.

40 Spangler, G.E. (2006) *Anal. Chem.*, **78**, 5205–5207.

41 Giddings, J.C. (1962) *Gas Chromatography* (eds N. Brenner, J.E. Callen, and M.D. Weiss), Academic Press, New York, pp. 57–77.

42 Krupčík, J., Repka, D., Hevesi, T., Nolte, J., Paschold, B., and Mayer, H. (1988) *J. Chromatogr.*, **448**, 203–218.

43 Maurer, T., Engewald, W., and Steinborn, A. (1990) *J. Chromatogr.*, **517**, 77–86.

44 Bautz, D.E., Dolan, J.W., Raddatz, L.R., and Snyder, L.R. (1990) *Anal. Chem.*, **62**, 1560–1567.

45 Dolan, J.W., Snyder, L.R., and Bautz, D.E. (1991) *J. Chromatogr.*, **541**, 21–34.

46 Bautz, D.E., Dolan, J.W., and Snyder, L.R. (1991) *J. Chromatogr.*, **541**, 1–19.

47 Castello, G., Vezzani, S., and Moretti, P. (1994) *J. Chromatogr. A*, **677**, 95–106.

48 Vezzani, S., Castello, G., and Pierani, D. (1998) *J. Chromatogr. A*, **811**, 85–96.

49 Lu, X., Kong, H., Li, H., Ma, C., Tian, J., and Xu, G. (1086) *J. Chromatogr. A*, **2005**, 175–184.

50 Wang, X., Stoll, D.R., Schellinger, A.P., and Carr, P.W. (2006) *Anal. Chem.*, **78**, 3406–3416.

51 Halász, I., Hartmann, K., and Heine, E. (1965) *Gas Chromatography 1964* (ed. A. Goldup), The Institute of Petroleum, London, pp. 38–61.

52 Blumberg, L.M. (1993) *J. Chromatogr.*, **637**, 119–128.

53 Lan, K. and Jorgenson, J.W. (2000) *Anal. Chem.*, **72**, 1555–1563.

54 Lan, K. and Jorgenson, J.W. (2001) *J. Chromatogr. A*, **905**, 47–57.

55 Martire, D.E. (1989) *J. Chromatogr.*, **461**, 165–176.

56 Martire, D.E., Riester, R.L., Bruno, T.J., Hussam, A., and Poe, D.P. (1991) *J. Chromatogr.*, **545**, 135–147.

57 James, A.T. and Martin, A.J.P. (1952) *Analyst*, **77**, 915–932.

58 Harris, W.E. and Habgood, H.W. (1966) *Programmed Temperature Gas Chromatography*, John Wiley & Sons, Inc., New York.

59 Ettre, L.S. and Hinshaw, J.V. (1993) *Basic Relations of Gas Chromatography*, Advanstar, Cleveland, OH.

60 Davankov, V.A. (1996) *Chromatographia*, **42**, 111.

61 Ettre, L.S. and Hinshaw, J.V. (1996) *Chromatographia*, **43**, 159–162.

62 Wu, Z. (1997) *Chromatographia*, **44**, 325.

63 Parcher, J.F. (1998) *Chromatographia*, **47**, 570.

64 Kurganov, A., Davankov, V.A., and Ettre, L.S. (2000) *Chromatographia*, **51**, 500–504.

65 Golay, M.J.E. (1961) *Gas Chromatography* (eds H.J. Noebels, R.F., Wall, and N. Brenner), Academic Press, New York, pp. 11–19.

66 Golay, M.J.E. (1963) *Nature*, **199**, 370–371.

67 Golay, M.J.E. (1968) *Anal. Chem.*, **40**, 382–384.

68 Golay, M.J.E. and Atwood, J.G. (1979) *J. Chromatogr.*, **186**, 353–370.

69 Golay, M.J.E. (1980) *J. Chromatogr.*, **196**, 349–354.

70 Giddings, J.C. (1965) *Dynamics of Chromatography*, Marcel Dekker, New York.

71 ASTM (1992) Subcommittee D02.04 on hydrocarbon analysis, in *Annual Book of ASTM Standards*, ASTM, West Conshohocken, PA Vol. 05.01, pp. 344–352.

72 Klee, M.S., Quimby, B.D., and Blumberg, L.M.(23 November 1999) Automated Retention Time Locking, USA Patent 5,987,959.

73 Blumberg, L.M. and Broske, A.D. (14 March 2000). Column Specific Parameters for Retention Time Locking in Chromatography, USA Patent 6,036,747.

74 Blumberg, L.M., Wilson, W.H., and Klee, M.S. (1999) *J. Chromatogr. A*, **842/1-2**, 15–28.

Part Three
Formation of Chromatogram

Temperature-Programmed Gas Chromatography. Leonid M. Blumberg

ISBN: 978-3-527-32642-6

8
Formation of Retention Times

8.1
Solute Mobility

From the point of view of the end result, a solute zone migrating along a GC column can be treated as a one-dimensional entity, Figure 1.3, described by distribution, $a(z)$, of specific amount of a solute (in units of mass/length, mole/length, and so forth) along the z-axis representing the distance from the column inlet. However, in order to identify the factors affecting the end result, a two-dimensional picture, such as the one, Figure 8.1, representing an *axial longitudinal plane* of a capillary column, is helpful.

Note 8.1

Generally, a distribution of a solute in a column can be a three-dimensional function. However, in a round column, all column and solute properties are typically the same for any main longitudinal plane regardless of its angle, and, therefore, there is no need to expand the scope of the inside view of a column to a three-dimensional picture. □

A fraction of a solute migrating in a column can be dissolved, Figure 8.1, in a carrier gas stream (a mobile phase) and be as mobile as the carrier gas itself, that is to migrate along the column with the same velocity as the velocity, u, of the gas. Another fraction of the same solute can be interacting with the column stationary phase being absorbed or adsorbed by it and remaining practically immobile (stationary) for the time of the interaction.

To be more specific, let us focus on wall coated capillary columns used in partition GC – the subject of primary attention in this book. In such columns, a process of absorption of a solute in a liquid stationary phase is the solute solvation in that phase. On the other hand, a process of desorption of a solute from the liquid phase is its evaporation from that phase.

Before a mobile molecule can become immobile, it has to reach the stationary phase. Due to the presence of a random component in the motions of molecules of a

Temperature-Programmed Gas Chromatography. Leonid M. Blumberg

ISBN: 978-3-527-32642-6

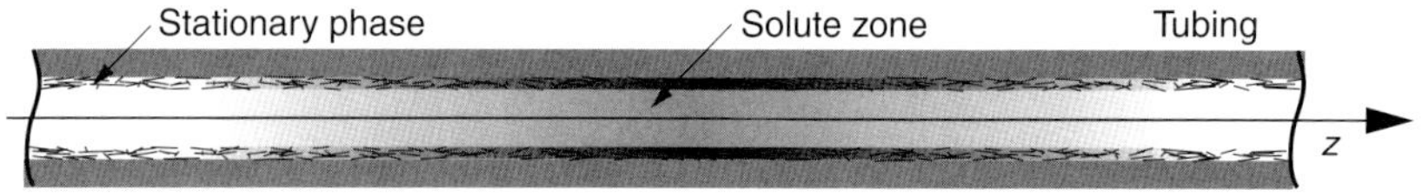

Figure 8.1 Solute zone in a carrier gas and in a stationary phase of a capillary column. The darkness of the solute represents its concentration (mass/volume, mole/volume and so forth). The higher the darkness the larger is the concentration. Typically, a solute concentration in a liquid stationary phase of a GC column is much larger than that in a carrier gas. At a radial equilibrium of a solute solvation/evaporation, the ratio of nonzero concentrations is the same at any coordinate *z*. The proportions of a solute zone shown here are not typical. The longitudinal width of a zone – its spread along the *z*-axis – is typically hundreds of times larger than the column internal diameter.

gas and the mobile molecules of a solute, there is a certain probability at any time that a mobile molecule of a solute can reach the stationary phase, be dissolved in it, and become immobile. At the same time, there is a certain probability that an immobile molecule can be evaporated from the stationary phase and become mobile. For conditions that are typical in GC there is equilibrium or near equilibrium between the two opposing process in any cross-section of a column.

Let m_{sol} and $m_{\text{sol,g}}$ be, respectively, Figure 8.1, the total amount of a solute and the amount of its mobile fraction

$$\mu = \frac{m_{\text{sol,g}}}{m_{\text{sol}}} \tag{8.1}$$

residing in the gas at the equilibrium of the solute solvation/evaporation in and out of a liquid stationary phase. To evaluate the effect of quantity μ on velocity of migration of a solute along a column, let us start with a simple case where decompression of a carrier gas along a column is weak and, therefore, the carrier gas velocity, u, is nearly uniform (nearly the same at any location along the column). While the entire solute zone remains at the equilibrium, any particular molecule in the zone does not remain permanently mobile or permanently immobile, but frequently changes its mobility status. As a result, the entire solute zone migrates along the column as a single entity. Since, at any time, a fraction μ of the solute molecules is dissolved in the carrier gas and migrates with the velocity, u, while the rest of the solute zone remains immobile, the entire zone appears to migrate with the zone velocity,

$$v = \mu u \tag{8.2}$$

The quantity μ is an important parameter of a solute interaction with a column stationary phase. Introduced in 1944 by Consden *et al.* [1, 2], it is a parameter of choice for the description of a solute migration in many studies [1–17]. As a fraction, Eq. (8.1), of a whole, μ can vary from zero to one, that is,

$$0 \leq \mu \leq 1 \tag{8.3}$$

and the larger μ is the higher is the solute velocity v. This suggests that μ can be treated as a solute mobility factor [4, 8] (briefly, solute mobility).

Note 8.2

In the original study [1], no name designation has been assigned to parameter μ (originally denoted as R_F). Later, several terms such as the *retardation factor* [2, 6, 7, 9–11, 17], the *retention ratio* [8, 12, 13, 16], the *frontal ratio* [14], the *propagation factor* [3], *migration rate* [19] and the *mobility factor* [4, 18] were used in the literature to designate this quantity.

The first two terms – the retardation factor and the retention ratio – that enjoy the widest acceptance (the retardation factor is also recommended by IUPAC [9]) are counterintuitive. They assign the lowest retardation factor and the lowest retention ratio to the most retarded (least mobile) and the most retained solutes. On the other hand, these terms assign the highest retardation factor and the highest retention ratio to the least retarded (most mobile) and the nonretained solutes. This obstructs interpretation of many relations involving μ. The term mobility factor evolved [4, 18] from these and similar considerations. Earlier, Connors used the term mobility [5] and the traditionally nameless parameter R_f as what appears to be the designations for the same concept. □

8.2 Solute–Column Interaction and Solute Migration

8.2.1 Parameters of a Solute–Column Interaction

To find a connection between a solute mobility, μ, and the level of a solute interaction with a liquid stationary phase, one can start with the expression

$$m_{sol} = m_{sol.g} + m_{sol.f} \tag{8.4}$$

where $m_{sol.f}$ is the immobile mass of a solute – its mass that is dissolved in a stationary phase liquid film, Figure 8.1. Comparing Eq. (8.4) with Eq. (5.4), one can find that μ can be expressed via the solute retention factor, k, as $\mu = 1/(1 + k)$ [1, 2, 14].

The parameters k and μ describe the same solute–liquid interaction from different points of view. Because there is a one-to-one relationship between μ and k, either of these parameters carries the same information. However, some aspects of a solute–liquid interaction, such as its dependence on T, can be simpler and more transparently expressed via k (Chapter 5) while others, such as velocity of a solute migration – via μ. Sometimes, the choice depends on a researcher's personal preferences. Giddings noticed [13] (page 233, different symbols were used instead of k and μ in the original text), "most equations describing chromatography are simpler when expressed in terms of μ rather than k."

There is yet another metric,

$$\omega = \frac{m_{sol.f}}{m_{sol}} \tag{8.5}$$

of the solute–liquid interaction that, sometimes, leads to the most transparent expressions [4, 18]. It follows directly from the definitions of k, ω and μ that

$$\mu = 1/(1+k) = 1-\omega \tag{8.6}$$

$$\omega = k/(1+k) = 1-\mu \tag{8.7}$$

$$k = \omega/(1-\omega) = (1-\mu)/\mu \tag{8.8}$$

Similar to the solute mobility factor (μ), the quantity ω is confined between 0 and 1, that is

$$0 \le \omega \le 1 \tag{8.9}$$

However, as follows from Eqs. (8.6) and (8.7), the meaning of the quantity ω is opposite to that of the quantity μ so that ω can be interpreted as a solute immobility factor (briefly, immobility). Indeed, when a solute has zero mobility ($\mu = 0$, zero fraction of the solute resides in the mobile phase) it has the highest immobility, $\omega = 1$ (100% of the solute reside in the stationary phase). On the other hand, a solute with the highest mobility ($\mu = 1$, 100% of the solute reside in the mobile phase) has the lowest immobility, $\omega = 0$ (zero fraction of the solute resides in the stationary phase).

The quantity ω can also be interpreted as a measure of the level of the solute–liquid interaction [4], or a solute engagement with liquid phase. The latter interpretation suggests that the quantity μ can be interpreted as a measure of a solute disengagement from the phase.

With some clarification, the parameters k, μ and, ω are valid not only for solute–liquid interactions but also for the interaction of solutes with all stationary phases, including the solid ones, and for all columns, including the packed ones. For that reason, it is convenient to use a broader term *solute sorption* instead of a more specific term *solute–liquid interaction* used in Chapter 5. It should be pointed out, however, that the solute–liquid interactions – the type of a solute sorption in partition GC (also known as *gas–liquid chromatography* or GLC) – remain the main focus of this book.

Example 8.1

Those who are familiar with the formulae for resolution, R_s, of two close peaks in static analysis, know that R_s is proportional to the ratio $k/(1 + k)$. Because $k/(1 + k) = \omega$, the dependence has a simple interpretation: R_s of two close peaks in static analysis is proportional to their immobility (normalized interaction level with the stationary phase, normalized engagement with it). □

8.2.2 Solute Mobility

The following discussion is based on material of Chapter 5.

Ideally, the dependence of the retention factor (k) on the temperature (T) is described by an ideal thermodynamic model in Eqs. (5.7) and (5.8) where entropy,

$\Delta\mathcal{G}_S$, and enthalpy, $\Delta\mathcal{G}_H$, are thermodynamic parameters of evaporation of a solute from liquid, φ is the stationary phase dimensionless film thickness (Eq. (2.7)), and $\mathcal{R} = 8.31447\ \mathrm{J\,K^{-1}mol^{-1}}$ is the molar gas constant. As an illustration, the parameters $\Delta\mathcal{G}_S$ and $\Delta\mathcal{G}_H$ for several n-alkanes are listed in Table 5.1.

The quantity ln k in Eq. (5.8) is a linear function, Figure 5.2, of $1/T$, and, therefore, four parameters, $\Delta\mathcal{G}_H$, $\Delta\mathcal{G}_S$, $\mathcal{R}$, and β, in Eqs. (5.7) and (5.8) can be reduced to two parameters – the intercept $-g_S$, Eq. (5.10), and the slope T_H, Eq. (5.9). This transforms Eqs. (5.7) and (5.8) into equivalent Eqs. (5.11) and (5.12). As an example, the quantities g_S and T_H corresponding to $\Delta\mathcal{G}_S$ and $\Delta\mathcal{G}_H$ in Table 5.1 are listed in Table 5.2. Equations (5.7), (5.8), (5.11) and (5.12) are equivalent expressions for the same ideal thermodynamic model of the dependence of k on T, $k(T)$. It is convenient to view the parameters $\Delta\mathcal{G}_H$, $\Delta\mathcal{G}_S$, g_S, and T_H as thermodynamic parameters of a solute sorption.

Substitution of Eqs. (5.7) and (5.11) into Eq. (8.6) yields two equivalent expressions of the ideal thermodynamic model for μ:

$$\mu = \frac{1}{1 + 4\varphi \cdot \exp\left(-\frac{\Delta\mathcal{G}_S}{\mathcal{R}} + \frac{\Delta\mathcal{G}_H}{\mathcal{R}T}\right)} \quad \text{(ideal thermodynamic model)} \tag{8.10}$$

$$\mu = \frac{1}{1 + \exp(-g_S + T_H/T)} \quad \text{(ideal thermodynamic model)} \tag{8.11}$$

There are problems with the use of these models. As illustrated in Chapter 5, the parameters in these models, be it the quantities $\Delta\mathcal{G}_S$ and $\Delta\mathcal{G}_H$ or the quantities g_S and T_H, do not have direct chromatographic meaning. More meaningful are characteristic parameters – characteristic temperature, T_{char}, and characteristic thermal constant, θ_{char}, of a solute.

The quantity T_{char} is the temperature that causes a given solute in a given column to have $k = 1$. As shown later in this chapter, T_{char} provides a reasonable presentation of a solute elution temperature in a temperature-programmed analysis. Also, as a general trend, solutes in isothermal and temperature-programmed analysis elute in the order of the increase in their characteristic temperatures.

The quantity θ_{char} – typically valued between 20 and 40 °C (Figure 5.6) – is an increase in T in the vicinity of T_{char} that causes k to decline by a factor of e (≈ 2.718). If, for example, $T_{char} = 500$ K and $\theta_{char} = 30$ °C, then a temperature increase from 470 to 500 K changes k from about 2.72 to 1. The quantity θ_{char} can also be described as an inverse of the rate, Eq. (5.20), of a relative decrease in k with an increase in T in the vicinity of T_{char}. In the previous example, that rate was $1/\theta_{char} = 1/30\,°\mathrm{C} \approx 0.033\,°\mathrm{C}^{-1}$, indicating that in the vicinity of 500 K, each 1 °C increase in T causes about 3.3% decline in k. Relationships between the parameters T_{char} and θ_{char} on the one hand and the parameters g_S and T_H on the other are described in Eqs. (5.28), (5.29), (5.31) and (5.32). They allow one to express Eq. (8.11) via the characteristic parameters T_{char} and θ_{char} as

$$\mu = \frac{1}{1 + \exp\left(\frac{T_{char}}{\theta_{char}}\left(\frac{T_{char}}{T} - 1\right)\right)} \quad \text{(ideal thermodynamic model)} \tag{8.12}$$

The same result can be obtained from substitution of Eq. (5.33) into Eq. (8.6). At $T = T_{char}$, the last formula yields $\mu = 0.5$ (as expected from Eq. (8.6) at $k = 1$).

While expressing μ via chromatographically meaningful parameters, Eq. (8.12) retains one substantial shortcoming of the ideal thermodynamic model. The argument of the exponential term in Eq. (8.12) is inversely proportional to T. This significantly complicates mathematical treatment of conventional linear or piecewise linear temperature programming [20, 21].

A mathematically simpler linearized model of $\mu(T)$ can be obtained from the linearized model of $k(T)$ in Eqs. (5.35) and (5.36,) which is an approximation of the ideal thermodynamic model in the vicinity of $T = T_{char}$. Substitution of Eq. (5.36) into Eq. (8.6) yields

$$\mu = \frac{1}{1 + \exp\frac{T_{char} - T}{\theta_{char}}} \quad \text{(linearized model)} \tag{8.13}$$

The exponential term in this expression is a linear function of T. This model has been used for theoretical evaluation of the key performance characteristics of temperature-programmed GC [4, 18, 22]. The model is similar to other linearized models known in GC [23–26] and to the *linear solvent strength* (*LSS*) model in LC [24, 27–34]. In the latter case, a solvent relative composition plays the role of temperature, T, in Eq. (5.15). Because it can describe a diverse class of interactions of solutes with the column inner materials (stationary and mobile), Eq. (5.15) can be viewed as a model of a broader class of *solute–column interactions* rather than only of the solute–liquid interactions.

The temperature that causes a predetermined μ can be found from Eq. (8.13) as

$$T = T_{char} - \theta_{char} \ln\frac{1-\mu}{\mu} \tag{8.14}$$

8.2.3
Generic Solutes

The parameters T_{char} and θ_{char} of interaction of a particular solute with a particular liquid stationary phase film are independent of each other. However, there is a general trend (Figure 5.6, Eqs. (5.46) and (6.43)) suggesting that the solutes with higher molecular weights that have generally higher T_{char} have also higher θ_{char}.

As described earlier (Chapters 2 and 5) this book is concerned with the general properties of GC analyses (such as the effect of the heating rate on retention times, elution order, widths, and other parameters of all peaks). This is different from specific properties concerned with prediction of retention times of particular solutes and with other similar specifics.

To study the general properties, it is convenient to deal with generic rather than actual solutes (Chapter 5). For generic solutes, Eq. (6.43) is not just a trend, but an exact relationship between T_{char} and θ_{char}, that is

$$\theta_{char} = \theta_{char,st}(T_{char}/T_{st})^{\xi} \quad \text{(generic solutes)} \tag{8.15}$$

where $T_{st} = 273.15$ K, $\theta_{char,st}$ (Eq. (5.47)) is θ_{char} at $T_{char} = T_{st}$, and ξ is a gas parameter listed in Table 6.12. For any generic solute with a given characteristic temperature (T_{char}) quantity θ_{char} in Eqs. (8.12), (8.13) and others can be found from Eq. (8.15). This means that characteristic temperature of a generic solute (the temperature for which $k = 1$) provides a complete description of its interaction with the stationary phase film.

In reality, there are probably very few, if any, actual solutes that interact with stationary phases as generic ones. The concept of generic solutes is simply a conceptual approach to the study of general trends. On the other hand, parameters θ_{char} of overwhelming majority of real solutes fall within a relatively narrow range (Figure 5.6j) around θ_{char} predicted from Eq. (8.15). In that sense, almost all solutes behave more or less as generic ones.

As shown later, generic solutes always elute in the order of the increase in their T_{char} and the difference in their elution temperatures is close to the difference in the values of their T_{char}.

8.2.4 Velocity of a Solute Zone

Let us now return to evaluation of the velocity of a solute migration along a column.

At the first glance, Eq. (8.2) appears to be intuitively clear. It suggests that a solute velocity, v, can be found as a product of the velocity, u, of a carrier gas, and, the solute mobility, μ. A closer look reveals, however, that some questions regarding the meaning of Eq. (8.2) still remain.

In line with a general concept of velocity (Chapter 4), the velocity, v, of a solute zone can be defined as, Eq. (4.2),

$$v = \frac{dz}{dt} \tag{8.16}$$

where z is a coordinate of the zone – its distance from a column inlet – and t is the zone migration time – the time since the injection of the solute.

At any time, t, a solute zone occupies a certain segment along a column. As a one-dimensional entity, the zone is described by the distribution, $a = a(z,t)$, Figure 1.3, of its material along the z-axes at a given time t. The quantity a is a specific amount (mass/length, mole/length, and so froth) of a solute zone – the sum of the fraction of a solute residing in a carrier gas at a given location, z, and the fraction dissolved in a stationary phase at the same z.

Equation (8.16) indicates that, in order to speak of the migration velocity, v, of a distributed solute, its location – a coordinate, z, of the solute zone as a single entity – at any time of its migration has to be clearly identified. In an ideal world of symmetric zones, a zone location could be identified with a coordinate, z, of the zone apex which also is a coordinate of the zone centroid (Chapter 3).

Unfortunately, the world of GC is not always ideal. A zone can be asymmetric, Figure 8.2, and its very shape can change during migration. It is not known how to predict the coordinate of the apex of asymmetric zone from the properties of the medium where the zone migrates. Another choice is to treat a coordinate, z, of a zone

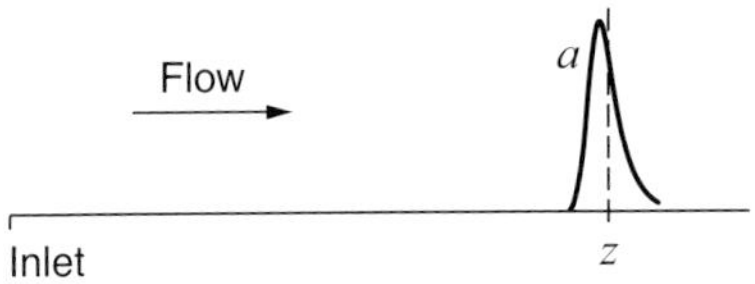

Figure 8.2 Distribution, a, of material in the asymmetric zone. The zone apex is trailing the zone centroid at z.

centroid (the first normalized mathematical moment of the zone, Chapter 3) as the coordinate of the zone itself. Because a zone centroid can be predicted in a broad class of *mass-conserving* migrations of solutes [35–38], it will be assumed from now on that the longitudinal coordinate, z, of the centroid of a solute zone is the coordinate of the zone itself. In that case, a zone velocity, v, is the same as the velocity, v, Eq. (8.16), of displacement of the zone centroid. Equation (8.2) provides a sufficiently accurate description of that velocity. Additional comments on that subject can be found in Appendix 10.A.1 (Chapter 10).

Note 8.3

One aspect of location of a solute zone needs additional clarification. During its elution, when a part of a solute is out of a column and a part of the solute still remains in the column, the *elution-related aberrations* take place. Strictly speaking, it is incorrect to say, for example, that a solute zone is located at the column end since the coordinate of the centroid of whatever is left of a zone in a column is always within the column, but not at its end. To avoid the logical difficulties associated with the aberrations, one can think of a column as being sufficiently longer than its actual length, L, is. Along with that, one can think of a column outlet as just a mark at a distance $z = L$ along the column. □

8.3
General Equations of a Solute Migration and Elution

A dual view (Chapter 4) on variables like the distance (z) from the column inlet, time (t) since the injection of a sample, gas velocity (u), solute mobility (μ) and velocity (v), and others is adopted in this book. On the one hand, these quantities can be viewed as the variables of a solute zone migrating in a medium (a column). On the other hand, they can be viewed as variables and parameters of the medium itself.

A location, z, of a solute and the time, t, of its travel to that location are mutually dependent variables. One can express z as a function, $z = z(t)$, of t. One can also express t as a function, $t = t(z)$, of z. In both cases,

$$z = 0 \text{ implies } t = 0, \text{ and } t = 0 \text{ implies } z = 0 \tag{8.17}$$

From another perspective, the solute migration variables such as the gas velocity (u) solute mobility (μ) and its velocity (v) can be treated as properties of a medium (a column) where the solute migrates. In that case, the variables can be expressed as the functions $u = u(z,t)$, $\mu = \mu(z,t)$ and $v = v(z,t)$ of z and t. Thus, in a typical temperature-programmed, isobaric (constant pressure) analysis, v is a strong function of t. Due to the dependence of gas viscosity on temperature, u is also a function of t. If, in addition to the temperature programming, a significant decompression of a carrier gas along a column takes place, then both u and v become functions of z and t. In all these cases, z and t are mutually independent variables of the medium, rather than a coordinate and a migration time of a particular solute. For example, the function $v(z,t)$ tells what would be a solute velocity, v, if, at an *arbitrary* time, t, the solute was at an arbitrary location, z, regardless of whether or not the solute can actually arrive to that location at a given time t. If a solute velocity, $v = v(z,t)$, is known as a function of mutually independent variables z and t, then the time, $t = t(z)$, it takes for a solute to arrive to an arbitrary location, z, can generally be found, Example 4.2, as a solution to Eqs. (8.16) and (8.17).

Let $t(z)$ be the time of arrival of a solute to an arbitrary location, z. Then the retention time (elution time), t_R, of a peak corresponding to the solute is the time, $t(L)$, of arrival of a solute to the end of the column, that is

$$t_R = t(L) \tag{8.18}$$

Note 8.4

The following is another aspect of elution-related aberrations, Note 8.3. During its elution, a portion of a solute that still remains in the column continues to disperse. As a result, even under the uniform static conditions, a symmetrically distributed solute yields an asymmetric peak whose retention time, t_R, is different from the time, $t(L)$, of the solute arrival to the column end. However [39–42], the relative difference $(t_R - t(L))/t(L)$ is in the order of $1/N$ where N is the column plate number – a quantity in the order of L/d_c. This, due to Eq. (2.9), allows one to ignore the difference between t_R and $t(L)$ as is done in Eq. (8.18). It is important to recognize however that Eq. (8.18) is a convenient and, typically, sufficiently accurate approximation for t_R, but not its definition. □

In static analysis, t and t_R can be expressed as

$$t = \frac{t_g}{\mu} \text{ (static conditions)} \tag{8.19}$$

$$t_R = \frac{t_M}{\mu} \quad \text{(static conditions)} \tag{8.20}$$

where t_g and t_M are, respectively, the gas propagation time and the hold-up time – the times (Chapter 7) that it takes for a gas packet to travel to an arbitrary location, z, and to the end of the column, respectively, assuming that

$$t_g = 0 \quad \text{when} \quad t = 0 \tag{8.21}$$

Due to Eq. (8.6), Eq. (8.20) can also be expressed as

$$t_R = (1 + k)t_M \quad \text{(static conditions)} \tag{8.22}$$

In a dynamic analysis, column temperature and/or pneumatic conditions of a carrier gas flow can be programmable functions of time. As a result, the expression for a solute migration time, t, as a function, $t = t(z)$, of its location, z can become more complex compared to Eq. (8.19). However, that formula can serve as a prototype for a solution for $t = t(z)$ in a dynamic medium.

Following a general definition (Eq. (4.2)) of velocity of a moving object, one can define a carrier gas velocity, u, in a general case of a dynamic nonuniform medium as the velocity of a narrow gas packet, that is

$$u = \frac{dz}{dt_g} \tag{8.23}$$

where dt_g – the elementary time required for a narrow gas packet to travel from z to $z + dz$ – is an elementary increment in a gas propagation time, t_g, considered in Chapter 7 for static conditions. At any time, t, when a given solute is located at z, the time dt_g is different from the elementary time, dt, required for the solute to traverse the same distance. It follows from Eqs. (8.16),(8.23) and (8.2) that

$$dt_g = \mu dt \tag{8.24}$$

which is a differential form of Eq. (8.19). A discussion of the properties of the quantity t_g in a dynamic medium is provided shortly. Till then, t_g is treated just as a variable formally defined in Eqs. (8.23) and (8.24).

Two parameters, $u(z,t)$ and $\mu(z,t)$, of the medium define the system of Eqs. (8.23) and (8.24). Due to the initial conditions described in Eqs. (8.17) and (8.21), the system can be expressed for the time t required for a solute zone to arrive to location z as

$$\int_0^{t(z)} \mu(z(t), t)dt = t_g \tag{8.25}$$

$$t_g = t_g(z) = \int_0^z \frac{dz}{u(z, t(z))} \tag{8.26}$$

Mathematically, both equations in this system are equally important because they are necessary for finding $t(z)$. From a chromatographic point of view, the key equation in the system is the solute migration equation, Eq. (8.25), which shows how a solute mobility, μ, links $t(z)$ with $t_g(z)$ for an arbitrary location, z, of the solute. Equation (8.26), showing how the gas propagation time, $t_g(z)$, depends on the gas velocity, u, at any z, plays a conceptually supportive role.

At the end of a column where $z = L$, Eqs. (8.25) and (8.26) become

$$\int_0^{t_R} \mu(z(t), t)dt = t_m \tag{8.27}$$

$$t_{\mathrm{m}} = \int_0^L \frac{\mathrm{d}z}{u(z, t(z))} \tag{8.28}$$

where t_{R} is the *solute retention time* defined in Eq. (8.18), Eq. (8.27) is a solute elution equation, and

$$t_{\mathrm{m}} = t_{\mathrm{g}}(L) \tag{8.29}$$

is *dynamic hold-up time* corresponding to that solute.

The system of Eqs. (8.25) and (8.26) as well as its special case in Eqs. (8.27) and (8.28) describes formation of retention times under a broad set of conditions that can combine dynamic operations (such as simultaneous pressure and temperature programming) in a nonuniform medium (gas decompression along a column, serial assembly of different columns, and so forth). As a result of their breadth, the systems cannot be directly used for practical description of column operation. Nevertheless, by providing compact descriptions of a solute migration, the systems serve useful purposes. They reveal the key factors affecting retention time formation and allow one to make several far-reaching conclusions that will be formulated later. The systems can also serve as a starting point for further simplifications.

A numerical algorithm (Appendix 8.A.1) for solving Eqs. (8.25) and (8.26) for one important special case provides additional insight into mechanics of Eqs. (8.25)–(8.28).

8.3.1 Dynamic Gas Propagation Time

An important concept identified in Eqs. (8.25) and (8.26) is that of the gas propagation time, t_{g}, in analysis under dynamic nonuniform conditions.

Generally, t_{g} is defined in Eqs. (8.21) and (8.24). Equation (8.21) together with the formula

$$\mathrm{d}t_{\mathrm{g}} = \frac{\mathrm{d}z}{u} \tag{8.30}$$

which follows from Eq. (8.23), can be treated as an equivalent alternative definition of t_{g}.

Note 8.5

Equation (8.30) includes the inverse velocity, $1/u$, of a carrier gas. Many formulae including inverse velocities, $1/v$ and $1/u$, could be simpler and more transparently expressed [36, 43–47] if the quantities $t' = 1/v = \mathrm{d}t/\mathrm{d}z$ and $t'_{\mathrm{g}} = 1/u = \mathrm{d}t_{\mathrm{g}}/\mathrm{d}z$ were used instead of, respectively, $1/v$, $1/u$. The root cause for the advantage of this approach comes from the fact that, in chromatography, all solutes travel through the same distance – column length, L, – before their elution, but it takes different time for each solute to make that trip. As a result, a distance, z, along a column rather than time, t, of a solute migration is a better choice for an independent variable in many equations describing a solute migration (otherwise, it would be desirable to express Eq. (8.30) as $dz = u dt_g$).

The quantities t' and t'_g are measured in units of time per unit of length (such as s/m) and represent the time, t, it takes for a solute or for a gas packet to travel along a unit of distance. One can also interpret t' and t'_g as the times that a solute or a gas packet resides within a unit of distance. This interpretation was a reason behind naming t' and t'_g as, respectively, the solute and the gas residency [43]. Further exploration of the quantities t' and t'_g is outside of the scope of this book. □

Consider an arbitrary location, z, along a column and an elementary increment dz. Suppose that, for solute A, it takes time t_A to reach location z. Generally, u at z can be a function, $u = u(z,t)$, of time, t. From Eq. (8.30), one has for the elementary increment, $dt_{g,A}$, in a gas propagation time at z and t_A: $dt_{g,A} = dz/u(z,t_A)$. This is the time it takes for a gas packet to travel from z to $z + dz$ at the time, t_A, when solute A is at z. For a different solute, say solute B, it might take different time, t_B, to arrive to location z. The time, $dt_{g,B}$, it takes for a gas packet to travel from z to $z + dz$ at the time t_B when solute B is at z can be found as $dt_{g,B} = dz/u(z,t_B)$. If, due to the dynamic nature of a chromatographic analysis, $u(z,t_B) \neq u(z,t_A)$, then dt_g at an arbitrary location, z, becomes a function of time required for a corresponding solute to arrive to z.

Note 8.6

The gas velocity, u, at a fixed location along a column can be a function of time in pressure-programmed analysis. In isobaric temperature-programmed GC analysis, the changes in u with time come from the temperature dependence of gas viscosity. □

If u at some or all locations along a column is a function of t, then the accumulation, $t_g = t_g(z)$, of all time increments $dt_g = dz/u$ in Eq. (8.26) can become different for different solutes. In other words,

Statement 8.1

In dynamic GC analysis, the gas propagation time, t_g, can be a different function of z for different solutes.

Note 8.7

Not always $t_g(z)$ in a dynamic analysis is different for different solutes, but only when gas velocity, u, is a function of t. In a temperature-programmed GC analysis with a negligible decompression of a carrier gas along a column, a fixed value of u can be obtained by pressure programming in order to compensate for the change in gas viscosity caused by the change in a column temperature. If the decompression of a carrier gas is significant, the compensation for the temperature-dependent changes in the gas viscosity simultaneously everywhere along a column is hardly possible. □

Let again t_g be a gas propagation time in a static as well as in a dynamic analysis. A *static gas propagation time* can be defined as t_g in the analysis where the gas velocity, u, is static. A *dynamic gas propagation time* can be defined as t_g in the analysis where the gas velocity, u, is dynamic. According to these definitions, t_g in a static analysis is always static, but t_g in a dynamic analysis can also be static if u in that analysis is static.

It follows from Eq. (8.30) that

Statement 8.2

The static gas propagation time is the same function of z for all solutes.

Particularly, a static gas propagation time corresponding to any solute is the same as the gas propagation time corresponding to an unretained solute. For the latter, $\mu = 1$ implying, according to Eq. (8.24), that

Statement 8.3

At any location z, the static gas propagation time, $t_g(z)$, is the same as the migration time, $t(z)$, of an unretained solute. The quantity $t_g(z)$ can also be interpreted as the migration time of a *single* gas packet.

One can conclude that, as mentioned earlier, the static gas propagation time is the same as the gas propagation time discussed in Chapter 7.

As for the dynamic gas propagation time, not only it can be different for different solutes (Statement 8.1), but it might also be different from the migration time (t) of an unretained solute.

Indeed, it follows from Eqs. (8.2) and (8.3) that, typically, a solute migrates along a column slower (frequently, much slower) than does a packet of a carrier gas molecules. Therefore, during its migration from one dz-long segment of a column to the next segment, a solute is being accompanied (and bypassed) by different narrow gas packets. Suppose that, in order to traverse a sequence of dz-long segments, it takes for a solute a sequence of elementary times dt_1, dt_2, dt_3, ..., while, for the sequence of the accompanying gas packets, it takes a sequence of elementary times dt_{g1}, dt_{g2}, dt_{g3}, ... According to Eq. (8.24), the last sequence can be expressed as $\mu_1 dt_1$, $\mu_2 dt_2$, $\mu_3 dt_3$. Equation (8.25) describes $t_g(z)$ as the accumulation of these elementary time intervals. In other words,

Statement 8.4

Equation (8.25) describes $t_g(z)$ corresponding to a given solute as the accumulation of consecutive elementary times dt_g each of which is the elementary time of propagation of a gas packet that bypasses the solute located at z and which travels through location z $1/\mu$ times faster than the solute does. If $\mu < 1$, then different packets bypass the same solute at different locations z.

The model of a gas propagation time as the accumulation of the elementary gas propagation times of different gas packets applies to a dynamic and a static analysis. However, due to Statements 8.2 and 8.3, the static gas propagation time, $t_g(z)$, is the same as the migration time, $t(z)$, *of an* unretained solute, and can be conveniently identified with it. This means, that, at least in principle, the static gas propagation time can be *measured* in an experiment that allows us to *observe* (directly or indirectly) the migration of an unretained solute within a column.

On the other hand, dynamic $t_g(z)$ cannot be identified with the migration time, $t(z)$, of an unretained solute in a given dynamic analysis. The resulting lack of a model for a simple experiment (or, at least, a simple thought experiment) allowing us to measure dynamic $t_g(z)$ causes an uncomfortable void in the interpretation of that important quantity. To address and, eventually, to resolve the problem, it is important to face the fact that, a dynamic $t_g(z)$ corresponding to one solute can be different from $t_g(z)$ corresponding to other solutes.

Example 8.2

Consider two solutes, "1" and "2", migrating with $\mu_1 = 0.01$ and $\mu_2 = 0.005$ along a column during a pressure-programmed, isothermal analysis. Let $t_1(z)$ and $t_2(z)$ be the times of arrival of solutes "1" and "2," respectively, to some location, z, along a column. Due to its two times lower mobility, solute "2" travels behind solute "1." In isobaric analysis, it would take for solute "2" twice as longer than for solute "1" to arrive to any given location. However, in the case of increasing inlet pressure, $t_2/t_1 < 2$.

Let us assume that the pressure program is such that, for some particular $z = z_0$, $t_1(z_0) = 10$ min and $t_2(z_0) = 15$ min. Whenever it takes Δt time units for a solute to advance from some location z to location $z + \Delta z$, it takes only $\Delta t_g = \mu \Delta t$ time units for a gas packet to advance from $z + \Delta z$. As a result, $t_{g1} = t_g(z_0, t_1(z_0)) = \mu_1 t_1(z_0) = 0.01 \times 10\,\text{min} = 6\,\text{s}$, $t_{g2} = t_g(z_0, t_2(z_0)) = \mu_2 t_2(z_0) = 0.005 \times 15\,\text{min} = 4.5\,\text{s}$. In other words, at $z = z_0$, the gas propagation times, t_{g1} and t_{g2}, corresponding to solutes "1" and "2" are, respectively, $t_{g1} = 6\,\text{s}$, $t_{g2} = 4.5\,\text{s}$. □

Let $t_g(z)$ be a dynamic gas propagation time corresponding to a particular solute with $\mu < 1$. Because the solute migrates slower than the carrier gas, the net time of migration of any given gas packet through the entire column is smaller than the net time, t_R, of migration of the solute. As a result, no single gas packet and no single unretained solute can reflect the changes in the gas velocity, u that took place during the time t_R. Also, according to Eq. (8.25), $t_g(z) < t(z)$ at any z.

Example 8.3

While t_{g1} and t_{g2} of Example 8.2 are only several second long (6 s and 4.5 s), they are affected by the programmable changes in pressure that took place during the time measured in minutes (10 min and 15 min, respectively). Due to the gradually increasing pressure (at fixed temperature), the longer a solute retained (and, therefore, the longer it migrates to a given location), the shorter is the gas propagation time, t_g, corresponding to the solute. (The latter effect does not occur in a temperature-programmed isobaric analysis.) □

The dilemma of the fact that a relatively short $t_g(z)$ depends on the changes that take place during a longer (frequently, much longer) solute migration time, $t(z)$, is not a contradiction. Only $t(z)$ is the actual time in the actual analysis while $t_g(z)$, when considered in respect to the actual analysis, is not an actual time, but a computed quantity. This observation not only explains the dilemma, but also provides a hint to making $t_g(z)$ an observable and a measurable quantity by constructing an accelerated experiment where the necessary changes in gas velocity u evolve much faster than they do in an actual analysis.

Consider an actual dynamic analysis, and an accelerated experiment constructed to measure a gas propagation time, $t_g(z)$, corresponding to a given solute in the actual analysis. Let t be the time in the actual analysis, $\mu(t)$ be the mobility of the solute in the actual analysis, and t_a be the time in the accelerated experiment. If all timed events (such as the pressure and temperature programs) in the accelerated experiment run $1/\mu(t)$ times faster than they do in the actual analysis, then the relationship between t_a and t can be described as

$$dt_a = \mu(t)dt \quad \text{with} \quad t_a = 0 \quad \text{at} \quad t = 0 \tag{8.31}$$

This means that any change that in the actual analysis takes place during an elementary time dt occurs in the accelerated experiment during the shorter time $\mu(t)dt$. As a result, Eq. (8.24) for the accelerated experiment becomes $dt_g = dt_a$, or, reproducing the form of Eq. (8.24), results in

$$dt_g = \mu dt_a \quad \text{where} \quad \mu = 1 \tag{8.32}$$

Equations (8.31) and (8.32) imply two things. It follow from Eq. (8.31) that dt_g in Eq. (8.32) is exactly the same as dt_g in Eq. (8.24), and, therefore, Eq. (8.32) describes exactly the same $t_g(z)$ – the gas propagation time corresponding to the given solute in the actual analysis – as Eq. (8.24) does. On the other hand, because μ in Eq. (8.32) is equal to unity, the dynamic gas propagation time, $t_g(z)$, described in Eq. (8.32) corresponds to an unretained solute and is equal to the migration time, $t_a(z)$, of that solute in the accelerated experiment. This means that the accelerated experiment allows us to measure a dynamic gas propagation time corresponding to a given solute in an actual analysis by observing an unretained solute in the accelerated experiment.

It should be stressed again that, because two different solutes in an actual analysis could at the same time, t, have different mobilities, $\mu(t)$, the time scales, t_a, in respective accelerated experiments could also be different.

Note 8.8

It is not unusual for μ^{-1} to be a large number measured in thousands. Therefore, the μ^{-1}-fold accelerated programming of pressure and/or temperature in an accelerated experiment could be outside of the laws of physics or it could be beyond the reach of a current state of the art. It is very likely that these factors might make it impossible to *physically* conduct an accelerated experiment. In this case, the experiment can be viewed only as a thought exercise. □

Note 8.9

An accelerated experiment can be constructed for the measurement of a dynamic as well as a static t_g. However, there is no need for the latter, according to Statements 8.2 and 8.3. □

The existence of the models for the experimental observation and measurement (at least, in principle) of any gas propagation time (static or dynamic) helps to clarify one of the key concepts of dynamic chromatography.

The system of Eqs. (8.25) and (8.26) might or might not have a closed form solution for migration of an arbitrary or a particular solute. However, whether it has a closed form or only a numerical (Appendix 8.A.1) solution, the system offers a clear and simple conceptual description of the timing of a solute migration. It has been mentioned earlier that the migration equation, Eq. (8.25), is the key equation in the system. It shows that migration of a solute in any chromatographic analysis can be described via only two quantities – the solute mobility, μ, and the gas propagation time, t_g. In other words,

Statement 8.5

All particulars of a GC analysis – column dimensions, carrier gas type and its (possibly programmable) flow, stationary phase type and thickness, (possibly programmable) column temperature, and so forth – affect the time, $t(z)$, of migration of a given solute to an arbitrary location, z, only as much as they affect two parameters: the solute mobility, μ, and the gas propagation time, $t_g(z)$.

This means that the quantities t_g and μ are suitable as the focal points in the studies of a solute migration in chromatography.

8.3.2
Solute Retention Time and Dynamic Hold-up Time

At the end of an L-long column, the solute migration time, $t(z)$, becomes its retention time, t_R, Eq. (8.18), and the dynamic gas propagation time, $t_g(z)$, becomes the dynamic hold-up time, t_m, Eq. (8.29). The system of Eqs. (8.25) and (8.26) becomes the system of Eqs. (8.27) and (8.28).

When it is necessary to distinguish a conventional [9] hold-up time, t_M, – the hold-up time in analysis with a static gas velocity (Chapter 7) – from the hold-up time, t_m, Eq. (8.29), in an analysis with dynamic gas velocity, let us refer to the former as to the static hold-up time. The quantities t_M and t_m relate to each other in the same way as do the static and the dynamic gas propagation times, t_g. The only difference between the two pairs is that the gas propagation times correspond to an arbitrary location along the column while the hold-up times correspond to the column's end.

How different can a dynamic and a static hold-up time be? A clear answer to this question can be obtained in the case of a conventional isobaric (constant pressure) temperature-programmed analysis. Due to the temperature increase in such analysis, the gas velocity, u, at any given z gradually decreases with time due to the increase in the temperature-dependent gas viscosity. It follows from Eq. (8.28), therefore, that t_m corresponding to a given solute is bound by the inequality

$$t_{M,init} < t_m < t_{M,R} \quad \text{(isobaric analysis)} \tag{8.33}$$

where $t_{M,init}$ and $t_{M,R}$ are static hold-up times corresponding to, respectively, the initial temperature, T_{init}, of the temperature program and the elution temperature, T_R, of the solute – the column temperature at the solute elution time, t_R. As static hold-up time is proportional, Eq. (7.96), to gas viscosity, the ratio $t_{M,R}/t_{M,init}$ can be estimated as, Eq. (6.20), Table 6.3, $t_{M,R}/t_{M,init} \approx (T_R/T_{init})^{0.7}$. One can conclude that, even when the temperature program in an isobaric analysis covers a wide temperature range from T_{init} to $T_R = 2T_{init}$, the quantity t_m departs from either of its bounds in Eq. (8.33) by less than a factor of 2. Later we will see that whenever there is a significant difference between $t_{M,init}$ and $t_{M,R}$ in a typical isobaric temperature-programmed analysis, t_m is decisively closer to $t_{M,R}$ than it is to $t_{M,init}$ so that it can be reasonably assumed that

$$t_m \approx t_{M,R} \tag{8.34}$$

It has been mentioned earlier that the elution equation, Eq. (8.27), is the key equation in the system of Eqs. (8.27) and (8.28). Equation (8.27) leads to the following interpretation of Statement 8.5.

Statement 8.6

All particulars of a GC analysis – column dimensions, carrier gas type and its (possibly programmable) flow, stationary phase type and thickness, (possibly programmable) column temperature, and so forth – affect the retention time, t_R, of a solute only as much as they affect two parameters: the solute mobility factor, μ, and the dynamic hold-up time, t_m, of the gas.

Several modifications of Eq. (8.27) can be found in the literature [3, 21, 24, 25, 34, 47–50]. Some of them are approximate and hardly suitable for the pressure-programmed and other similar conditions. Others are as exact as Eq. (8.27), but have a limited scope (of, for example, being valid only for a constant pressure GC analysis). None of them provides the exact description of a solute migration in arbitrary dynamic nonuniform medium as Eq. (8.27) does.

The concept of dynamic hold-up time, t_m, plays a key role not only in descriptions of a solute migration, but also in descriptions of broadening of solute zones. Admittedly, the concept is not as simple as the concept of a static hold-up time,

t_M. However, it serves a useful purpose of providing the means for an exact and a compact expression of equation of a solute migration, Eq. (8.27), in the most general dynamic and nonuniform conditions. It also serves as a basis for further simplification of Eq. (8.27). To a certain degree, the rest of this chapter consists of a search for simpler substitutes (exact and approximate) for t_m in order to simplify Eq. (8.27) and future formulae utilizing t_m.

8.4
Uniform Solute Mobility in Isobaric Analysis

As mentioned earlier, the general elution equation, Eq. (8.27), provides a clear conceptual description of formation of a solute retention time under a broad set of time-varying nonuniform conditions. However, neither Eq. (8.27) alone nor the system of Eqs. (8.27) and (8.28) has a general closed form solution. As a result, neither the equation nor the system provides practically oriented description of retention time formation. The mathematical complexity of the system comes from the fact the integrands in the equations comprising the system include an unknown function, $t(z)$ and its inversion, $z(t)$. It is important to find such constraints to the conditions of a GC analysis that can allow one to eliminate the system complexity while retaining its suitability for a broad class of practically important techniques of GC analyses.

8.4.1
Uniform Mobility

A condition of *uniform mobility*, μ, of a solute can be expressed as

$$\frac{\partial \mu}{\partial z} = 0 \tag{8.35}$$

It follows from Eq. (8.6) that μ is uniform if the *retention factor, k*, is uniform. In a column with uniform dimensions, k and, therefore, μ is uniform for all solutes if physicochemical properties of a stationary phase are uniform and the column is uniformly heated. All these conditions include the overwhelming majority of practical GC analyses, and are listed in the outline (Chapter 2) of the scope of this book. In other words, the scope of this book is limited to the condition of uniform μ.

It should also be added that, if all other conditions are the same, then a column with a nonuniform μ *cannot* outperform a column with a uniform μ [44, 51–54]. Golay predicted [55] the superior performance of the uniform heating as early as in 1962 [56]. Giddings has shown [44] that, in the case of a column assembled of segments with different stationary phases, each transition from one stationary phase to another can only reduce the overall column performance. Not only the operations with nonuniform μ cannot outperform the operations with uniform μ, but also the implementations of a controllable nonuniformity of μ in GC, such as a nonuniform column heating [57–62] or creation of serial column assemblies with a real-time tunable solute mobility [63–65], are more complex.

Note 8.10

In its key aspects, temperature programming in GC is equivalent to programming of a composition of a mobile phase in gradient-elution LC. Both result in the time-dependent mobilities, μ, for some or all solutes. However, the specifics of programming of μ in LC – the modification of the mobile phase composition at the column inlet and gradual propagation of the changed composition toward the outlet – causes μ to be nonuniform. Although this could reduce the performance of an actual gradient-elution LC analysis compared to what could have been (a hardly possible task of) a uniform programming of μ, the losses could be practically insignificant because only a significant nonuniformity of μ can cause a significant deterioration of a column performance [36, 43]. □

Uniform $\mu(z,t)$ in Eqs. (8.25) and (8.27) is independent of z. This eliminates the need for relying on an unknown function $z(t)$ in the integrands of Eqs. (8.25) and (8.27), thus substantially simplifying mathematical treatment of the equations. Thus Eq. (8.27) – the elution equation – becomes

$$\int_0^{t_R} \mu \mathrm{d}t = t_m \tag{8.36}$$

where $\mu = \mu(t) = \mu(0,t)$ is a known function of t as long as μ is a known function of temperature, T, and a column temperature program is known.

In the case of a uniform heating of a column, that is when $\partial T/\partial z = 0$, a column temperature, T, at any instant, t, of time is the same at all locations along a column. The uniform heating is the only type of a column heating considered in the book.

As an example, a computer-generated chromatogram and functions $z(t)$ and $z(T)$ are shown in Figures 8.3 and 8.4, respectively. All three were found from the numerical solution (Appendix 8.A.1) of Eq. (8.25) for a single-ramp uniform heating and constant pressure.

Comparing chromatograms in Figure 8.3, one can notice that the linearized model tends to yield about 0.4 min shorter retention times or (equivalently) about 10 °C lower elution temperatures for all peaks. Although this shift does not significantly affect parameters of a column general performance (such, for example, as analysis time), it might be useful to know its source. A likely source of the shift is the approximations where linearized model for each solute is a tangent line to its ideal thermodynamic model (Figure 5.5). This approximation tends to reduce retention factors in the linearized model for all temperatures other than T_{char}. A more careful linear fit to idealized model would certainly reduce the shift. This might be considered as a way of further fine tuning of the linearized model if it becomes necessary or desirable.

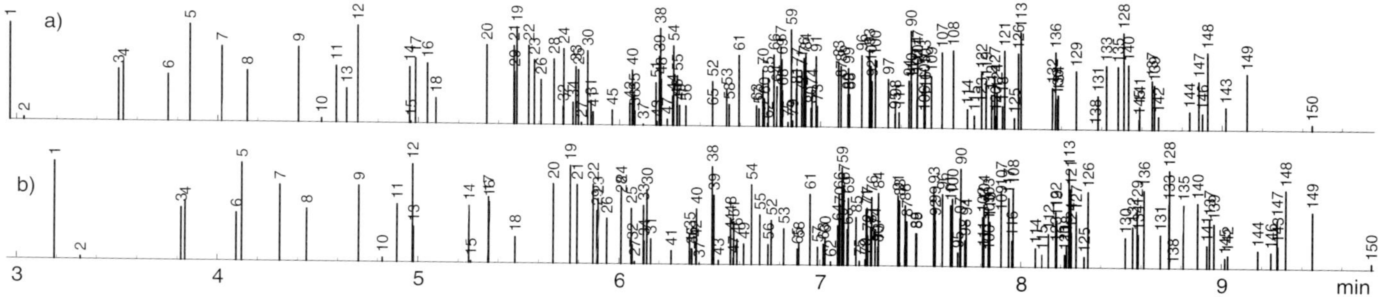

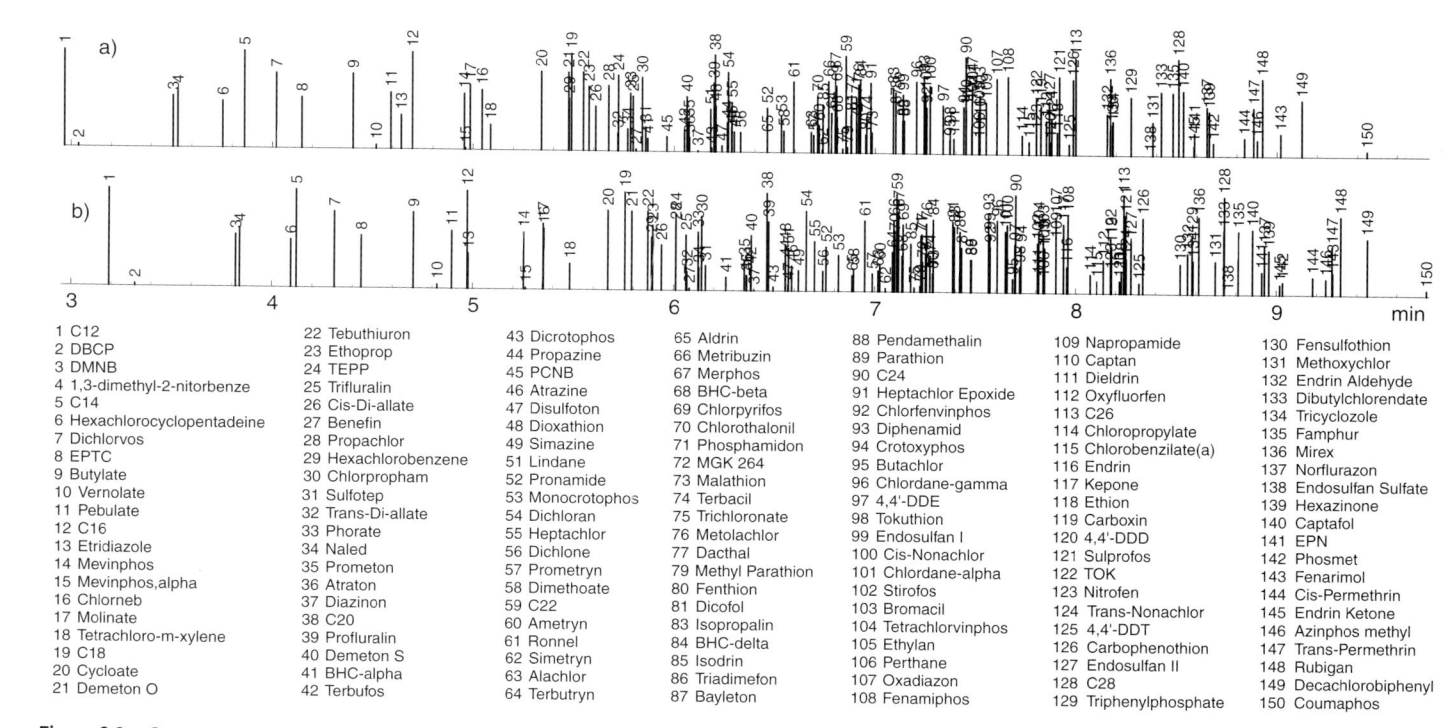

1 C12
2 DBCP
3 DMNB
4 1,3-dimethyl-2-nitorbenze
5 C14
6 Hexachlorocyclopentadeine
7 Dichlorvos
8 EPTC
9 Butylate
10 Vernolate
11 Pebulate
12 C16
13 Etridiazole
14 Mevinphos
15 Mevinphos,alpha
16 Chlorneb
17 Molinate
18 Tetrachloro-m-xylene
19 C18
20 Cycloate
21 Demeton O
22 Tebuthiuron
23 Ethoprop
24 TEPP
25 Trifluralin
26 Cis-Di-allate
27 Benefin
28 Propachlor
29 Hexachlorobenzene
30 Chlorpropham
31 Sulfotep
32 Trans-Di-allate
33 Phorate
34 Naled
35 Prometon
36 Atraton
37 Diazinon
38 C20
39 Profluralin
40 Demeton S
41 BHC-alpha
42 Terbufos
43 Dicrotophos
44 Propazine
45 PCNB
46 Atrazine
47 Disulfoton
48 Dioxathion
49 Simazine
51 Lindane
52 Pronamide
53 Monocrotophos
54 Dichloran
55 Heptachlor
56 Dichlone
57 Prometryn
58 Dimethoate
59 C22
60 Ametryn
61 Ronnel
62 Simetryn
63 Alachlor
64 Terbutryn
65 Aldrin
66 Metribuzin
67 Merphos
68 BHC-beta
69 Chlorpyrifos
70 Chlorothalonil
71 Phosphamidon
72 MGK 264
73 Malathion
74 Terbacil
75 Trichloronate
76 Metolachlor
77 Dacthal
79 Methyl Parathion
80 Fenthion
81 Dicofol
83 Isopropalin
84 BHC-delta
85 Isodrin
86 Triadimefon
87 Bayleton
88 Pendamethalin
89 Parathion
90 C24
91 Heptachlor Epoxide
92 Chlorfenvinphos
93 Diphenamid
94 Crotoxyphos
95 Butachlor
96 Chlordane-gamma
97 4,4'-DDE
98 Tokuthion
99 Endosulfan I
100 Cis-Nonachlor
101 Chlordane-alpha
102 Stirofos
103 Bromacil
104 Tetrachlorvinphos
105 Ethylan
106 Perthane
107 Oxadiazon
108 Fenamiphos
109 Napropamide
110 Captan
111 Dieldrin
112 Oxyfluorfen
113 C26
114 Chloropropylate
115 Chlorobenzilate(a)
116 Endrin
117 Kepone
118 Ethion
119 Carboxin
120 4,4'-DDD
121 Sulprofos
122 TOK
123 Nitrofen
124 Trans-Nonachlor
125 4,4'-DDT
126 Carbophenothion
127 Endosulfan II
128 C28
129 Triphenylphosphate
130 Fensulfothion
131 Methoxychlor
132 Endrin Aldehyde
133 Dibutylchlorendate
134 Tricyclozole
135 Famphur
136 Mirex
137 Norflurazon
138 Endosulfan Sulfate
139 Hexazinone
140 Captafol
141 EPN
142 Phosmet
143 Fenarimol
144 Cis-Permethrin
145 Endrin Ketone
146 Azinphos methyl
147 Trans-Permethrin
148 Rubigan
149 Decachlorobiphenyl
150 Coumaphos

Figure 8.3 Computer-generated line chromatogram (all peaks have zero widths) of 147 pesticides and even-numbered *n*-alkanes C_{12} through C_{28}. Amounts of pesticide have random distribution. Amounts of all *n*-alkanes are the same and are 10% larger than the largest pesticide amount. Column: 10 m × 0.1 mm × 0.1 μ, stationary phase – type 1701 polymer (Table 5.4). Source of thermodynamic data: commercial software [66], Figure 5.6e. Gas: helium, isobaric, $p_o = 1$ atm, initial flow 0.8 mL/min ($t_{M,init} = 0.32$ min). Temperature program: $T = 300\,K + (30\,°K/min)t$ ($R_T = 30\,°C/min = 9.63\,°C/t_{M,init}$). Models of the solute–column interaction: (a) linearized (Eq. (8.13)), (b) ideal thermodynamic (Eq. (8.12)). (Limit to stationary phase thermal stability in real columns is ignored).

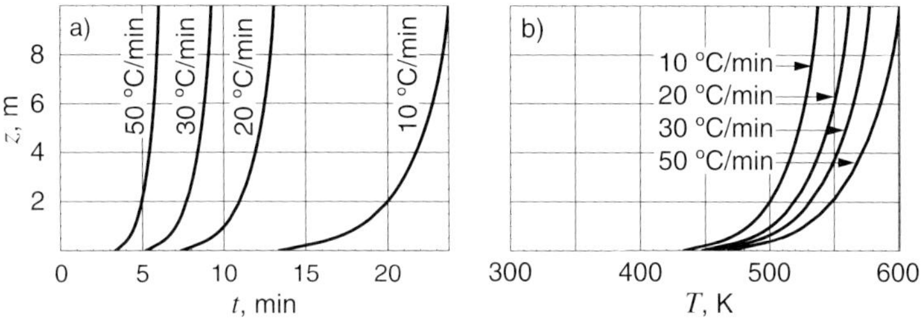

Figure 8.4 Distance, z, of a solute zone from inlet vs. (a) time, t, since the solute injection, and (b) column temperature, T, for several heating rates, R_T. Column: 10 m × 0.1 mm, gas: helium, constant pressure, initial flow 0.8 mL/min, temperature program: $T = 300\,\text{K} + R_T t$, ideal thermodynamic model of the solute–column interaction (Eq. (8.12)) with $T_{char} = 600\,\text{K}$, $\theta_{char} = 40\,^\circ\text{C}$.

8.4.2
Isobaric (Constant Pressure) GC Analysis

In the case of isobaric (constant pressure) GC analysis and in other special cases (Appendix 8.A.2) the need for the unknown function, $t(z)$, in Eq. (8.28) can be removed, and the system of simplified Eqs. (8.27) and (8.28) can be combined in one equation. In view of the wide use of isobaric analyses and comparative ease of mathematical treatment of equations describing the technique, it is convenient to approach it as a benchmark GC analysis and treat the study of it as a case study. The key factors of a column performance in pressure-programmed analyses can be deduced from their comparison with an isobaric analysis.

The migration equation for a solute with a uniform mobility in an isobaric GC analysis can be described as [3] (Appendix 8.A.2)

$$\int_0^t \frac{\eta_{ref}}{\eta}\mu \mathrm{d}t = t_{g,ref} \quad \text{(isobaric analysis)} \tag{8.37}$$

where η and η_{ref} are the viscosities of a carrier gas at an arbitrary temperature, T, and at a fixed predetermined reference temperature, T_{ref}, respectively. In a temperature-programmed analysis, η can be a function of time. The quantity $t_{g,ref}$ is a (static) gas propagation time (Chapter 7) at T_{ref} and at (fixed) actual pressure. At the end of a column, Eq. (8.37) becomes [3]

$$\int_0^{t_R} \frac{\eta_{ref}}{\eta}\mu \mathrm{d}t = t_{M,ref} \quad \text{(isobaric analysis)} \tag{8.38}$$

where $t_{M,ref}$ is a reference hold-up time – a conventional (static) hold-up time measured at T_{ref} and an at actual (fixed) pressure.

An important observation, similar to Statement 8.6, follows from Eq. (8.38).

Statement 8.7

In an isobaric analysis, the net effect of column dimensions as well as a carrier gas type and its initial conditions on a solute retention time (t_R) can be expressed via a single parameter – the static hold-up time ($t_{M,ref}$) measured at a predetermined temperature (T_{ref}) and at actual column pressure in the analysis.

One implication of this observation deserves a special notice. Let us first highlight the relevant difference between the general system of Eqs. (8.27) and (8.28) on the one hand and Eq. (8.38) on the other. Because, in the former case, both integrands depend on the functions $t(z)$ and $z(t)$ describing the process of migration of a solute, the properties of the dependence of the solute retention time, t_R, on u and μ are affected by specifics of a solute migration. Particularly, the degree of the decompression of a carrier gas along a column might affect the value of t_R. In Eq. (8.38), on the other hand, decompression or not, $\mu = \mu(t)$ is the same function of time, t, and $t_{M,ref}$ is a fixed quantity. Nothing in Eq. (8.38) reflects the decompression of the carrier gas which, therefore, has no effect on t_R and on the elution temperature, T_R, for a given $\mu(t)$ and $t_{M,ref}$. In other words [3],

Statement 8.8

Other than through its effect on the hold-up time ($t_{M,ref}$) at some fixed temperature (T_{ref}), a degree of a carrier gas decompression along a column in isobaric temperature-programmed analysis has no effect on the retention time (t_R) and elution temperature (T_R) of any solutes.

Example 8.4

Consider two 40-m-long capillary columns with the same type of stationary phase, the same dimensionless film thickness, and helium as a carrier gas. Let the internal diameters of the columns be 0.53 mm and 0.1 mm, and their gauge pressures be, respectively, 25 kPa (3.63 psi) and 852 kPa (123.6 psi). At these conditions, both columns have the same hold-up time, t_M, at the same temperature ($t_M = 2.46$ min at 25 °C). In two analyses utilizing the columns with their respective pneumatic conditions and having the same temperature program, the retention time, t_R, and elution temperature, T_R, of any solute will be the same. As a result, any mixture will yield the same set of retention times in both analyses although there is weak gas decompression in one column and strong decompression in another. □

Statement 8.8 suggests that, as far as the solute migration in isobaric GC analysis is concerned, the decompression of a carrier gas along a column and the subsequent nonuniformity of a carrier gas velocity is no more than a nuisance. While complicating the study of a solute migration, it does not affect the key parameters of the outcome such as the ratio $t_R/t_{M,ref}$, the solute elution temperature, its elution mobility, and so forth. Therefore, one can ignore the gas decompression in the studies of a solute migration that are concerned with the end result rather than with the migration process itself. The following transformations of Eq. (8.38) offer the ways to bypass the intermediate effects of the decompression.

Due to Eqs. (7.96) and (7.117), Eq. (8.38) can be expressed as

$$\int_0^{t_R} \frac{\mu dt}{t_M} = 1 \quad \text{(isobaric analysis)} \tag{8.39}$$

$$\int_0^{t_R} \mu \bar{u} dt = L \quad \text{(isobaric analysis)} \tag{8.40}$$

where the temperature-dependent parameters t_M and $\bar{u}$ are static hold-up time and average gas velocity measured at the conditions existing at the time t in a temperature-programmed analysis. Equation (8.39) is known from the literature [24, 25, 48, 50], although it is not always stressed that it is valid only for isobaric conditions. More relevant to current discussion is Eq. (8.40). It is a simplification of a more general (and mathematically unmanageable) form [47, 49] involving u instead of $\bar{u}$. Equation (8.40) shows that a way to bypass the very fact of the gas decompression along the column in an isobaric temperature-programmed analysis is to ignore the fact that actual gas velocity, u, is a nonuniform (coordinate-dependent) quantity and to treat it as a uniform quantity equal to $\bar{u}$ (at any coordinate) This substitute will have no effect on t_R, T_R, and other related parameters.

8.4.3
Two Factors Affecting Retention Time

The time-dependent quantity η/η_{ref} in Eq. (8.38) represents a relative change in gas viscosity (η) due to the time-dependent change in T. From Eq. (6.20), one has

$$\eta/\eta_{ref} = (T/T_{ref})^{\xi} \tag{8.41}$$

where (Table 6.12) $\xi \approx 0.7$.

Similar to a solute mobility (μ) in a given analysis, the quantity η/η_{ref} depends only on a column temperature (T). However, the temperature dependence of η/η_{ref} is the same for all solutes while μ is a solute-dependent function of T. In view of these considerations, one can think of the quantity $\mu\eta_{ref}/\eta$ in Eqs. (8.37) and (8.38) as of a solute effective mobility [3],

$$\mu_{eff} = \frac{\mu\eta_{ref}}{\eta} \tag{8.42}$$

Further interpretation of the quantity μ_{eff} comes from using it in Eq. (8.2). Writing it as a function,

$$v = \mu_{\text{eff}} u \tag{8.43}$$

of μ_{eff} representing all temperature-dependent properties of isobaric analysis with uniform μ by a single temperature-dependent parameter μ_{eff} and treating gas viscosity as a temperature-independent quantity equal to η_{ref}.

Substitution of Eq. (8.42) into Eqs. (8.37) and (8.38) yields

$$\int_0^t \mu_{eff}\,\mathrm{d}t = t_{\text{g,ref}} \quad \text{(isobaric analysis)} \tag{8.44}$$

$$\int_0^{t_{\text{R}}} \mu_{eff}\,\mathrm{d}t = t_{M,\text{ref}} \quad \text{(isobaric analysis)} \tag{8.45}$$

It follows from Eq. (8.45) that

Statement 8.9

All parameters (column dimensions, carrier gas, stationary phase, and so forth) of isobaric analysis with a uniformly heated column affect the retention times through only two factors. One of them is temperature-dependent effective solute mobility (μ_{eff}, Eq. (8.42)) representing a normalized effect of column temperature on solute velocity. Another is isothermal hold-up time ($t_{\text{M,ref}}$) representing the effect of column dimensions and carrier gas type on the absolute values of solute velocities.

Equations (8.44) and (8.45) are based only on two assumptions:

1) uniform solute mobility (in practical terms, this means that the stationary phase and column temperature are uniform along a column);
2) analysis is isobaric (pressure does not change during the analysis).

Notice that Eqs. (8.44) and (8.45) do not rely on any particular model of the solute–column interaction and are valid for any temperature program. Thus, solute mobility (μ) can be any positive function of the column temperature (T), temperature program does not have to be a linear or piecewise linear function of time, and T does not have to rise with time, but it can fall or change direction several times during the analysis.

8.4.4 Approximate Forms of Migration and Elution Equations

When, due to temperature programming, T changes from, say, 300 to 600 K covering a substantial portion of the entire practically available temperature range, η increases by a factor of 1.6. At the same time, μ in Eq. (8.38) can increase by several orders of magnitude.

Example 8.5

Consider a solute that has $T_{char} = 600$ K. According to Eq. (8.15), a typical value of θ_{char} for this solute is 38 °C. Let $\theta_{char} = 40$ °C. From Eq. (8.12) one has $\mu \approx 0.3 \times 10^{-6}$ at $T = 300$ K, and $\mu \approx 0.5$ at $T = 600$ K. □

Giddings noticed that "nearly all migration [of a solute] occurs at the higher temperatures, near the elution temperature" [19] of that solute. Therefore, only the temperatures, T, that are close to the solute elution temperature, T_R, significantly affect the solute retention time, t_R.

Consider a solute eluting near the beginning of a heating ramp. No significant change in T and, hence, in η takes place during the solute migration. Therefore, the quantity η can be treated as a fixed quantity without adding a noticeable error to the solution of Eq. (8.38) for t_R.

A solute that elutes close to the end of the ramp covering a wide temperature range does so due to its low mobility at the beginning of the ramp and during its major portion preceding the solute elution. Only shortly before its elution, such a solute begins a noticeable accelerated migration toward the column outlet, Figure 8.4. How was the very low velocity changing during the time when the solute remained near the column inlet with almost no movement is irrelevant to its retention time.

Under typical conditions, all but almost unretained solutes elute at the temperatures that are close [22, 67] to their characteristic temperatures, T_{char}. Therefore, in a typical temperature-programmed analysis, it seems acceptable to approximate the actual dynamic (time-dependent) viscosity, η, in Eq. (8.38) with the static (fixed in time) characteristic viscosity, that is

$$\eta \approx \eta_{char} \tag{8.46}$$

where

$$\eta_{char} = \eta_{st}(T_{char}/T_{st})^{\xi} \tag{8.47}$$

Equations (8.42) and (8.44) become

$$\mu_{eff} = \frac{\mu\eta_{ref}}{\eta_{char}} \tag{8.48}$$

$$\int_0^t \frac{\eta_{ref}}{\eta_{char}} \mu dt = t_{g,ref} \quad \text{(isobaric analyses)} \tag{8.49}$$

The reference temperature, T_{ref}, corresponding to η_{ref} and to $t_{g,ref}$ can be any temperature, including the characteristic temperature, T_{char}, of a solute whose migration this formula describes. In the latter case,

$$\eta_{ref} = \eta_{char}, \; t_{g,ref} = t_{g,char} \tag{8.50}$$

where $t_{g,char}(z)$ is the characteristic gas propagation time corresponding to a solute – the time that it takes for a gas packet to travel from a column inlet to z when the column temperature is fixed at the solute characteristic temperature, T_{char}. Substitution of Eq. (8.50) into Eq. (8.49) yields

$$\int_0^t \mu dt = t_{g,char} \quad \text{(isobaric analyses)} \tag{8.51}$$

Comparison of this equation with Eq. (8.25) which is a general migration equation valid for *any* model of μ in an *arbitrary* pressure- and/or temperature-programmed GC analysis suggests that, in the case of constant pressure and uniform solute mobility, the transition from the most general case to Eq. (8.51) is equivalent to the assumption that the dynamic gas propagation time, $t_g(z,t(z))$, in Eq. (8.25) can be approximated by the static gas propagation time, $t_{g,char}(z)$.

At the end of a column where $z = L$, one has

$$t_{g,char}(L) = t_{M,char} \tag{8.52}$$

where $t_{M,char}$ is the *characteristic hold-up time* corresponding to a given solute – the static hold-up time measured at a solute characteristic temperature, T_{char}, and at an actual (fixed) column pressure. Equations (8.18) and (8.52) allow one to express Eqs. (8.51) at $z = L$ as

$$\int_0^{t_R} \mu dt = t_{M,char} \quad \text{(isobaric analyses)} \tag{8.53}$$

Comparison of Eqs. (8.51) and (8.25) also suggests that $t_m \approx t_{M,char}$, and, because the solute elution temperature (T_R) is typically close to its characteristic temperature, $t_{M,R} \approx t_{M,char}$ and, therefore, $t_m \approx t_{M,R}$ as earlier suggested in Eq. (8.34).

8.5
Scalability of Retention Times in Isobaric Analyses

The time (t) since injection of a sample can be expressed in terms of its relation to the hold-up time (t_M).

The *dimensionless time*, τ, *and dimensionless retention time*, τ_R, defined as [3]

$$\tau = \frac{t}{t_{M,ref}}, \quad \tau_R = \frac{t_R}{t_{M,ref}} \tag{8.54}$$

transform Eq. (8.45) (elution equation) into the *dimensionless elution equation*

$$\int_0^{\tau_R} \mu_{ref} d\tau = 1 \tag{8.55}$$

where μ_{eff} is a function of τ. To find an expression for μ_{eff} as a function of τ, recall that the quantities μ and η in Eq. (8.42) defining μ_{eff} are known as the functions of T. Therefore, μ_{eff} can also be known as the function of T. Let us denote it as $\mu_{eff,T}(T)$. If $T(t)$ is a known temperature program, then μ_{eff} as a function, $\mu_{eff}(\tau)$, of τ can be defined as $\mu_{eff}(\tau) = \mu_{eff,T}(T(\tau t_{M,ref}))$.

Equation (8.55) is a basis for the GC method translation [3] for isobaric analyses. Its extension to nonisobaric analyses has been described elsewhere [68]. Only the isobaric analyses are assumed in the discussion of the method translation below.

Columns A and B are translations of each other if both have the same type of stationary phase and the same dimensionless film thickness.

The temperature program $T_B(t)$ in method B of GC analysis is the translation (rescaling along time axis) of the temperature program $T_A(t)$ in method A if the only difference between $T_A(t)$ and $T_B(t)$ is the time scale so that

$$T_B(t) = T_A(Gt) \tag{8.56}$$

where a fixed dimensionless quantity G – the speed gain in program B relative to program A – is defined as

$$G = \frac{t_{M.ref.A}}{t_{M.ref.B}} \tag{8.57}$$

Equation (8.56) is suitable for any *temperature program*, $T(t)$. The latter can be described by the *heating rate*,

$$R_T(t) = \frac{dT(t)}{dt} \tag{8.58}$$

and initial temperature, T_{init}, – a column temperature at $t = 0$ – or by other means.

In the case of a piecewise linear temperature program, Figure 8.5, Eq. (8.56) means that, for each temperature plateau at some temperature in A there is a plateau in B at the same temperature, the *durations* ($T_{p,A}$ and $T_{p,B}$) of respective plateaus relate as $T_{p,B} = T_{p,A}/G$, and respective heating rates ($R_{T,A}$ and $R_{T,B}$) relate as $R_{T,B} = GR_{T,A}$. If $T_B(t)$ is the translation of $T_A(t)$, then $T_A(t)$ is the translation of $T_B(t)$.

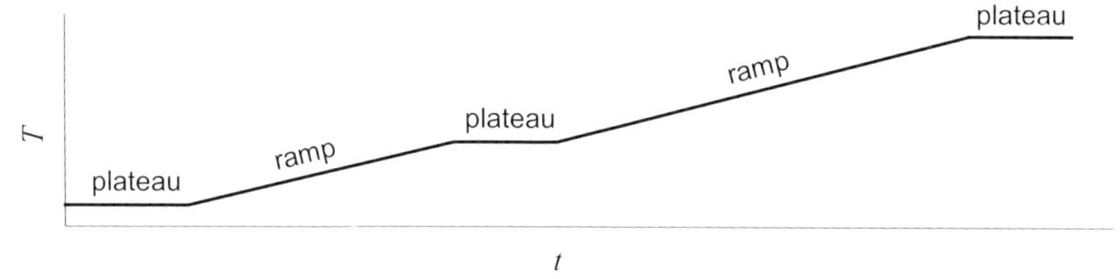

Figure 8.5 Piecewise linear temperature program – sequence of linear heating ramps. A temperature plateau can be viewed as a ramp with zero rate ($R_T = 0$).

Example 8.6

Two temperature programs,

$$T_A(t) = \begin{cases} 36\,°\text{C}, & \text{if} \quad 0 \leq t \leq 12\,\text{min} \\ \dfrac{3\,°\text{C}}{\text{min}}t, & \text{if} \quad 12\text{min} < t \leq 120\,\text{min} \end{cases},$$

$$T_B(t) = \begin{cases} 36\,°\text{C}, & \text{if} \quad 0 \leq t \leq 1.2\,\text{min} \\ \dfrac{30\,°\text{C}}{\text{min}}t, & \text{if} \quad 1.2\,\text{min} < t \leq 12\,\text{min} \end{cases}$$

satisfy Eq. (8.56). Indeed, $T_B(t) = T_A(10t)$. Therefore, $T_B(t)$ is 10 times faster than A and the speed gain in analysis B relative to analysis A is 10. On the other hand, $T_A(t) = T_B(0.1t)$. $T_A(t)$ is 10 times slower than $T_B(t)$ and the speed gain in A relative to B is 0.1. □

Method B is the *translation* of method A if column B and temperature program $T_B(t)$ are translations of column A and temperature program $T_A(t)$, respectively. If method B is the translation of method A, then method A is the translation of method B. Two methods that are translations of each other can use the same or different columns, the same or different carrier gases, the same or different flow rates, and the same or different outlet pressures (for example, ambient pressure in one method and vacuum in its translation).

Two GC analyses of the same mixture are *translations* of each other if their methods are translations of each other.

The following are the properties of method translation that establish the *scalability* of GC analyses.

If two GC analyses are *translations* of each other, then for any solute in both analyses, the quantity μ_{eff} in the dimensionless elution equation (Eq. (8.55)) is the same functions of the dimensionless time (τ). Therefore,

Statement 8.10

In all translations of isobaric GC analysis, a given solute elutes at the same temperature.

For a given chromatogram, a set, $(\tau_{R,a}, \tau_{R,b}, \ldots)$, of normalized retention times ordered according to a predetermined sequence, (solute1, solute2, . . .) of solutes is a *retention pattern* of the chromatogram.

Example 8.7

In the chromatogram of Figure 8.3a, retention times of the first six solutes (#1, #2, #3, #4, #5, #6) are 2.968 min, 3.037 min, 3.507 min, 3.53 min, 3.861 min, 3.753 min, respectively. At $t_{M,ref} = t_{M,init} \approx 0.32$ min, the first six components of the corresponding retention patter are (9.276, 9.492, 10.961, 11.0352, 12.068, 11.731, ...). Notice that solute #6 elutes before solute #5. However, this does not affect the order of listing of components in the retention pattern. □

Statement 8.11

Two analyses that are *translations* of each other yield for the same mixture the same retention pattern.

The only difference between retention times in these analyses is in the absolute scale. Thus, the retention time, $t_{R,B}$, of a solute in analysis B relates to the retention time, $t_{R,A}$, of the same solute in analysis A as

$$t_{R,B} = \frac{t_{R,A}}{G} \tag{8.59}$$

where the speed gain (G, Eq. (8.57)) is the same for all solutes.

An important concept of scalability of GC analyses is that of the normalized temperature program.

Substitution of Eqs. (8.54) and (8.57) into Eq. (8.56) yields

$$T_B(t_{M,ref,B}\tau) = T_A(t_{M,ref,A}\tau) \tag{8.60}$$

Considered as functions of dimensionless time (τ), these programs can be viewed as normalized temperature programs, $T_{A,norm}(\tau)$ and $T_{B,norm}(\tau)$ where $T_{A,norm}(\tau) = T_A(t_{M,ref,A}\tau)$ and $T_{B,norm}(\tau) = T_B(t_{M,ref,B}\tau)$. It follows from Eq. (8.60) that

$$T_{B,norm}(\tau) = T_{A,norm}(\tau) = T_{norm}(\tau) \tag{8.61}$$

Statement 8.12

Two temperature programs that are translations of each other have the same normalized form.

Example 8.8

Suppose that, in Example 8.6, $t_{M,ref,A} = 3$ min and $t_{M,ref,B} = 0.3$ min. Then substitution of t from Eq. (8.54) into $T_A(t)$ and $T_B(t)$ of Example 8.6 (substitution of $t = (3\text{ min})\,\tau$ in $T_A(t)$ and $t = (0.3\text{ min})\,\tau$ in $T_B(t)$) yields

$$T_{B,norm}(\tau) = T_{A,norm}(\tau) = T_{norm}(\tau) = \begin{cases} 36\,^\circ\text{C}, & \text{if} \quad 0 \le \tau \le 4 \\ 9\,^\circ\text{C}\cdot\tau, & \text{if} \quad 4 < \tau \le 40 \end{cases}$$

— equal normalized temperature programs in both cases. □

Equation (8.61) has other important implications. It implies that what affects the retention pattern is not the absolute temperature program in terms of T as a function of time in absolute units (minutes, seconds, and so forth), but only its normalized form where time is expressed in units of the hold-up time (t_M) measured at some predetermined temperature for all analyses under consideration. This suggests that t_M plays a role of a fundamental time unit in chromatography.

A convenient choice for the reference hold-up time ($t_{M,ref}$) in considerations of the retention times scalability is the initial hold-up time, $t_{M,init}$ – the hold-up time at the beginning of each analysis, that is

$$t_{M,ref} = t_{M,init} \tag{8.62}$$

This leads to several interesting concepts.

Temperature increment, ΔT, during time increment, Δt, in a linear heating ramp with the rate R_T can be expressed as

$$\Delta T = R_T \Delta t = \Delta T_{M,init} \Delta\tau = R_{T,norm} \Delta\tau \tag{8.63}$$

where $\Delta\tau = \Delta t / t_{M,init}$ is the dimensionless time increment (Eq. (8.54)) and

$$\Delta T_{M,init} = R_{T,norm} = R_T t_{M,init} \tag{8.64}$$

is the hold-up temperature (dead temperature [20], void temperature [3]) – the temperature increment during the $t_{M,init}$-long time interval.

The hold-up temperature, $\Delta T_{M,init}$, can also be viewed as the normalized heating rate, $R_{T,norm}$, – the heating rate in units of temperature per initial hold-up time. In the future, both notations, $\Delta T_{M,init}$ and $R_{T,norm}$, will be used depending on the need to emphasize different aspects of the same quantity in Eq. (8.64).

It follows from Eq. (8.61) that

Statement 8.13

The heating ramps that are mutual translations of each other have equal normalized heating rates.

Example 8.9

For the data in Example 8.6, $R_{T,A} = 3\,°C/min$, $R_{T,B} = 30\,°C/min$. Suppose that $t_{M,init,A} = 3\,min$ and $t_{M,init,B} = 0.3\,min$. Then $R_{T,norm,A} = t_{M,init,A} R_{T,A} = 3\,min \times 3\,°C/min = 9\,°C$, $R_{T,norm,B} = 0.3\,min \times 30\,°C/min = 9\,°C$. Therefore, $R_{T,norm,A} = R_{T,norm,B}$. □

According to Eq. (8.64), the actual (absolute) heating rate, R_T, in a particular analysis can be found as $R_T = R_{T,norm}/t_{M,init}$.

Example 8.10

In both ramps of Example 8.9, $R_{T,norm} = 9\,°C$. Therefore, $R_T = 9\,°C/t_{M,init}$. For the hold-up times in Example 8.9, this yields $R_{T,A} = 9\,°C/t_{M,init,A} = 3\,°C/min$, and $R_{T,B} = 9\,°C/t_{M,init,B} = 30\,°C/min$. These are the rates in Example 8.6. □

Generally, two heating rates, $R_{T,A}$ and $R_{T,B}$, can be different by several orders of magnitude. And yet, if these are the heating rates in two GC analyses that are translations of each other, they yield the same retention pattern. This implies that it is not the absolute heating rate, R_T (in units of temperature per minute, per second, and so forth), but the normalized heating rate, $R_{T,norm}$ (in units of temperature per initial hold-up time) that affects the retention pattern in a chromatogram.

In closing, it can be mentioned that, in computerized data analysis systems, the requirement for the same retention pattern as a basis for the method translation can be replaced with a more relaxed requirement for a predictable retention pattern. As a result, the use of the mutually translatable columns – the columns with the same type and dimensionless thickness of stationary phase film – becomes the only requirement for the method translation [68]. Fixed pressure is no longer required. However, the retention pattern still remains to be the result of the relationship between the temperature program and the hold-up time rather than of the temperature program alone.

8.6 Dimensionless Parameters

8.6.1 General Considerations

Discussing retention time scalability (Section 8.5), we found that using dimensionless time, $\tau = t/t_{M,ref}$, where $t_{M,ref}$ is the hold-up time at a predetermined reference temperature leads to the form of elution equation (Eq. (8.55)) that does not depend on column dimensions and pneumatic conditions (outlet pressure and flow) of a carrier gas. This and other studies [3, 4, 18, 22, 67, 69–73] suggest that the hold-up time (t_M) is a natural choice for a time unit in chromatography. Many chromatographic formulae become simpler and their scope becomes more universal if all time-related quantities, such as retention time, heating rate, and so forth, are expressed in a dimensionless form where t_M is used as the time-normalization factor.

In the study of retention time scalability, the same reference temperature (T_{ref}) has been assumed for all solutes. However, in the studies of other aspects of temperature-programmed GC, the choice of its own T_{ref} for each solute, like choosing $T_{ref} = T_{char}$ for Eq. (8.51), is more beneficial.

In addition to time (t), column temperature (T) is another important *variable* in GC and dimensionless temperature is another important dimensionless variable. When choosing the solute-specific parameter $t_{M,char}$ as the time-normalizing factor, it is convenient to also choose the solute-specific characteristic thermal constant (θ_{char}) of that solute as the *temperature-normalizing factor.*

The quantities

$$\tilde{\tau} = \frac{t}{t_{M,char}}, \quad \tilde{\tau}_{g,char} = \frac{t_{g,char}}{t_{M,char}}, \quad \tilde{\tau}_R = \frac{t_R}{t_{M,char}}, \quad \Theta = \frac{T}{\theta_{char}}, \quad \Theta_{char} = \frac{T_{char}}{\theta_{char}}, \quad \zeta = \frac{z}{L} \tag{8.65}$$

are, respectively, dimensionless time, dimensionless characteristic gas propagation time, dimensionless retention time, dimensionless temperature, dimensionless characteristic temperature and dimensionless distance from the column inlet. The variables $\tilde{\tau}_{g,char}$ and ζ are the normalized ones. They change between 0 and 1. Other dimensionless variables in this collection can be larger than 1. Notations such as $\tilde{\tau}$ and $\tilde{\tau}_R$ distinguish dimensionless times with solute-specific normalizations in Eq. (8.65) from dimensionless times, τ and τ_R, with solute-independent normalization in Eq. (8.54).

One can write Eqs. (8.13), (8.12), (8.51) and (8.53) as

$$\mu = 1/(1 + e^{\Theta_{char} - \Theta}) \quad \text{(linearized model)} \tag{8.66}$$

$$\mu = \frac{1}{1 + e^{(\Theta_{char}/\Theta - 1)\Theta_{char}}} \quad \text{(ideal thermodynamic model)} \tag{8.67}$$

$$\int_0^{\tilde{\tau}} \mu d\tilde{\tau} = \tilde{\tau}_{g,char} \quad \text{(isobaric analyses)} \tag{8.68}$$

$$\int_0^{\tilde{\tau}_R} \mu d\tilde{\tau} = 1 \quad \text{(isobaric analyses)} \tag{8.69}$$

As mentioned earlier, a pair $(T_{char}, \theta_{char})$ of characteristic parameters of a solute uniquely represents the properties of its thermodynamic interaction with a column. Although θ_{char} is independent of T_{char}, the former generally falls within a relatively narrow range, Figure 5.6, determined by T_{char}. The general trend of the relation between T_{char} and θ_{char} is well represented by generic solutes for which θ_{char} is a function, Eq. (8.15), of T_{char}. Due to Eq. (8.15) and (5.50), Θ_{char} of a generic solute can be described as

$$\Theta_{char} = \frac{T_{char}}{\theta_{char}} = \frac{T_{char}}{\theta_{char,st}} \left(\frac{T_{st}}{T_{char}}\right)^{\xi} = \Theta_{st} \cdot \left(\frac{T_{char}}{T_{st}}\right)^{1-\xi} \tag{8.70}$$

Since $\theta_{char,st}$ is a weak function (Eq. (5.47)) of relative thickness (φ, Eq. (2.7)) of stationary phase film, the quantity Θ_{st} in the last formula is also a weak function (Eq. (5.50)) of φ. Due to Eq. (5.50), the boundaries of the quantity Θ_{char} for generic solutes can be estimated as

$$12.8(10^3\varphi)^{0.09} \le \Theta_{char} \le 15.7(10^3\varphi)^{0.09} \quad \text{when} \quad 300\ \text{K} \le T_{char} \le 600\ \text{K} \tag{8.71}$$

The last inequality allows one to conclude that, for a fixed film thickness, Θ_{char} is confined within a relatively narrow range. For a typical film thickness ($\varphi = 0.001$),

$$12.8 \le \Theta_{char} \le 15.7 \quad (300\ \text{K} \le T_{char} \le 600\ \text{K}, \varphi = 0.001) \tag{8.72}$$

Use of dimensionless parameters leads to several interesting observations.

In the case of a weak gas decompression along a column, gas propagation time, $t_{g,char}$, at a fixed temperature, $T = T_{char}$, can be expressed, according to Eq. (7.189), as $t_{g,char} = t_{M,char}z/L$. In dimensionless form, this becomes

$$\tilde{\tau}_{g,char} = \zeta \quad \text{(isobaric analyses, } \Delta p \ll p_o) \tag{8.73}$$

This allows one to rewrite Eq. (8.68) as

$$\int_0^{\tilde{\tau}} \mu d\tilde{\tau} = \zeta \quad \text{(isobaric analyses, } \Delta p \ll p_o) \tag{8.74}$$

which implies that

$$\frac{d\zeta}{d\tilde{\tau}} = \mu \quad \text{(isobaric analyses, } \Delta p \ll p_o) \tag{8.75}$$

Another observation follows from Eq. (8.66). Let us denote the temperature difference, $T - T_{char}$, in Eq. (8.13) and its dimensionless equivalent, $\Theta - \Theta_{char}$, as

$$\Delta T_\mu = T - T_{char} \tag{8.76}$$

$$\Delta\Theta_\mu = \Theta - \Theta_{char} = \frac{\Delta T_\mu}{\theta_{char}} \tag{8.77}$$

According to Eq. (8.13), the solute mobility, μ, monotonically increases with an increase in ΔT_μ. This justifies referring to ΔT_μ as to mobilizing temperature increment for a solute. Accordingly, $\Delta\Theta_\mu$ is its dimensionless mobilizing temperature increment. Substitution of Eq. (8.77) into Eq. (8.66) yields

$$\mu = 1/(1 + e^{-\Delta\Theta_\mu}) \quad \text{(linearized model)} \tag{8.78}$$

indicating that, in the case of linearized model, a single quantity $\Delta\Theta_\mu$ can describe the dependence of a solute mobility, μ, on temperature. When $\Delta\Theta_\mu \leq -3$, that is when

$$T_{char} - T \geq 3\theta_{char} \tag{8.79}$$

μ in Eq. (8.78) becomes $\mu \approx 1/\exp(\Delta\Theta_\mu)$ – a small quantity that exponentially depends on $\Delta\Theta_\mu$. On the other hand, at $\Delta\Theta_\mu = 0$ ($T = T_{char}$), $\mu = 0.5$ leaving a little room for further significant growth in μ as Eq. (8.3) indicates. An average value of θ_{char} can be estimated as 30 °C, Figure 5.6.

The solutes that are highly retained, Figure 5.4, at the beginning of a heating ramp are especially important for the study of a column operation in temperature-programmed GC [18]. Let us refer to these solutes as to the *highly interactive* ones. For practical considerations, it can be assumed that

$$\text{a solute is highly interactive if its} \quad T_{char} > T_{init} + 3\theta_{char} \tag{8.80}$$

It follows from Eq. (8.13) that for such solute, $\mu_{init} < 0.05$.

One can make the following observation.

Statement 8.14

Departure of a highly interactive solute from the column inlet becomes noticeable only when the column temperature reaches the level close to $T_R - 3\theta_{char}$, where T_R is the solute elution temperature.

It follows from Eq. (8.15) that, for the generic solutes, the $3\theta_{char}$ margin can be estimated as $3\theta_{char} = (1000\varphi)^{0.09}(T_{char}/273\text{ K})^{0.7}66\,°\text{C}$. For a typical film thickness ($\varphi = 1000$), it becomes $3\theta_{char} = (T_{char}/273\text{ K})^{0.7}66\,°\text{C}$. For example, for a solute with $T_{char} = 500$ K, $3\theta_{char} \approx 100$ K. Therefore, during a typical heating ramp, a noticeable departure of a solute from the column inlet begins to develop only when the column temperature exceeds about 400 K.

Sometimes, it might be necessary to reverse Eqs. (8.67) and (8.66) in order to express T as a function μ. To simplify the formulae, it is convenient to express T as a function of k and use Eq. (8.8) to express k via μ. From Eqs. (5.36) and (5.34), one has

$$\Theta = \Theta_{char} - \ln k \quad \text{(linearized model)} \tag{8.81}$$

$$\Theta = \Theta_{char}\left(1 - \frac{\ln k}{\Theta_{char} + \ln k}\right) \quad \text{(ideal thermodynamic model)} \tag{8.82}$$

Both formulae yield nearly the same results when $k \approx 1$.

8.6.2
Linear Heating Ramp in an Isobaric Analysis

It is known from experimental observations that all peaks in a typical temperature-programmed analysis have roughly the same width. This means (Chapter 10) that all solutes elute with roughly the same elution mobility, μ_R, that depends on the heating rate, R_T. The following is a quantitative evaluation [4, 18–20, 67, 70, 71, 74] of this observation.

A linear heating ramp starting at an initial temperature, T_{init}, and proceeding with a fixed rate, R_T, can be described as

$$T = T_{init} + R_T t \tag{8.83}$$

To express this formula in a dimensionless form, it is convenient to first define the characteristic heating rate,

$$R_{T,char} = \frac{\theta_{char}}{t_{M,char}} \tag{8.84}$$

corresponding to a solute. As described at its introduction in Eq. (8.52), the quantity $t_{M,char}$ is the static hold-up time at $T = T_{char}$ where T_{char} is the solute characteristic temperature.

Generally, several solutes with the same T_{char} can have different θ_{char}, Figure 5.6. Although the difference is contained within a relatively narrow range, nevertheless, it means that the quantity $R_{T,char}$ can be different for different solutes having the same

T_{char}. However, in isobaric analysis, $R_{T,char}$ is the same for *all* generic solutes, not only for those that have the same T_{char}. Indeed, in isobaric analysis, t_M is proportional to the gas viscosity η (Eq. (7.96)). As a result, it follows from Eqs. (8.15) and (6.20) that the ratio $\theta_{char}/t_{M,char}$ is the same for all generic solutes. Equation (8.84) can be modified as

$$R_{T,char} = \frac{\theta_{char,init}}{t_{M,init}} \quad \text{(generic solutes)} \tag{8.85}$$

where $t_{M,init}$ is the (static) hold-up time measured at the conditions existing at the beginning of the heating ramp, and

$$\theta_{char,init} = (T_{init}/T_{st})^{\xi}\theta_{char,st} \tag{8.86}$$

is θ_{char} of a generic solute with $T_{char} = T_{init}$.

Since the characteristic heating rate, $R_{T,char}$, is the same for all generic solutes, it is convenient, when dealing with the generic solutes, to treat $R_{T,char}$ as a method rather than a solute parameter. Equation (8.84) can be expressed as

$$R_{T,char} = \frac{\theta_T}{t_M} \quad \text{(generic solutes)} \tag{8.87}$$

where θ_T is not the characteristic thermal constant of a solute, but just a parameter – equivalent thermal constant – calculated as

$$\theta_T = (T/T_{st})^{\xi}\theta_{char,st} \tag{8.88}$$

for the same column temperature (T) as the temperature in the measurement of t_M. Specifically, Eq. (8.87) can be expressed as

$$R_{T,char} = \frac{\theta_{T,init}}{t_{M,init}} \quad \text{(generic solutes)} \tag{8.89}$$

where

$$\theta_{T,init} = (T_{init}/T_{st})^{\xi}\theta_{char,st} \tag{8.90}$$

Comparison of Eqs. (8.86) and (8.90) implies that, θ_T is equal to θ_{char} of a generic solute with $T_{char} = T$. Particularly,

$$\theta_{T,init} = \theta_{char,init} \tag{8.91}$$

In the dimensionless form, Eq. (8.83) can be expressed as

$$\Theta = \Theta_{init} + r_T\tilde{\tau} \tag{8.92}$$

where

$$\Theta_{init} = \frac{T_{init}}{\theta_{char}} \tag{8.93}$$

$$r_T = \frac{R_T}{R_{T,char}} \tag{8.94}$$

are, respectively, the dimensionless initial temperature and dimensionless heating rate. Due to Eqs. (8.84), (8.89) and (8.64), r_T can also be expressed as

$$r_T = \frac{R_T}{R_{T,char}} = \frac{R_T t_{M,char}}{\theta_{char}} \tag{8.95}$$

$$r_T = \frac{R_T}{R_{T,char}} = \frac{R_T t_{M,init}}{\theta_{T,init}} \quad \text{(generic solutes)} \tag{8.96}$$

$$r_T = \frac{R_{T,norm}}{\theta_{T,init}} = \frac{\Delta T_{M,init}}{\theta_{T,init}} \quad \text{(generic solutes)} \tag{8.97}$$

Example 8.11

Suppose that $R_T = 10\,°\text{C/min}$ and, for some solute, $t_{M,char} = 1$ min, $\theta_{char} = 30\,°\text{C}$. The characteristic and the dimensionless heating rates corresponding to this solute are $R_{T,char} = \theta_{char}/t_{M,char} = 30\,°\text{C/min}$, and $r_T = R_T/R_{T,char} \approx 0.33$. □

Generally, because it is a function of solute parameters, r_T can be different for different solutes eluting during *the same* linear heating ramp. The difference, however, is not relatively large. There is a strong correlation, Eqs. (5.40) and (5.46), Figure 5.6, between the characteristic parameters, T_{char} and θ_{char}, of different solutes. As a result, the solutes with larger T_{char} tend to have larger θ_{char}. This tends to diminish the difference between characteristic heating rates, $R_{T,char}$, Eq. (8.84), corresponding to different solutes. Let $t_{M,ref}$ be the hold-up time at some reference temperature, T_{ref}, in a given isobaric analysis. Because, being proportional to gas viscosity, t_M is proportional to the quantity T^{ξ} (Chapter 6) where $\xi \approx 0.7$ and T is the temperature at which t_M is measured, the quantity $t_{M,char}$ in Eq. (8.95) can be expressed as $t_{M,char} = (T_{char}/T_{ref})^{\xi} t_{M,ref}$. This, due to Eq. (5.46), allows us to estimate Eq. (8.95) as

$$r_T \approx \left(\frac{T_{st}}{T_{ref}}\right)^{\xi} \frac{R_T t_{M,ref}}{\theta_{char,st}} \tag{8.98}$$

where $\theta_{char,st}$ is a fixed quantity described in Eq. (5.47).

Example 8.12

Let $\theta_{char,st} = 22\,°\text{C}$. If $R_T = 10\,°\text{C/min}$ and $t_M = 1$ min at 300 K, then, for any solute in a column, r_T can be estimated as $(273/300)^{0.7} 10\,°\text{C/min} \times 1\,\text{min}/(22\,°\text{C}) \approx 0.43$. □

There are no parameters of a particular solute on the right-hand side of Eq. (8.98). This indicates that, within the accuracy of the estimations in Eqs. (5.40) and (5.46), Figure 5.6,

Statement 8.15

Migration of all solutes during the same heating ramp is governed by roughly the same dimensionless heating rate.

Because, in isobaric analysis, all solutes eluting during the same heating ramp have the same or nearly the same dimensionless heating rate, the latter can be treated as a method rather than a solute parameter.

The quantity r_T for a generic solute can be found from Eqs. (8.96) and (8.97). There are no solute parameters in these formulae. Therefore, all generic solutes eluting during the same heating ramp in isobaric analysis have the same dimensionless heating rate.

If the dimensionless heating rate, r_T, is known, then the actual heating rate, R_T, and normalized heating rate, $R_{T,norm}$, can be obtained as

$$R_T = \frac{r_T \theta_{char}}{t_{M,char}} \tag{8.99}$$

$$R_{T,norm} = \Delta T_{M,init} = r_T \theta_{T,init} = r_T \theta_{char,st}(T_{init}/T_{st})^{\xi} \approx$$
$$\approx 22\,^{\circ}C\,(10^3\varphi)^{0.09}(T_{init}/T_{st})^{0.7} r_T \tag{8.100}$$

The former follows from Eq. (8.95) while the latter from Eqs. (8.97), (8.90), and (5.47).

Example 8.13

According to Eqs. (8.90) and (5.47), at $\varphi = 0.001$ and $T_{init} = 50\,^{\circ}C$ (323 K), $\theta_{T,init} \approx 25^{\circ}C$. Under these conditions, the optimal heating rate of $10\,^{\circ}C/t_{M,init}$ [67] ($R_{T,norm} = \Delta T_{M,init} = 10\,^{\circ}C$) corresponds to $r_T = R_{T,norm}/\theta_{T,init} \approx 0.4$. For a column with $t_{M,init} = 1$ min, R_T becomes $10\,^{\circ}C/min$. For a column with 10 times shorter $t_{M,init}$, the same $R_{T,norm}$ and r_T yield $R_T = 100\,^{\circ}C/min$. □

8.6.3
Analytical Solutions for the Linearized Model and Linear Heating Ramp

It follows from Eq. (8.66) that, for the linearized model of the solute–column interaction and linear heating ramp, the evolution of solute mobility (μ) as a function of the dimensionless time, $\tilde{\tau}$, of a solute migration along a column can be described as, Figure 8.6a,

$$\mu = \frac{1}{1 + e^{-\Delta\Theta_{\mu,init} - r_T \tilde{\tau}}} \quad \text{(linearized model)} \tag{8.101}$$

where

$$\Delta\Theta_{\mu,init} = \Theta_{init} - \Theta_{char} = \frac{T_{init} - T_{char}}{\theta_{char}} \tag{8.102}$$

is the dimensionless initial mobilizing temperature increment.

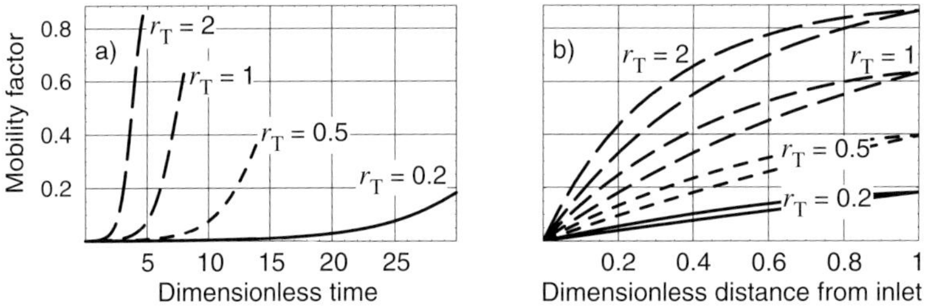

Figure 8.6 Mobility factors (μ) of highly interactive solutes for several dimensionless heating rates (r_T). (a) μ of a solute with $\Delta\Theta_{\mu,\text{init}} = -7.5$ as a function, Eq. (8.101), of its dimensionless migration time ($\tilde{\tau}_R$). (b) μ of *any* highly interactive solute ($\Delta\Theta_{\mu,\text{init}} < -3$) as a function, Eqs. (8.105), of its dimensionless distance (ς) from the inlet. Dimensionless time at the end of each curve in (a) is the dimensionless retention time ($\tilde{\tau}_R$) found from Eq. (8.104) at $\varsigma = 1$. Each pair of similarly dashed curves in (b) corresponds to the same r_T. In each pair, the lower curve corresponds to weak decompression ($\Delta p \ll p_o$) while the upper one corresponds to strong decompression ($\Delta p \gg p_o$). The graphs in (b) confirm that the degree of the decompression has no effect on the elution mobility, μ_R (μ at $\varsigma = 1$).

Example 8.14

Let $T_{\text{char}} = 600$ K, $\theta_{\text{char}} = 40\,^\circ$C. Then $\Theta_{\text{char}} = 600/40 = 15$. If $T_{\text{init}} = 300$ K, then $\Theta_{\text{init}} = 300/40 = 7.5$, $\Delta\Theta_{\mu,\text{init}} = 7.5 - 15 = -7.5$ and, from Eq. (8.101), $\mu_{\text{init}} = 0.00\,055$. □

Equation (8.80) can be rephrased as

$$\text{a solute is highly interactive if its} \quad \Delta\Theta_{\mu,\text{init}} < -3 \tag{8.103}$$

The linearized model and linear heating ramp make it possible to find the analytical (closed form) solution for migration integral. Substitution of Eq. (8.101) into Eq. (8.68) yields

$$\frac{1}{r_T}\ln\frac{1+e^{\Delta\Theta_{\mu,\text{init}}+r_T\tilde{\tau}}}{1+e^{\Delta\Theta_{\mu,\text{init}}}} = \frac{1-(1-j_2\zeta)^{3/2}}{1-(1-j_2)^{3/2}}, \quad j_2 = \frac{p^2-1}{p^2} \tag{8.104}$$

The right-hand side of this equation comes from Eq. (7.109) with notations in Eq. (8.65). Solving Eq. (8.104) for $\tilde{\tau}$ yields the expression $\tilde{\tau}(\zeta)$ for $\tilde{\tau}$ as a function of the dimensionless distance (ζ) traveled by a solute, solving Eq. (8.104) for ζ yields the expression $\zeta(\tilde{\tau})$ for ζ as a function of $\tilde{\tau}$, and substitution of the expression $\tilde{\tau}(\zeta)$ into Eq. (8.101) yields the following function of ζ, Figure 8.6b:

$$\begin{aligned}\mu &= 1-\frac{1}{1+e^{\Delta\Theta_{\mu,\text{init}}}}\exp\left(r_T\frac{(1-j_2\zeta)^{3/2}-1}{1-(1-j_2)^{3/2}}\right) = \\ &= 1-\frac{1}{1+e^{\Delta\Theta_{\mu,\text{init}}}}\begin{cases} e^{-r_T\zeta}, & \Delta p \ll p_o \\ e^{-r_T\cdot(1-(1-\zeta)^{3/2})}, & \Delta p \gg p_o \end{cases}\end{aligned} \tag{8.105}$$

Figure 8.6 shows the evolution of the mobility of a highly interactive solute from two perspectives – as a function $\mu(\tilde{\tau})$ (Eq. (8.101), Figure 8.6a) of the solute dimensionless migration time ($\tilde{\tau}$), and as a function $\mu(\zeta)$ ((8.106), Figure 8.6b) of its dimensionless distance (ζ) from the inlet. The function $\mu(\tilde{\tau})$ strongly depends on initial conditions represented by the parameter $\Delta\Theta_{\mu,\text{init}}$. The function $\mu(\zeta)$, on the other hand, is the same for *all* highly interactive solutes.

At $\zeta = 1$, solute mobility (μ) becomes its elution mobility, μ_R, while dimensionless time ($\tilde{\tau}$) becomes dimensionless retention time ($\tilde{\tau}_R$). Equations (8.104) and (8.105) offer quantitative confirmations of Statement 8.8 which suggests that the gas decompression has no effect on $\tilde{\tau}_R$ or μ_R.

At weak gas decompression ($\Delta p \ll p_o$), Eq. (8.105) has the simplest form

$$\mu = 1 - \frac{e^{-r_T\zeta}}{1 + e^{\Delta\Theta_{\mu,\text{init}}}} \tag{8.106}$$

Several factors suggest that this formula can be used at any gas decompression. First, the difference between the functions $\mu(\zeta)$ at two extremes of gas decompression (weak and strong) is not large (Figure 8.6b). What is more, the decompression has no effect on the final outcome – the elution mobility (μ_R) and related parameters (Statement 8.8).

Equation (8.106) becomes more transparent when expressed as a dependence of a solute immobility, ω, on ζ. Due to Eq. (8.7), one has [4] from Eq. (8.106), Figure 8.7,

$$\omega = \frac{e^{-r_T\zeta}}{1 + e^{\Delta\Theta_{\mu,\text{init}}}} = \omega_{\text{init}}\omega_a \tag{8.107}$$

where

$$\omega_{\text{init}} = \omega|_{\zeta=0} = 1 - \mu|_{\zeta=0} = \frac{1}{1 + e^{\Delta\Theta_{\mu,\text{init}}}} = \frac{1}{1 + e^{(T_{\text{init}} - T_{\text{char}})/\theta_{\text{char}}}} \tag{8.108}$$

$$\omega_a = \omega|_{\Delta\Theta_{\mu,\text{init}} = -\infty} = \omega|_{\omega_{\text{init}} = 1} = e^{-r_T\zeta} \tag{8.109}$$

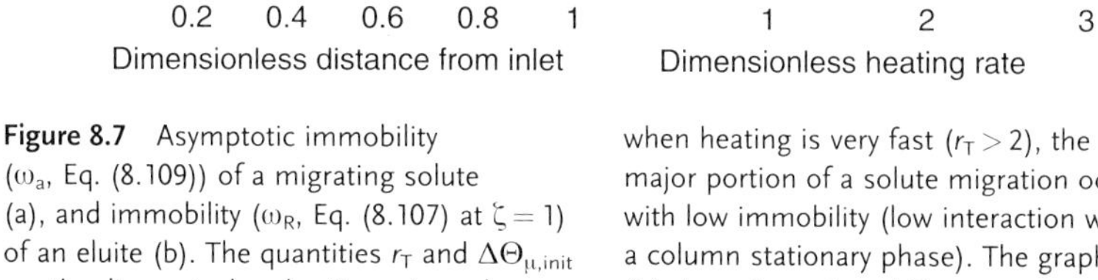

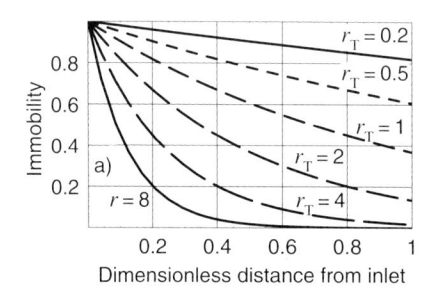

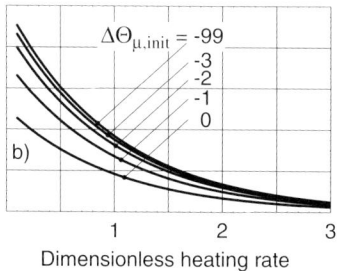

Figure 8.7 Asymptotic immobility (ω_a, Eq. (8.109)) of a migrating solute (a), and immobility (ω_R, Eq. (8.107) at $\zeta = 1$) of an eluite (b). The quantities r_T and $\Delta\Theta_{\mu,\text{init}}$ are the dimensionless heating rate and initial mobilizing temperature increment, respectively. The graphs in (a) show that, when heating is very fast ($r_T > 2$), the major portion of a solute migration occurs with low immobility (low interaction with a column stationary phase). The graphs in (b) show that, when $\Delta\Theta_{\mu,\text{init}} \leq -3$, further lowering of T_{init} has a minor effect on ω_R for any r_T.

The quantities ω_{init} and ω_a are, respectively, the initial and asymptotic immobilities of a solute. It follows from Eq. (8.107) that ω approaches its asymptotic level when ω_{init} approaches unity (at the beginning of the heating ramp, nearly 100% of the solute is dissolved in a stationary phase). Earlier, these solutes were defined as the highly interactive ones. According to Eq. (8.108), a solute is highly interactive if $-\Delta\Theta_{\mu,init}$ is sufficiently large ($T_{char} - T_{init} \gg \theta_{char}$). It follows from Eqs. (8.107) and (8.109) combined with Statement 8.15 that

Statement 8.16

In isobaric temperature-programmed analysis, the immobilities of all highly interactive solutes have nearly the same profiles as a function of the distance, z, of their departure from the column inlet. This profile approaches the asymptotic level, ω_a, Eq. (8.109), Figure 8.7a, that depends only on the dimensional heating rate, r_T. Similar conclusions are valid for solute mobilities and other related parameters.

Equation (8.107) allows one to see the earlier discussed factors affecting a solute migration from a different angle.

First, as shown in Figure 8.7a, ω almost linearly declines with the distance from the column inlet when the dimensionless heating rate, r_T, is relatively small ($|r_T| \ll 1$), and exponentially declines with the distance when r_T is large ($r_T \gg 1$). Thus, at $r_T = 8$, for example, even the highly interactive solutes have almost no interaction with a column stationary phase along the major portion of the solute migration. In other words, Figure 8.7a and Eq. (8.107) quantify a well-known fact that, when heating is too fast, the solutes interact with the stationary phase only within a small portion of a column near its inlet. The rest of the column acts as an inert tube with no solute retention, and, therefore, with no additional contribution to the solute separation.

Second, according to Eq. (8.108), $\omega_{init} \geq 0.95$ when $\Delta\Theta_{\mu,init} \leq -3$. This means that, for the conditions expressed in Eq. (8.79), ω is nearly the same, Figure 8.7b, as its asymptotic level, ω_a, confirming the view of Eq. (8.79) as of a qualifier for a solute to be highly retained at T_{init}, and, therefore, as a highly interactive one.

Example 8.15

Consider a solute that has $T_{char} = 373$ K (100 °C). According to Eq. (5.46), a typical value of θ_{char} for this solute is 27.4 K. The column temperature (T) which is $3\theta_{char}$ below T_{char} for this solute can be estimated as $T = T_{char} - 3\theta_{char} \approx 373\text{ K} - 82\text{ K} \approx 291$ K (18 °C). This means that, at typical room temperatures, all solutes whose T_{char} is 373 K (100 °C) or greater are highly retained. It also means that during a linear heating ramp starting at room temperature, all these solutes behave as the highly interactive ones. □

According to Eq. (8.109), ω_a is ω of a solute that, at $T = T_{init}$ has $\omega_{init} = 1$. Due to Eqs. (8.6) and (8.8), the latter is equivalent to $\mu_{init} = 0$, $k_{init} = \infty$. Equation (8.109),(8.6) and (8.8) yield

$$\mu_a = \mu|_{\mu_{init}=0} = 1 - e^{-r_T\zeta} \tag{8.110}$$

$$k_a = k|_{k_{init}=\infty} = 1/(e^{r_T\zeta} - 1) \tag{8.111}$$

The last formula allows one to find a solute *asymptotic temperature*, T_a, – the column temperature at the time when a highly interactive solute is located at ζ. From Eq. (8.81), one has

$$T_a = \lim_{k_{init}\to\infty} T = T_{char} + \theta_{char}\ln(e^{r_T\zeta} - 1) \tag{8.112}$$

At the time of its elution ($\zeta = 1$), the solute elution immobility, ω_R (its elution interaction level), is, according to Eq. (8.107),

$$\omega_R = \omega|_{\zeta=1} = \omega_{init}\omega_{R,a} \tag{8.113}$$

where, Figure 8.8,

$$\omega_{R,a} = \omega_a|_{\zeta=1} = e^{-r_T} \tag{8.114}$$

is the solute's asymptotic elution immobility [4].

Similarly, Eqs. (8.10)–(8.112) yield for the asymptotic elution mobility factor, $\mu_{R,a}$, asymptotic elution retention factor, $k_{R,a}$, and asymptotic elution temperature, $T_{R,a}$, – the elution parameters of highly interactive solutes [4], Figure 8.9, Figure 8.8,

$$\mu_{R,a} = \mu_a|_{\zeta=1} = 1 - e^{-r_T} \tag{8.115}$$

$$k_{R,a} = k_a|_{\zeta=1} = 1/(e^{r_T} - 1) \tag{8.116}$$

$$T_{R,a} = T_a|_{\zeta=1} = T_{char} + \theta_{char}\ln(e^{r_T} - 1) \tag{8.117}$$

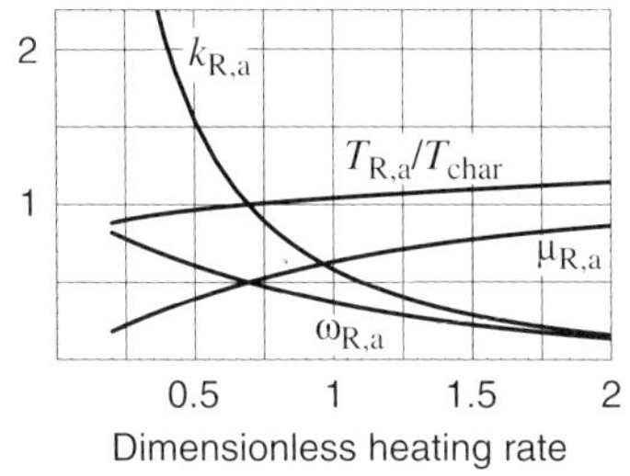

Figure 8.8 Elution parameters, Eqs. (8.114), (8.115), (8.116) and (8.124), of a highly interactive solute in isobaric temperature-programmed GC analysis. No visible change in the graphs for $T_{R,a}/T_{char}$ can be noticed when they are obtained from Eq. (8.123) for $300\,\text{K} \le T_{char} \le 600\,\text{K}$, $10^2 \le \varphi \le 10^4$.

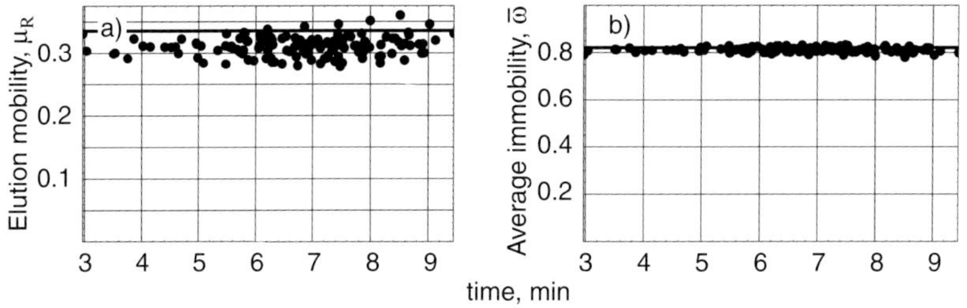

Figure 8.9 Elution mobility (μ_R) and average immobility ($\bar{\omega}$) of the pesticides under conditions in Figure 8.3. The dots represent computer-generated data for the conditions in Figure 8.3. The horizontal solid lines represent $\mu_{R,a}$ and $\bar{\omega}_a$ in Eqs. (8.115) and (8.122) for the same conditions.

Statement 8.17

In isobaric temperature-programmed analysis, elution parameters of a highly interactive solute depend *only* on the dimensionless heating rate, r_T, as shown in Eqs. (8.114),(8.115),(8.116) and (8.117).

This together with Statement 8.15 reflects a widely known fact that all highly interactive solutes elute with roughly the same mobilities (and, therefore, the same immobilities and retention factors).

Example 8.16

Let $r_T = 0.5$. Then $\omega_{R,a} = 0.61$, $\mu_{R,a} = 0.39$, $k_{R,a} = 1.54$, and, for the solute in Example 8.14 ($T_{char} = 600$ K, $\theta_{char} = 40\,°C$), $T_{R,a} = 582.7$ K. The same result for $T_{R,a}$ follows from Eq. (8.14) for $\mu = 0.39$. Solving Eq. (8.12) for T, one can find that, for $\mu = \mu_{R,a}$, the ideal thermodynamic model yields $T_{R,a} = 583.2$ K. □

The elution parameters, μ_R and k_R, of a solute with arbitrary (not necessarily high) initial retention, can be different from their respective asymptotic counterparts, $\mu_{R,a}$ and $k_{R,a}$. They can be found as

$$\mu_R = \mu_{init} + \mu_{R,a} - \mu_{init}\mu_{R,a} \tag{8.118}$$

$$k_R = \frac{k_{init} k_{R,a}}{1 + k_{init} + k_{R,a}} \tag{8.119}$$

These formulae are not as simple and transparent as Eqs. (8.107) and (8.113) are. This suggests that, of the three parameters, k, μ, and ω, the solute immobility (its engagement, interaction level with the column), ω, is the most suitable metric of interaction of a migrating and eluting solute with the column in a temperature-programmed analysis. On the other hand, neither of these parameters is the best choice for all situations. Thus, μ is convenient for the description of a solute velocity (Eq. (8.2)) while k is the most suitable for the expression of the dependence of solute retention on temperature and vice versa (Eqs. (5.33) and (5.36)).

Expressing all parameters in Eq. (8.118) via the parameters T_{init}, T_{R} and $T_{\text{R,a}}$ in Eq. (8.13) and solving the result for T_{R} yields

$$T_{\text{R}} = T_{\text{R,a}} + \theta_{\text{char}} \ln\left(1 + \frac{e^{r_{\text{T}}} + \Delta\Theta_{\mu,\text{init}}}{e^{r_{\text{T}}} - 1}\right) \tag{8.120}$$

When T_{init} is significantly lower than T_{char}, T_{R} asymptotically approaches $T_{\text{R,a}}$ in Eq. (8.117).

An important factor of a column operation is the distance-averaged ω. Let us denote it as $\bar{\omega}$. It follows from Eqs. (8.107), (8.109) and (8.115) that, Figure 8.9,

$$\bar{\omega} = \omega_{\text{init}} \bar{\omega}_{\text{a}} \tag{8.121}$$

where

$$\bar{\omega}_{\text{a}} = \frac{1}{L}\int_0^L \omega_{\text{a}}\, dz = \frac{1}{L}\int_0^L e^{-r_{\text{T}} z/L}\, dz = \int_0^1 e^{-r_{\text{T}}\zeta}\, d\zeta = \frac{1 - e^{-r_{\text{T}}}}{r_{\text{T}}} = \frac{\mu_{\text{R,a}}}{r_{\text{T}}} \tag{8.122}$$

The elution parameters of a solute generally depend on both of its characteristic parameters (T_{char} and θ_{char}). For generic solutes, θ_{char} is a unique function (Eq. (8.15)) of T_{char}. This makes it possible to reduce the number of parameters of a solute–column interaction to one. Thus substitution of Eqs. (8.15) and (5.47) into Eq. (8.117) yields, Figure 8.8,

$$T_{\text{R,a}} = T_{\text{char}}\left(1 + \frac{(10^3\varphi)^{0.09} \ln(e^{r_{\text{T}}} - 1)}{12.4 (T_{\text{char}}/T_{\text{st}})^{1-\xi}}\right) \tag{8.123}$$

where φ is the dimensionless film thickness. A simpler estimate, Figure 8.8,

$$T_{\text{R,a}} \approx T_{\text{char}}/k_{\text{R,a}}^{0.08} \approx T_{\text{char}} \cdot (e^{r_{\text{T}}} - 1)^{0.08} \quad (r_{\text{T}} \leq 2.4) \tag{8.124}$$

can be obtained from Eqs. (5.52) and (8.116).

Let us define the *benchmark heating rate* as the dimensionless heating rate, r_{T0}, corresponding to $k_{\text{R,a}} = 1$ (and, hence, to $\mu_{\text{R,a}} = \omega_{\text{R,a}} = 1/2$). From Eq. (8.116), one has

$$r_{\text{T0}} = r_{\text{T}}\big|_{k_{\text{R,a}}=1} = \ln 2 \approx 0.7 \tag{8.125}$$

which is comparable with the optimal dimensionless heating rate of about 0.4 (Example 8.13).

It follows from Eqs. (8.117), (8.123), and (8.124) that

$$T_{\text{R},a} = T_{\text{char}} \quad \text{when} \quad r_{\text{T}} = r_{\text{T0}} = \ln 2 \tag{8.126}$$

For heating rates that are different from r_{T0}, Eq. (8.117) yields

$$T_{\text{R,a}} \approx T_{\text{char}} + \theta_{\text{char}} \cdot \begin{cases} \ln r_{\text{T}}, & r_{\text{T}} < 0.3 \\ r_{\text{T}}, & r_{\text{T}} > 2 \end{cases} \tag{8.127}$$

A moderate heating rate is the one that yields a moderate $k_{\text{R,a}}$, that is, Figure 5.4, $0.1 \leq k_{\text{R,a}} \leq 10$. It follows from Eq. (8.116) that

$$0.1 \leq r_T \leq 2.4 \quad \text{(moderate heating rate)} \tag{8.128}$$

Equations (8.124) and (8.128) yield

$$0.83 \leq T_{R,a}/T_{char} \leq 1.2 \quad (0.1 \leq r_T \leq 2.4) \tag{8.129}$$

Statement 8.18

In an isobaric temperature-programmed GC analysis with a moderate heating rate, a highly interactive solute elutes at the temperature (T_R) that is close to its characteristic temperature (T_{char}).

Being a function of its characteristic parameters (T_{char} and θ_{char}), a solute elution temperature (T_R) can be expressed via other thermodynamic parameters that can be found from T_{char} and θ_{char}. For example, due to Eqs. (8.120) and (8.108), T_R can be express as a function of ω_{init}. Sometimes, it can be necessary to reverse these expressions in order to find thermodynamic parameters like ω_{init} from the measured (observed) parameters like T_R.

Solving together Eqs. (8.120), (8.117) and (8.108) yields ω_{init} as a function, Figure 8.10,

$$\omega_{init} = \frac{e^{(T_R - T_{init})/\theta_{char}} - e^{r_T}}{e^{(T_R - T_{init})/\theta_{char}} - 1} \quad (T_R \geq \theta_{char} r_T + T_{init}) \tag{8.130}$$

of T_R. The requirement $T_R \geq \theta_{char} r_T + T_{init}$ prevents negative ω_{init} by recognizing that, as a result of the temperature rise during the hold-up time, the first eluite cannot emerge from the column before the column temperature departs from T_{init} by, at least, $r_T\theta_{char}$ (which, according to Eq. (8.95), is equal to $R_T t_{M,char}$).

Equations (8.130),(8.6) and (8.7) make it possible to express μ_R in Eq. (8.118) as a function,

$$\mu_R = \frac{\mu_{R,a}}{1 - e^{-(T_R - T_{init})/\theta_{char}}} \quad (T_R \geq \theta_{char} r_T + T_{init}) \tag{8.131}$$

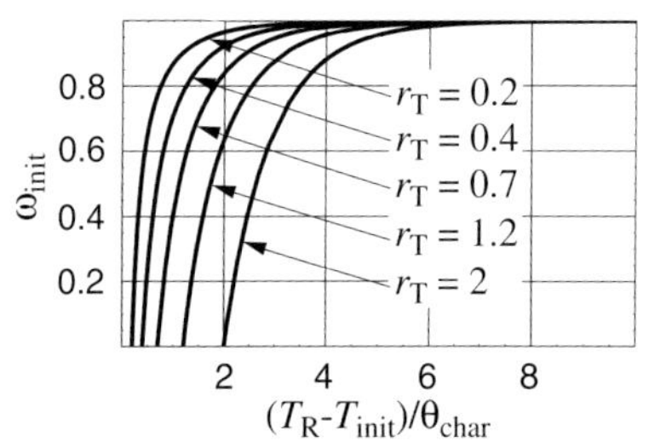

Figure 8.10 A solute initial immobility (ω_{init}, Eq. (8.130)) reconstructed from the dimensionless heating rate (r_T) and dimensionless departure, $(T_R - T_{init})/\theta_{char}$, of the solute elution temperature (T_R) from the initial temperature (T_{init}) of a heating ramp. At $r_T < 1$, all solutes with $T_R - T_{init} > 3\theta_{char}$ have $\omega_{init} > 0.95$. They were highly retained at T_{init}, and, therefore migrated with nearly asymptotic parameters.

of T_R. It was found earlier that highly interactive solutes – those that were highly retained at T_{init} – elute with almost the same μ_R approaching the asymptotic elution mobility $\mu_{R,a}$. Equation (8.131) highlights this fact from a different perspective. The larger the solute's T_R compared to T_{init}, the closer is its μ_R to $\mu_{R,a}$. Specifically, according to Eq. (8.131), μ_R of a solute that elutes at $T_R > T_{init} + 3\theta_{char}$ is within 5% range of $\mu_{R,a}$.

Equation (8.131) provides a good qualitative picture of the dependence of μ_R on T_R, but quantitatively it is not complete. During the same temperature program, solutes with different θ_{char} can elute at different temperatures. Therefore, μ_R in Eq. (8.131) is a function of two mutually dependent parameters, T_R and θ_{char}. For generic solutes, the dependence of μ_R on T_R can be found from solving together Eqs. (8.15), (8.120) and (8.131) in order to exclude θ_{char} from Eq. (8.131). Unfortunately, the solution is much too complex. Significantly simpler is an approximation, Figure 8.11a,

$$\mu_R = \frac{\mu_{R,a}}{1-e^{-(T_R-T_{init})/\theta_{T,init}}} = \frac{1-e^{-r_T}}{1-e^{-(T_R-T_{init})/\theta_{T,init}}} \qquad (T_R \geq T_{init} + R_T t_{M,init}) \tag{8.132}$$

where θ_{char} is replaced with the quantity $\theta_{T,init}$ defined in Eq. (8.90) as a simple function of T_{init}. Substitution of this formula into Eq. (8.8) yields, Figure 8.11b,

$$k_R = k_{R,a}\left(1-e^{r_T-(T_R-T_{init})/\theta_{T,init}}\right) = \frac{1-e^{r-(T_R-T_{init})/\theta_{T,init}}}{e^{r_T}-1} \qquad (T_R \geq T_{init} + R_T t_{M,init}) \tag{8.133}$$

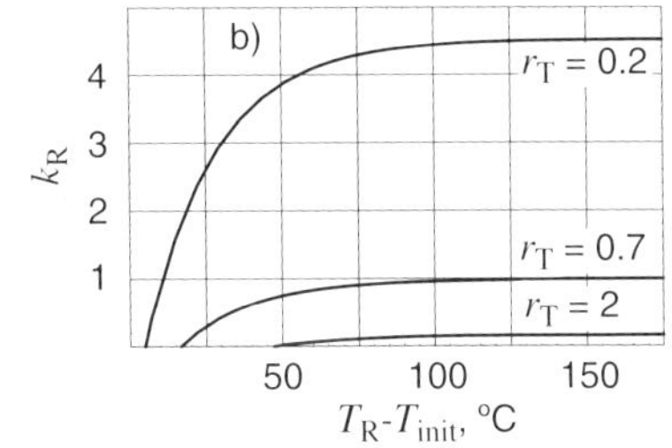

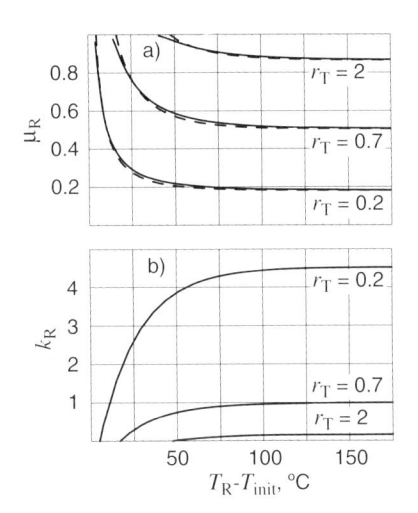

Figure 8.11 Elution mobility (μ_R) and retention factor (k_R, Eq. (8.133)) as functions of elution temperature (T_R) and dimensionless heating rate (r_T). T_{init} is initial temperature of heating ramp. Solid lines in (a) represent exact solutions of Eqs. (8.15), (8.120) and (8.131) while dashed lines are approximations in Eq. (8.132). In all cases, $\xi = 0.7$.

Replacement of θ_{char} with $\theta_{T,\text{init}}$ in Eq. (8.134) does not suggest that these quantities are always approximately equal to each other. Thus, according to Eqs. (5.47) and (8.90), if $T_{\text{init}} = 300$ K then $\theta_{T,\text{init}} \approx 23\,^\circ$C while, for a solute with $T_{\text{char}} = 600$ K, $\theta_{\text{char}} \approx 38\,^\circ$C. However, there is another important factor. The elution temperature (T_{R}) of a solute with $T_{\text{char}} = 600$ K is close to 600 K. As a result, the denominator of Eq. (8.135) is very close to unity regardless of the choice of θ_{char} as 38 °C or 23 °C. On the other hand, $\theta_{T,\text{init}}$ is close to θ_{char} where it counts, that is at the beginning of the heating ramp where, as a result of T_{R} being not very much higher than T_{init}, μ_{R} is sufficiently different from $\mu_{\text{R,a}}$. As a result, Eq. (8.132) is a close approximation (Figure 8.11a) to Eq. (8.136) for any θ_{char}.

An important feature of the transition of the curves in Eqs. (8.132) and (8.133) to their horizontal asymptotic levels ($\mu_{\text{R,a}}$ and $k_{\text{R,a}}$) can be observed in Figure 8.11. Let us notice first that the product $R_{\text{T}} t_{\text{M,init}}$ in the statements of constraints for T in Eqs. (8.132) and (8.133) is the hold-up temperature ($\Delta T_{\text{M,init}}$) defined in Eq. (8.64). This temperature is the intercept in Figure 8.11b. Each curve has its own intercept proportional to r_{T}. In Figure 8.11a, the quantity $\Delta T_{\text{M,init}}$ for each curve is the value of $T_{\text{R}} - T_{\text{init}}$ at the intersection with the upper bound of the graph (at $\mu_{\text{R}} = 1$). The graphs in Figure 8.11 show that

Statement 8.19

For any heating rate, the transition of μ_{R} to its asymptotic lever $\mu_{\text{R,a}}$ as well as the transition of k_{R} to its asymptotic lever $k_{\text{R,a}}$ takes place during the first 50 °C or so immediately following the hold-up temperature ($\Delta T_{\text{M,init}}$) for each curve.

In other words, the *transitional temperature increment*, $\Delta T_{\text{trans}} = T_{\text{R,trans}} - T_{\text{init}}$, for any heating rate can be estimated as $\Delta T_{\text{trans}} \approx 50\,^\circ\text{C} + R_{\text{T}} t_{\text{M,init}}$.

The functions of the elution temperature (T_{R}) in Eqs. (8.132) and (8.133) can be transformed into functions of dimensionless retention time.

Previously in this chapter, when the properties of migration of different solutes were considered, the dimensionless time ($\tilde{\tau}$) of each solute had its own solute-dependent normalization described in Eq. (8.65). To describe elution parameters as functions of time regardless of a particular solute, it is convenient to use the same normalization for all solutes as it has been done in the study of scalability of retention times. Let us denote a solute-independent dimensionless retention time as

$$\tau_{\text{R}} = \frac{t_{\text{R}}}{t_{\text{M,init}}} \tag{8.134}$$

Combination of this with Eqs. (8.83) and (8.96) yields $(T_{\text{R}} - T_{\text{init}})/\theta_{T,\text{init}} = r_{\text{T}} \tau_{\text{R}}$. Equations (8.132) and (8.133) become, Figure 8.12,

$$\mu_{\text{R}} = \frac{\mu_{\text{R,a}}}{1 - e^{-r_{\text{T}} \tau_{\text{R}}}} = \frac{1 - e^{-r_{\text{T}}}}{1 - e^{-r_{\text{T}} \tau_{\text{R}}}} \quad (\tau_{\text{R}} \geq 1) \tag{8.135}$$

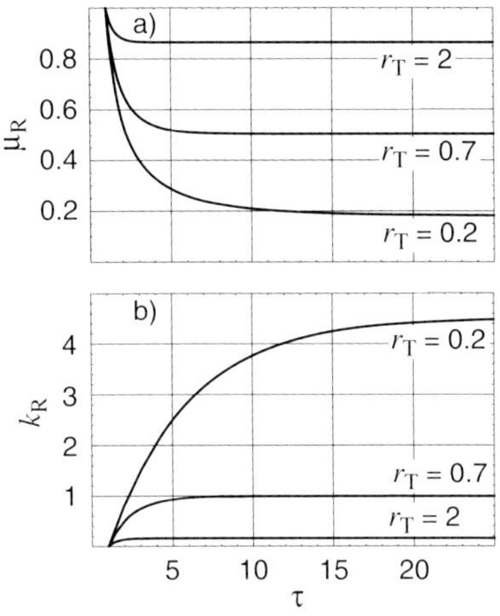

Figure 8.12 Elution mobility (μ_R) and retention factor (k_R) as functions of the dimensionless retention time (τ_R) and dimensionless heating rate (r_T).

$$k_R = k_{R,a}(1-e^{r_T(1-\tau_R)}) = \frac{1-e^{r_T\tau_R}}{e^{r_T}-1} \quad (\tau_R \geq 1) \tag{8.133}$$

The dimensionless transitional time, τ_{trans}, during which the transition of the curves $\mu_R(\tau_R)$ and $k_R(\tau_R)$ to their asymptotic levels ($\mu_{R,a}$ and $k_{R,a}$) takes place can be estimated as $\tau_{trans} \approx 1 + 2/r_T$.

8.7
Boundaries of the Linearized Model

Closed form mathematical formulae for elution parameters found in the previous section were based on the following operational constraints and underlying approximations for temperature-programmed GC analyses.

Operational constraints:

- uniform (the same along the column) solute mobility (μ). In practical terms, this means uniform stationary phase film and uniform column heating;
- constant pressure.

Underlying approximations:

- linearized model of a solute–column interaction;
- during migration of each solute, gas viscosity remains fixed at the level corresponding to the solute characteristic temperature (T_{char}).

The accuracy of the found formulae for elution parameters depends on the accuracy of the underlying approximations.

Both models of a solute–column interaction – ideal thermodynamic and linearized – adopted in this book are based on respective models of a solute–liquid interaction. All numerical data and computer-simulated chromatograms are based on experimental data for the solute–liquid interaction (Chapter 5). How close is the linearized model of a solute–liquid interaction to physical reality? The answer depends on a particular kind of reality being considered.

A very real thing might not be relevant to a column performance under consideration. For example, prediction of peak retention times for the purpose of solute identification requiring a very accurate model might be of a great importance in some studies. However, the solute identification is outside of a subject of a column general operation – the topic of this book. The accuracy of a thermodynamic model required for the study of a column performance can be substantially lower than that suitable for addressing the needs of a solute identification.

The linearized model might be considered as an approximation to ideal thermodynamic model. Some aspects of the accuracy of the approximation were addressed in Chapter 5. However, the ideal thermodynamic model has its own limitations, and the difference between the two models becomes large exactly where the accuracy of the ideal thermodynamic model itself is questionable even within the scope of a column general performance.

Let us first highlight the conditions under which the accuracy of both models is questionable, and then outline the conditions where the linearized model is expected to be sufficiently accurate.

Example 8.17

For the solute of Example 8.14 ($T_{char} = 600$ K, $\theta_{char} = 40\,^{\circ}$C) at normal temperature (25 °C), the ideal thermodynamic model in Eq. (8.12) yields $\mu \approx 0.3 \times 10^{-6}$ while the linearized model in Eq. (8.13) yields $\mu \approx 0.55 \times 10^{-3}$. The three orders of magnitude relative difference between the μ-values in the two models is very large. In a typical case of $t_M \approx 1$ min ($L = 30$ m, $d_c = 0.25$ mm, helium), the solute elution times would, according to Eq. (8.20), be $t_R \approx 3.3 \times 10^6$ min ≈ 6.2 years (ideal thermodynamic model), and $t_R \approx 30$ h (linearized model). □

Suppose that the solute in this example is the last to elute. Its elution time, as a measure of the duration of isothermal analysis, can be treated as a measure of a column performance. Which answer is correct? $t_R \approx 6.2$ years? $t_R \approx 30$ h? Neither?

There are several ways to address this question.

First, there is hardly any practical reason for using very long isothermal analysis where the mobility of the last eluite is lower than 10^{-3} (solute elution velocity is 1000 lower than gas velocity). Not only the analysis is prohibitively long, but it might also be difficult to detect and quantify its very wide last peaks. Generally, the use of isothermal conditions is acceptable only in the analyses of relatively simple mixtures where the isothermal analysis time is not very long so that the time savings do not justify additional complexity of temperature-programmed conditions. Example 8.17 certainly does not represent that category and, therefore, does not represent a realistic column performance.

Second, the linearized model is advantageous in the studies of temperature-programmed conditions. However, at fixed temperature of isothermal analysis, there is no need to choose between the models because neither is too complex for the task at hand. As a result, the choice of a model for isothermal conditions can be based on the accuracy of the model rather than on its relative simplicity.

Finally, as mentioned in Chapter 5, majority of currently available data for solute–liquid interactions came from isothermal measurements with retention factors that do not depart very far from $k = 1$ – certainly not as far as in Example 8.17. Temperature-programmed analyses can also be used for these measurements [23–25]. However, again, elution retention factors close to 1 were used in these applications [24]. On the other hand, parameters of the ideal thermodynamic model change with temperature [75–78]. This implies that extrapolation of the ideal thermodynamic model for very large retention factors might be significantly inaccurate. As a result, not only it is practically irrelevant, but it is also unclear which of the two models produces more accurate results for very large retentions like those evaluated in Example 8.17.

One can conclude that both large retention times of Example 8.17 have questionable accuracy. However, both are also irrelevant to practical aspects of a column performance.

When it comes to the study of the performance of temperature-programmed GC analysis, finding a simple model for the description of a solute migration becomes imperative. Fortunately, there is one important property of temperature programming that makes its outcome dependent only on those properties of a solute retention that are very similarly described by both the thermodynamic and the characteristic models, and that rely on actual rather than extrapolated experimental data.

In a typical temperature-programmed GC, a solute departure from the inlet becomes significant only shortly before its elution, Figure 8.4. It is practically inconsequential whether the mobility of a solute residing near the inlet was 10^{-3}, 10^{-6} or lower. What counts most is the solute mobility (μ) – typically higher than 0.1 ($k < 10$), Figure 8.6a, – shortly before its elution. Furthermore, in a typical temperature program, solutes elute with $\mu < 0.9$ ($k > 0.1$), Figure 8.6. The region of $0.1 \leq k \leq 10$ ($0.1 \leq \mu \leq 0.9$) is exactly the one where the linearized model is close (Figure 5.5) to the ideal thermodynamic model (or to any slightly curved dependence of $\ln k$ on T), and where thermodynamic parameters can be found from experimental data without significant extrapolations.

Good correlation of linearized models (differently arranged than the linearized model in this book) with experimental data has been demonstrated in several studies [23–26]. However, the amount of data for each model were insufficient for quantifying the accuracy of the models.

For quantitative comparison of linearized and ideal thermodynamic models, let us recall that solute migration during temperature-programmed analysis depends not only on the temperature-dependent solute–column interaction, but also on the temperature-dependent gas viscosity (η).

The solute elution temperature (T_R) and elution mobility (μ_R) are the important parameters of a column performance in temperature-programmed analysis. These

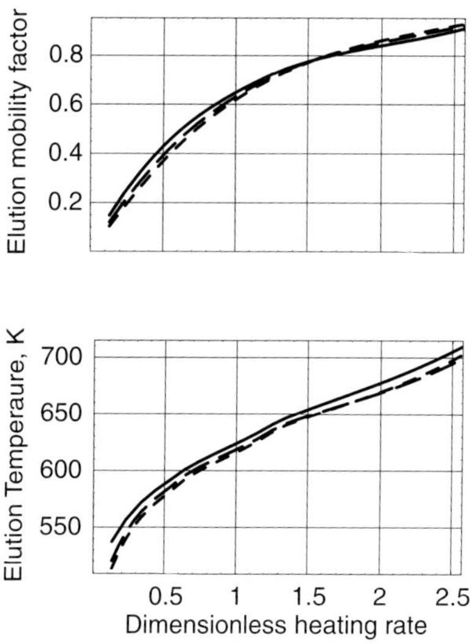

Figure 8.13 Elution mobility (μ_R) and temperature (T_R) of a solute as functions of dimensionless heating rate (r_T) in isobaric temperature-programmed analysis with linear heating ramp. *Solid* and *short-dashed* lines represent the numerical solution of Eq. (8.38) with μ from Eqs. (8.12) and (8.13), respectively. Long-dashed lines represent Eqs. (8.115) and (8.117) found from solving Eq. (8.69). In all cases, $\theta_{char} = 40\,^\circ C$, $T_{char} = 600$ K, $T_{init} = 300$ K. The systematic shift of the dashed lines from the solid one for T_R is likely to have the same roots as the systematic retention time shifts in Figure 8.3.

parameters found from three elution equations – relying on two models and accounting for or ignoring the temperature dependence of η – are compared in Figure 8.13. All three equations yield sufficiently close results for each parameter. The systematic downshift in T_R for linearized models relative to T_R for the ideal thermodynamic model in Figure 8.13 is consistent with retention time shifts in chromatograms of Figure 8.3.

In the future, unless contrary is explicitly stated, the linearized model of solute–-column interaction (Eq. (8.13)) is always assumed.

Appendix 8.A

8.A.1 Uniform Mobility – Numerical Solution of the Migration Equation

In a uniformly heated column, gas velocity, u, can be found from Eq. (7.67) for any z and t. This allows one to express Eq. (8.23) as

$$dt_g = \frac{\sqrt{1-z/L_{ext}}}{u_i} dz \tag{A1}$$

where the quantity L_{ext} is described in Eq. (7.64), and the gas inlet velocity (u_i) is described in Eqs. (7.50) and (7.38). When the inlet pressure (p_i) and/or outlet pressure (p_o) change with time, L_{ext} and u_i also become functions of time. The quantity u_i can also become a function of t due to the change in a column temperature (T) as a function of t.

Let us assume that the dependences ($\mu = \mu(T)$ and $\eta = \eta(T)$) of a solute mobility and carrier gas viscosity on temperature (T) are known. Also known are the column length (L) internal diameter (d), pressure program $p_i(t)$, and temperature program $T(t)$. Numerical calculation of $t(z_0)$ and $t_g(z_0,t(z_0))$ for a predetermined coordinate $z = z_0$ in Eq. (8.25) can proceed as follows.

Start at $z = 0$, $t_g = 0$, $t = 0$. Choose dz – a short increment in the distance traveled by the solute. For example, $\mathrm{d}z = L/100$.

Step 1: Compute $p_i = p_i(t)$, $P = p_i/p_o$, $\eta = \eta(T(t))$, and $\mu = \mu(T(t))$
Step 2: Compute L_{ext} and u_i
Step 3: Compute $z = z + \mathrm{d}z$
Step 4: Compute $\mathrm{d}t_g$, Eq. (A1), and $t_g = t_g + \mathrm{d}t_g$
Step 5: Compute $\mathrm{d}t$ as $\mathrm{d}t = \mathrm{d}t_g/\mu$, Eq. (8.25), and $t = t + \mathrm{d}t$,
Step 6: If $z < z_0$, then go back to Step 1, else output the results and stop.

8.A.2 Elution Equation under Uniform Mobility

The dynamic hold-up time, t_m, on the right-hand side of the elution equation (8.36) can be different for different solutes. Theoretically powerful and practically useful conclusions, such as *scalability* of chromatograms and *method translation* techniques, can be deduced from the elution equation if its right-hand side could be made equal for all solutes [3], that is when Eq. (8.36) is rearranged to allow replacing the *dynamic* hold-up time, t_m, with the *static* hold-up time, $t_{M,ref}$, measured at some predetermined fixed *reference inlet pressure*, $p_{i,ref}$, and *reference temperature*, T_{ref}.

Due to Eqs. (8.2) and (8.16), the differential equation of a solute migration in pressure- and temperature-programmed analysis with uniform solute mobility $\mu = \mu(t)$ can be expressed as $\mu(t)u(z,t)\mathrm{d}t = \mathrm{d}z$. In this formula $u(z,t)$ is the gas velocity that, due to the programming of pressure and/or temperature, can be a function of t, and, due to gas decompression along a column, could be a function of z. Due to relations between pneumatic parameters of ideal gas (Chapter 7), the differential equation of a solute migration can be arranged as

$$\frac{\bar{u}(t)j_{32}(t)\mu(t)\mathrm{d}t}{\bar{u}_{ref}} = t_{M,ref}P_\zeta(\zeta)\mathrm{d}\zeta, \quad j_{32}(t) = \frac{2j_3(t)}{3j_2(t)}, \quad P_\zeta(\zeta) = \sqrt{1 - \frac{P^2(t(\zeta))-1}{P^2(t(\zeta))}\zeta} \tag{A2}$$

where $\bar{u}_{ref}$ is the average gas velocity at $p_i = p_{i,ref}$ and $T = T_{ref}$; $\zeta = z/L$ is the dimensionless distance from the column inlet; $t(\zeta)$ is the time of migration of a solute to the location $z = \zeta L$; $P(t) = p_i(t)/p_o$, $\bar{u}(t)$, $j_2(t)$ and $j_3(t)$ are the time-dependent relative inlet pressure, average gas velocity and gas compressibility factors defined in Eqs. (7.65) and (7.102). It is assumed that the time-averaging in Eq. (7.116) for $\bar{u}(t)$ takes place under static (isothermal and isobaric) conditions existing in actual pressure- and temperature-programmed analysis at time t. According to Eqs. (7.66) and (7.64), the function $P_\zeta(\zeta)$ describes the normalized pressure profile, $P_\zeta(\zeta) = p_\zeta(\zeta, t(\zeta))/p_o$, along the path of a solute – the pressure at location ζ at the time when the solute is at ζ. In this definition, p_o is the outlet pressure and $p_\zeta(\zeta, t)$ is the pressure at arbitrary location, ζ, at arbitrary time, t.

In order to solve Eq. (A2), one needs to know the time, $t(\zeta)$, of arrival of a solute to an arbitrary location z along a column. Generally, $t(\zeta)$ is not known and, therefore, a close form integration of the right-hand side of Eq. (A2) is not possible. This problem does not exist in an isobaric analysis where $P(t)$ is a fixed quantity and, therefore, $P(t) = P(0) = P$, $j_{32}(t) = j_{32}(0) = j_{32}$. As a result,

$$\int_0^1 P_\zeta(\zeta)\,\mathrm{d}\zeta = \int_0^1 \sqrt{1-\frac{P^2-1}{P^2}\zeta}\,\mathrm{d}\zeta = j_{32} \tag{A3}$$

and integration of Eq. (A2) yields

$$\int_0^{t_R} \frac{\bar{u}(t)}{\bar{u}_{\mathrm{ref}}}\mu(t)\mathrm{d}t = t_{M,\mathrm{ref}} \tag{A4}$$

To find a suitable approximation for the general form of the right-hand side of Eq. (A2) leading to its simple solution, let us notice first that the difference (if any) between the right- and the left-hand sides of Eq. (A4) diminishes when gas decompression along the column is either weak or strong. At weak decompression, $P(t) \approx 1$, and, therefore, $P_\zeta(\zeta) \approx 1$. At strong decompression, $P(t) \gg 1$, and, therefore, $P_\zeta(\zeta) \approx (1-\zeta)^{1/2}$. When $P_\zeta(\zeta) = 1$ or $P_\zeta(\zeta) = (1-\zeta)^{1/2}$, the integration of $P_\zeta(\zeta)$ over ζ yields Eq. (A3) where j_{32} is a fixed quantity equal to 1 or to 2/3, respectively. The same is the value of $j_{32}(t)$ on the left-hand side of Eq. (A2). As a result, j_{32} in Eq. (A2) cancels out and integration of Eq. (A2) yields Eq. (A4).

One can conclude that the difference (if any) between the right- and the left-hand sides of Eq. (A4) should be the largest under the intermediate gas decompression – something corresponding to pressure programming where $1.5 \le P(t) \le 3$. Several observations help to evaluate the possible difference.

Integration of the left-hand side of Eq. (A2) can be treated as follows. According to the mean-value theorem of integration [79], there exists such t_1 between zero and t_R $(0 \le t_1 \le t_R)$ that

$$\int_0^{t_R} \frac{\bar{u}(t)j_{32}(t)}{\bar{u}_{\mathrm{ref}}}\mu(t)\,\mathrm{d}t = j_{32}(t_1)\int_0^{t_R} \frac{\bar{u}(t)}{\bar{u}_{\mathrm{ref}}}\mu(t)\,\mathrm{d}t \tag{A5}$$

The solution of Eq. (A2) can be described as

$$\int_0^{t_R} \frac{\bar{u}(t)}{\bar{u}_{ref}} \mu(t)\, dt = \frac{t_{M.ref}}{j_{32}(t_1)} \int_0^1 \sqrt{1 - \frac{P^2(t(\zeta)) - 1}{P^2(t(\zeta))}} \zeta d\zeta \tag{A6}$$

In these formulae, the quantity j_{32} is a weak function of P. Thus, the range $1 \leq P \leq \infty$ of variable P maps on the range $1 \leq j_{32} \leq 3/2$ of the quantity j_{32}. Furthermore, in a typical pressure- and temperature-programmed analysis, the change in $P(t)$ during the entire analysis is not very large. The most frequent use of the pressure programming in a temperature-programmed analysis is to maintain constant gas flow rate. During migration of a solute along the last 90% of a column length in a typical isobaric temperature-programmed analysis, the absolute temperature, T, of a column increases by less than 20%, Figure 8.4b. Roughly the same is true for the analysis with constant gas flow rate. When T increases by 20%, gas viscosity increases by about $1.2^{0.7} - 1 \approx 14\%$. At moderate gas decompression ($P \approx 2$), a smaller than 6% increase in P is sufficient to maintain the constant flow. At $P = 2$, this causes a smaller than 1.5% change in j_{32}. The quantity $j_{32}(t_1)$ is the area under one of the solid lines in Figure 8.A.1 while the integral on the right-hand side of Eq. (A6) is the area under the dashed line. The difference between these areas is very small (most likely smaller than 1%) and, therefore, the integral and $j_{32}(t_1)$ can be canceled out. Equation (A6) becomes Eq. (A4).

After substitution of Eqs. (7.132) and (7.38) into Eq. (A4), it becomes

$$\int_0^{t_R} \frac{\Delta\tilde{p}(t)}{\Delta\tilde{p}_{ref}} \cdot \frac{\eta_{ref}}{\eta(t)} \cdot \mu(t) dt = t_{M.ref} \tag{A7}$$

where η is the gas viscosity and $\Delta\tilde{p}$ is the virtual pressure drop, Eqs. (7.136) and (7.140). These, along with a solute mobility, μ, are the independent factors that affect

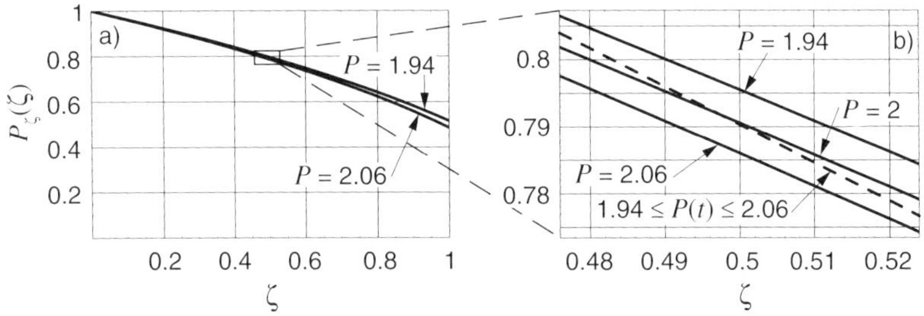

Figure 8.A.1 Normalized pressure profiles, $P_\varsigma(\varsigma)$, Eq. (A2), in $\pm 3\%$ inlet pressure region around $P = 2$ ($p_i = 2p_o$) – the pressure range during migration of a solute along 90% of a column length prior to the solute elution in a typical temperature-programmed analysis with constant flow and moderate gas decompression along the column. The variable $\varsigma = z/L$ is the dimensionless distance from the column inlet (L is the column length and z is the distance from the inlet). Solid lines represent the profiles at three fixed inlet pressures, $p_i = 1.94p_o$, $p_i = 2p_o$ and $p_i = 2.06p_o$. The dashed line shows the normalized pressure at ς at the time when the solute is at ς. Actual scale (a) shows a barely distinguishable difference between the profiles.

t_R in a pressure- and temperature-programmed analysis with a given $t_{M,ref}$. Equation (A7) is the basis for *universal method translation* [68] – translation of a pressure- and temperature-programmed analysis into another analysis with an arbitrarily chosen pressure program.

References

1 Consden, R., Gordon, A.H., and Martin, A.J.P. (1944) *Biochem. J.*, **38**, 224–232.
2 Keulemans, A.I.M. (1959) *Gas Chromatography*, 2nd edn, Reinhold Publishing Corp., New York.
3 Blumberg, L.M. and Klee, M.S. (1998) *Anal. Chem.*, **70**, 3828–3839.
4 Blumberg, L.M. and Klee, M.S. (2001) *J. Chromatogr. A*, **918/1**, 113–120.
5 Connors, K.A. (1974) *Anal. Chem.*, **46**, 53–58.
6 Cramers, C.A., Keulemans, A.I.M., and McNair, H.M. (1967) *Chromatography* (ed. E. Heftmann), Reinhold Publishing Corp., New York, pp. 182–209.
7 Dal Nogare, S. and Juvet, R.S. (1962) *Gas-Liquid Chromatography. Theory and Practice*, John Wiley & Sons, Inc., New York.
8 Dondi, F., Cavazzini, A., and Remilli, M. (1998) *Advances in Chromatography*, vol. **38** (eds P.R. Brown and E. Grushka), Marcel Dekker, New York, pp. 51–74.
9 Ettre, L.S. (1993) *Pure Appl. Chem.*, **65**, 819–872.
10 Ettre, L.S. and Hinshaw, J.V. (1993) *Basic Relations of Gas Chromatography*, Advanstar, Cleveland, OH.
11 Gaspar, G., Annino, R., Vidal-Madjar, C., and Guiochon, G. (1978) *Anal. Chem.*, **50**, 1512–1518.
12 Giddings, J.C. (1965) *Dynamics of Chromatography*, Marcel Dekker, New York.
13 Giddings, J.C. (1991) *Unified Separation Science*, Wiley, New York.
14 Guiochon, G. and Guillemin, C.L. (1988) *Quantitative Gas Chromatography for Laboratory Analysis and On-Line Control*, Elsevier, Amsterdam.
15 Klee, M.S. and Blumberg, L.M. (2001) 24th International Symposium on Capillary Chromatography and Electrophoresis, Abstracts of 24th International Symposium on Capillary Chromatography and Electrophoresis, Las Vegas, May 21-24, 2001, MeetingAbstracts.com Internet.
16 Knox, J.H. and Saleem, M. (1969) *J. Chromatogr. Sci.*, **7**, 614–622.
17 Littlewood, A.B. (1970) *Gas Chromatography: Principles, Techniques, and Applications*, 2nd edn, Academic Press, New York.
18 Blumberg, L.M. and Klee, M.S. (2001) *J. Chromatogr. A*, **933**, 13–26.
19 Giddings, J.C. (1962) *J. Chem. Educ.*, **39**, 569–573.
20 Giddings, J.C. (1962) *Gas Chromatography* (eds N. Brenner, J.E. Callen, and M.D. Weiss), Academic Press, New York, pp. 57–77.
21 Harris, W.E. and Habgood, H.W. (1966) *Programmed Temperature Gas Chromatography*, John Wiley & Sons, Inc., New York.
22 Blumberg, L.M. and Klee, M.S. (2000) *Anal. Chem.*, **72**, 4080–4089.
23 Abbay, G.N., Barry, E.F., Leepipatpiboon, S., Ramstad, T., Roman, M.C., Siergiej, R.W., Snyder, L.R., and Winniford, W.L. (1991) *LC-GC*, **9**, 100–114.
24 Bautz, D.E., Dolan, J.W., Raddatz, L.R., and Snyder, L.R. (1990) *Anal. Chem.*, **62**, 1560–1567.
25 Bautz, D.E., Dolan, J.W., and Snyder, L.R. (1991) *J. Chromatogr.*, **541**, 1–19.
26 Dolan, J.W., Snyder, L.R., and Bautz, D.E. (1991) *J. Chromatogr.*, **541**, 21–34.
27 Hao, W., Zhang, X., and Hou, K. (2006) *Anal. Chem.*, **78**, 7828–7840.
28 Kaliszan, R., Bączek, T., Buciński, A., Buszewski, B., and Sztupecka, M. (2003) *J. Sep. Sci.*, **26**, 271–282.

29 Scott, R.P.W. and Kucera, P. (1973) *Anal. Chem.*, **45**, 749–754.

30 Snyder, L.R. (1964) *J. Chromatogr.*, **13**, 415–434.

31 Snyder, L.R. (1992) *Chromatography, 5th Edition: Fundamentals and Applications of Chromatography and Related Differential Migration Methods. Part A: Fundamentals and Techniques* (ed. E. Heftmann), Elsevier, Amsterdam, pp. A1–A68.

32 Snyder, L.R. and Dolan, J.W. (2006) *High-Performance Gradient Elution: Practical Application of the Linear-Solvent-Strength Model*, John Wiley & Sons, Inc., Hoboken, NJ.

33 Snyder, L.R., Dolan, J.W., and Gant, J.R. (1979) *J. Chromatogr.*, **165**, 3–30.

34 Snyder, L.R. and Saunders, D.L. (1969) *J. Chromatogr. Sci.*, **7**, 195–208.

35 Blumberg, L.M. (1993) *J. Chromatogr.*, **637**, 119–128.

36 Blumberg, L.M. and Berger, T.A. (1992) *J. Chromatogr.*, **596**, 1–13.

37 Lan, K. and Jorgenson, J.W. (1999) *Anal. Chem.*, **71**, 709–714.

38 Lan, K. and Jorgenson, J.W. (2001) *J. Chromatogr. A*, **905**, 47–57.

39 Grubner, O. (1968) *Advances in Chromatography*, vol. 6 (eds J.C. Giddings and R.A. Keller), Marcel Dekker, New York, pp. 173–209.

40 Grushka, E. (1972) *J. Phys. Chem.*, **76**, 2586–2593.

41 Jönsson, J.A. (1987) *Chromatographic Theory and Basic Principles* (ed. J.A. Jönsson), Marcel Dekker, New York, pp. 27–102.

42 Kučera, E. (1965) *J. Chromatogr.*, **19**, 237–248.

43 Blumberg, L.M. (1993) *J. High Resolut Chromatogr.*, **16**, 31–38.

44 Giddings, J.C. (1963) *Anal. Chem.*, **35**, 353–356.

45 Lan, K. and Jorgenson, J.W. (1998) *Anal. Chem.*, **70**, 2773–2782.

46 Lan, K. and Jorgenson, J.W. (2000) *Anal. Chem.*, **72**, 1555–1563.

47 Schettler, P.D., Eikelberger, M., and Giddings, J.C. (1967) *Anal. Chem.*, **39**, 146–157.

48 Dose, E.V. (1987) *Anal. Chem.*, **59**, 2414–2419.

49 Giddings, J.C. (1960) *J. Chromatogr.*, **4**, 11–20.

50 Zhang, Y.J., Wang, G.M., and Qian, R. (1990) *J. Chromatogr.*, **521**, 71–87.

51 Blumberg, L.M. (1992) *Anal. Chem.*, **64**, 2459–2460.

52 Blumberg, L.M. (1994) *Chromatographia*, **39**, 719–728.

53 Blumberg, L.M. (1995) *Chromatographia*, **40**, 218.

54 Blumberg, L.M. (1997) *J. Chromatogr. Sci.*, **35**, 451–454.

55 Golay, M.J.E. (1962) *Gas Chromatography* (eds N. Brenner, J.E. Callen, and M.D. Weiss), Academic Press, New York, pp. xi–xv.

56 Ettre, L.S. (1987) *J. High Resolut. Chromatogr.*, **10**, 221–230.

57 Jain, V. and Phillips, J.B. (1995) *J. Chromatogr. Sci.*, **33**, 601–605.

58 Ohline, R.W. and DeFord, D.D. (1963) *Anal. Chem.*, **35**, 227–234.

59 Phillips, J.B. and Jain, V. (1995) *J. Chromatogr. Sci.*, **33**, 541–550.

60 Rubey, W.A. (1991) *J. High Resolut. Chromatogr.*, **14**, 542–548.

61 Rubey, W.A. (1992) *J. High Resolut. Chromatogr.*, **15**, 795–799.

62 Zhukhovitskii, A.A., Zolotareva, O.V., Sokolov, V.A., and Turkel'taub, N.M. (1951) *Doklady Akademii Nauk S.S.S.R.*, **77**, 435–438.

63 Akard, M. and Sacks, R. (1994) *Anal. Chem.*, **66**, 3036–3041.

64 McGuigan, M. and Sacks, R. (2001) *Anal. Chem.*, **73**, 3112–3118.

65 Veriotti, T., McGuigan, M., and Sacks, R. (2001) *Anal. Chem.*, **73**, 279–285.

66 Analytical Innovations, Inc . (2002) Pro EzGC, version 2.20 for Windows, Distributed by Restek Corp. (cat #21487), Bellefonte, PA.

67 Blumberg, L.M. and Klee, M.S. (2000) *J. Micro. Sep.*, **12**, 508–514.

68 Blumberg, L.M.(21 October 2003) Method Translation in Gas Chromatography, USA Patent 6,634,211.

69 Blumberg, L.M. and Klee, M.S. (2001) *Anal. Chem.*, **73**, 684–685.

70 Cramers, C.A. and Leclercq, P.A. (1999) *J. Chromatogr. A*, **842**, 3–13.

71 Klee, M.S. and Blumberg, L.M. (2002) *J. Chromatogr. Sci.*, **40**, 234–247.

72 Schutjes, C.P.M., Vermeer, E.A., Rijks, J.A., and Cramers, C.A. (1982) *J. Chromatogr.*, **253**, 1–16.

73 Snyder, W.D. and Blumberg, L.M. (1992) Fourteenth International Symposium on Capillary Chromatography, Proceedings of the Baltimore, May 25–29, 1992, ISCC92, Baltimore (eds P. Sandra and M.L. Lee), pp. 28–38.

74 Schutjes, C.P.M., Leclercq, P.A., Rijks, J.A., Cramers, C.A., Vidal-Madjar, C., and Guiochon, G. (1984) *J. Chromatogr.*, **289**, 163–170.

75 Beens, J., Tijssen, R., and Blomberg, J. (1998) *J. Chromatogr. A*, **822**, 233–251.

76 Girard, B. (1996) *J. Chromatogr.*, **721**, 279–288.

77 González, F.R. and Nardillo, A.M. (1997) *J. Chromatogr. A*, **779**, 263–274.

78 Héberger, K., Görgényi, M., and Kowalska, T. (2002) *J. Chromatogr. A*, **973**, 135–142.

79 Korn, G.A. and Korn, T.M. (1968) *Mathematical Handbook for Scientists and Engineers*, McGraw-Hill Book Company, New York.

9
Formation of Peak Spacing

When two solutes, say "1" and "2," simultaneously injected in a column migrate along the column with different velocities, v_1 and v_2, their *longitudinal distance,*

$$\Delta z = z_1 - z_2 \tag{9.1}$$

from each other, and *temporal distance* – the difference,

$$\Delta t = t_2 - t_1 \tag{9.2}$$

in the time of arriving to any predetermined location, z – gradually increases. As a result, the solutes arrive to the column outlet at different times, t_{R2} and t_{R1}, so that the temporal spacing [1–4] of the peak becomes

$$\Delta t_R = t_{R2} - t_{R1} \tag{9.3}$$

Temporal spacing (Δt_R) of peaks "1" and "2" eluting during a linear heating ramp can be found as

$$\Delta t_R = (T_{R2} - T_{R1})/R_T = \Delta T_R / R_T \tag{9.4}$$

where T_{R1} and T_{R2} are the elution temperatures of solutes "1" and "2," R_T is the column heating rate, and

$$\Delta T_R = T_{R2} - T_{R1} \tag{9.5}$$

is thermal spacing of the peaks.

Generally, a closed form solution for Δt_R cannot be obtained. A numerical solution can be found from subtraction of numerical solutions of Eqs. (8.27) and (8.28) for t_{R1} and t_{R2}. In several practically important special cases, substantial simplifications in evaluations of Δt_R or ΔT_R can be made.

Temperature-Programmed Gas Chromatography. Leonid M. Blumberg

ISBN: 978-3-527-32642-6

9.1
Static GC Analysis

It follows from Eq. (8.22) that, in static (isothermal and isobaric) analysis, Δt_R in Eq. (9.3) can be found as

$$\Delta t_R = t_M \Delta k \tag{9.6}$$

where

$$\Delta k = k_2 - k_1 = (\alpha - 1)k_1 = \frac{\alpha - 1}{\alpha} k_2 \tag{9.7}$$

The quantity, $\alpha = k_2/k_1$ in this formula is the relative retention [5–9] (relative volatility [10], separation factor [9, 11–13], selectivity (factor) [9, 14–19], relative selectivity [20]) of two solutes in static analysis.

9.2
Closely Migrating Solutes in Dynamic Analysis

Finding Δt_R in dynamic (temperature- and/or pressure-programmed) analysis is not as simple as it is in static analysis and it might require a numerical integration of Eqs. (8.27) and (8.28) for t_{R1} and t_{R2} followed by the calculation of Δt_R as a difference (Eq. (9.3)) of the two solutions. A very high accuracy of numerical integrations might be required when Δt_R is a small fraction of t_{R1} and t_{R2}. However, a special treatment of closely migrating solutes – the ones that remain close to each other at any time of their migration – can eliminate the need for the high accuracy.

Let t be an arbitrary time, t_R be the retention time of an arbitrarily chosen solute – briefly, a t_R-solute – and $t_{R,1} = t_R + \Delta t_R$ be the retention time of its neighbor. Let also $\Delta z = z_1 - z$ and $\Delta z_R = z_{1,R} - L$ be the longitudinal distance between the t_R-solute and its neighbor at the times t and t_R, respectively. For closely migrating solutes, $|\Delta z| << L$ at any t. As a result, Δt_R can be found as

$$\Delta t_R = -\Delta z_R / v_{oR}, \ |\Delta z| \ll L \tag{9.8}$$

where v_{oR} is the outlet velocity of the t_R-solute at the time t_R of its elution.

The last formula is based on two assumptions. First, the change in Δz_R during the time interval Δt_R is negligible compared to Δz_R, and, second, the difference between v_{oR} and the outlet velocity, $v_{oR,1}$, of the neighboring solute is negligible compared to v_{oR}. Both the assumptions are direct consequence of the properties of the solutes that make them to migrate close to each other. The negative sign in the formula reflects the fact that $t_{R,1} > t_R$ when the t_R-solute arrives to the column outlet before its neighbor, and therefore, $\Delta z_R = z_{1,R} - L < 0$.

An infinitesimally small change, $d(\Delta z)$, in Δz that takes place during infinitesimally small time interval, dt, when the t_R-solute and its neighbor advance from arbitrary locations, z and z_1 to locations $z + dz$ and $z_1 + dz_1$, respectively, can be found as $d(\Delta z) = dz_1 - dz = (dz_1/dz - 1)dz = (v_1/v - 1)dz = (\Delta v/v)dz$. Combining this result with Eq. (9.8), one can find Δt_R as

$$\Delta t_R = -\frac{1}{v_{oR}}\int_0^L \frac{\Delta v}{v}dz, \quad |\Delta v| \ll v \tag{9.9}$$

The relatively small value of Δv is the only condition for the last formula. Thus, the formula is valid for any model of a solute–column interaction for the analyses with simultaneous pressure and temperature programming, and for any degree of gas decompression along the column. Generally, both v and Δv can be functions of time, and, therefore, can change with z.

In the case of a uniform column heating combined with either weak gas decompression and arbitrary pressure program or any gas decompression and constant pressure (isobaric analysis), Eq. (9.9) can be made more specific.

According to Eq. (8.2), when gas decompression along the uniformly heated column is weak, the only source of the difference, Δv, in the solute velocities is the difference, $\Delta\mu = \mu_1 - \mu$ in their mobilities [21–23]. In addition to that, negligible decompression implies that v_{oR} in a L-long column can be found from Eq. (8.2) as $v_{oR} = \mu_R u_{oR} = \mu_R L/t_{M,R}$, where μ_R and $t_{M,R}$ are the solute mobility and static hold-up time measured at the conditions existing at the time t_R. As a result, Eq. (9.9) becomes

$$\Delta t_R = -\frac{t_{M,R}}{\mu_R}\int_0^1 \frac{\Delta\mu}{\mu}d\zeta, \quad |\Delta\mu| \ll \mu \tag{9.10}$$

where $\zeta = z/L$.

Being valid for any pressure program when the decompression is weak, Eq. (9.10) is valid when the pressure is fixed. In other words, it is valid for weak gas decompression in isobaric analysis. Being isobaric, this analysis can be translated (Section 8.5) into another isobaric analysis with arbitrary gas decompression. The translation does not affect μ_R and the ratio $\Delta t_R/t_{M,R}$ for any pair of peaks. Therefore, it does not affect the integral in Eq. (9.10). One can conclude that validity of Eq. (9.10) for arbitrary pressure program at weak gas decompression implies its validity for isothermal conditions with any gas decompression.

No particular model of solute–column interaction and no particular temperature program were used in deriving Eq. (9.10) or in establishing scalability of GC analyses in Section 8.5. This implies that

Statement 9.1

In isobaric temperature-programmed analysis with uniform heating of uniform column (the conditions that are always assumed in this book), Eq. (9.10) is valid for any model of solute–column interaction and for any temperature program. The requirement for isobaric conditions can be removed if gas decompression along the column is weak.

9.3 Isobaric Linear Heating Ramp and Highly Interactive Solutes

Previously we evaluated the difference, Δt_R, in retention times of two closely migrating solutes in an arbitrary dynamic analysis except, in the case of pressure programming, the gas decompression in the column had to be weak. The solutes considered here do not have to migrate close to each other. However, additional restrictions apply. Only the highly interactive solutes migrating during a linear heating ramp in an isobaric GC analysis are considered.

Although the linearized model of a solute–column interaction has already been chosen as a default, it is worth mentioning that, as long as a model of a column–solute interaction – be it a ideal thermodynamic, linearized, or any other model – is described by two parameters, Δt_R can be attributed only to the difference in those two parameters. The solute mobility, μ, in linearized and in ideal thermodynamic model can be expressed, Eqs. (8.13) and (8.12), via characteristic temperature, T_{char}, and characteristic thermal constant, θ_{char}. Therefore, regardless of the model, Δt_R is a function of one or both the differences,

$$\Delta T_{char} = T_{char,2} - T_{char,1} \tag{9.11}$$

$$\Delta \theta_{char} = \theta_{char,2} - \theta_{char,1} \tag{9.12}$$

Figure 9.1 shows that not always two different models – linearized or ideal thermodynamic – predict the same solute elution order at the same conditions. However, for the two different models, the difference in the time spacing of two peaks corresponding to the same pair of solutes seldom exceeds 0.1 min or about 1% of the entire analysis time shown in Figure 8.3. It means that, although the two different models fail to predict the same elution order, they fail to do so only for very close peaks. This observation appears to be in agreement with experimental data confirming that the ideal thermodynamic model [24–27] and several linearized models [2, 3, 28, 29] provide a good prediction of a general elution order of all peaks. It is not clear, however, which of the two models leads to a better prediction of elution order in most cases. A comparative study [30] seems to indicate an inferiority of both models compared to other models in the study in some cases. A more extensive comparative study that includes other models [31–36] can resolve the issue. This, however, together with the whole issue of prediction of a solute elution order is outside the scope of this book.

For the study of a column operation in this book, it is only important that, although the difference in the solute elution order predicted from different models appears to be rather chaotic, there is a high degree of consistency in one important aspect of the predictions – the sensitivity of the peak spacing to changes in heating rate. As will be shown shortly, although two different models might predict different spacing and even different elution order of two peaks corresponding to the same pair of solutes, both models predict similar change in the spacing

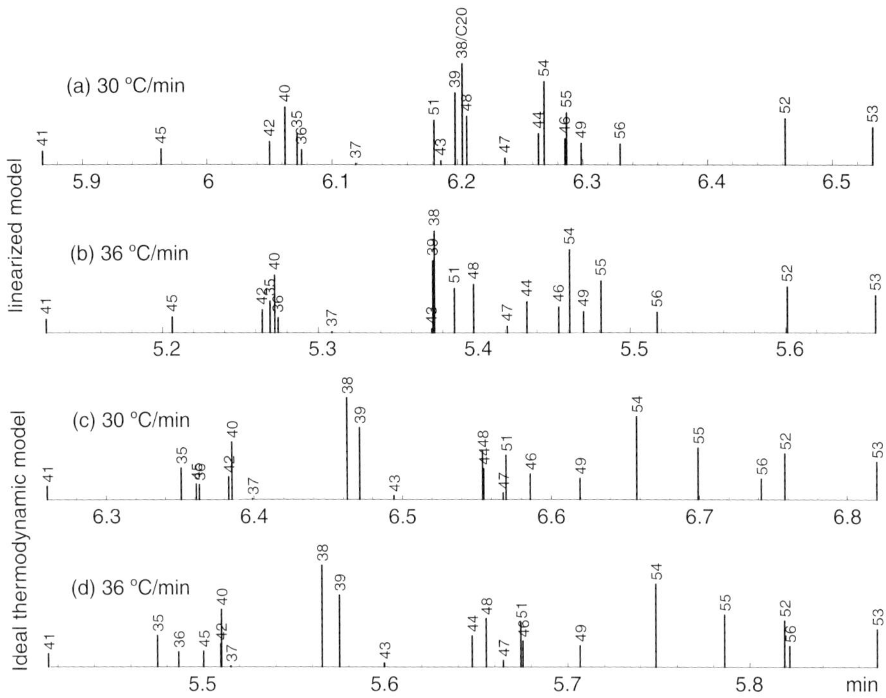

Figure 9.1 C_{20} region of line chromatograms (all peaks have zero widths) of a pesticide mixture in Table 5.5. Except for the heating rates of 36 °C/min in (b) and (d), all conditions are the same as in Figure 8.3.

resulting from the same change in the heating rate. This suggests that both models are likely to give a correct prediction of sensitivity of the peak spacing to changes in the heating rate in actual GC analyses. Evaluation of this sensitivity together with evaluation of the effect of the difference in T_{char} and θ_{char} on the peak spacing is the main focus of this chapter.

In the case of a linear heating ramp in a temperature-programmed isobaric analysis, a closed form solution for the elution interval, Δt_{R}, can be obtained. The solution becomes simpler if both solutes are highly interactive (highly retained at the beginning of the heating ramp) – the condition that is assumed through the rest of this chapter.

The retention parameters – $k_{\text{R,a}}$, $\mu_{\text{R,a}}$, $\omega_{\text{R,a}}$ and $T_{\text{R,a}}$ – of highly interactive solutes can be found from Eqs. (8.114)–(8.117). Since only the highly interactive solutes are considered in this chapter, they are denoted here as k_{R}, μ_{R}, ω_{R}, and T_{R}, respectively, that is

$$k_{\text{R}} = k_{\text{R},a}, \quad \mu_{\text{R}} = \mu_{\text{R},a}, \quad \omega_{\text{R}} = w_{\text{R},a}, \quad T_{\text{R}} = T_{\text{R,a}} \tag{9.13}$$

These simplifications simplify the forthcoming expressions.

Thermal spacing (ΔT_R) of two peaks can be found from Eqs. (8.14) and (8.6) as

$$\Delta T_R = \Delta T_{char} - (\theta_{char,2} \ln k_{R,2} - \theta_{char,1} \ln k_{R,1}) \tag{9.14}$$

The effect of ΔT_{char} on ΔT_R follows directly from Eq. (9.14) while the effect of $\Delta\theta_{char}$ on ΔT_R is not as direct. The structure of Eq. (9.14) suggests that it might be convenient to separately evaluate the following two cases:

$$k_{R,2} = k_{R,1} = k_R \tag{9.15}$$

$$T_{char,2} = T_{char,1} = T_{char} \tag{9.16}$$

In the following evaluations, the main attention is given to moderate heating rates, Eq. (8.128).

9.3.1
Solutes Eluting with Equal Mobilities

In a study of a solute migration, it is convenient to focus on the solute mobility, μ, rather than on its retention factor, k. Due to one-to-one relation, Eq. (8.6), of k and μ, Eq. (9.15) implies

$$\mu_{R,2} = \mu_{R,1} \tag{9.17}$$

and vice versa. Because Eq. (9.17) implies Eq. (9.15), it allows one to express Eq. (9.14) as

$$\Delta T_R = \Delta T_{char} - \Delta\theta_{char} \ln k_R \tag{9.18}$$

where $\Delta\theta_{char}$ is described in Eq. (9.12). Due to Eq. (8.115), one can further conclude that, in order for Eq. (9.17) to exist, dimensionless heating rates for both solutes have to be equal, that is,

$$r_{T2} = r_{T1} = r_T \tag{9.19}$$

That, in turn, implies, according to Appendix 9.A.2, that parameters ($T_{char,1}$, $\theta_{char,1}$) and ($T_{char,2}$, $\theta_{char,2}$) of the solutes relate as, Figure 9.2a,

$$\theta_{char,2}/\theta_{char,1} = (T_{char,2}/T_{char,1})^{\xi} \tag{9.20}$$

where parameter ξ is listed in Table 6.12.

Statement 9.2

If two highly interactive solutes eluting with equal mobilities have different characteristic temperatures, $T_{char,1}$ and $T_{char,2}$, then they also have different characteristic thermal constants, $\theta_{char,1}$ and $\theta_{char,2}$, as described in Eq. (9.20).

It also follows from Appendix 9.A.2 that Eq. (9.20) implies Eq. (9.19). Therefore, conditions described in Eqs. (9.19) and (9.20) are equivalent to each other.

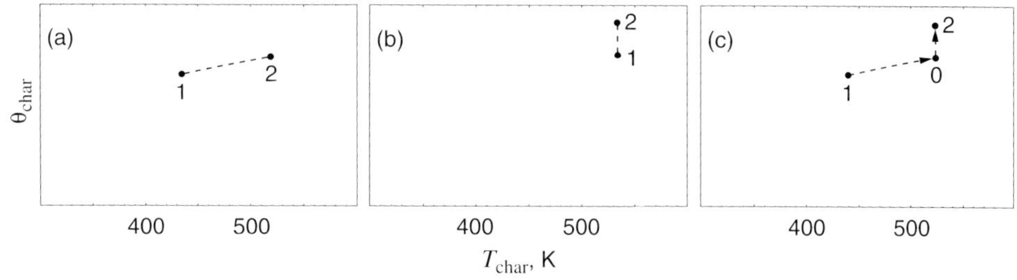

Figure 9.2 Location of points ($T_{char,1}$,$\theta_{char,1}$) and ($T_{char,2}$,$\theta_{char,2}$) for solutes "1" and "2." In (a) both solutes elute with equal mobilities. The dashed line connecting points "1" and "2" is described in Eq. (9.20). In (b), both solutes have the same T_{char}. An arbitrary location (c) of points "1" and "2" can be treated as a combinations of cases (a) – points "1" and "0", – and (b) – points "0" and "2." Point "0" is induced in the picture to facilitate the presentation of the case (c) as a combination of cases (a) and (b).

For moderate heating rates, Eq. (9.18) can be estimated as (Appendix 9.A.2)

$$\Delta T_R \approx \Delta T_{char} \tag{9.21}$$

Statement 9.3

Two highly interactive solutes eluting with equal mobilities elute in the order of increased characteristic temperatures. The difference, ΔT_R, in the elution temperatures of these solutes is close to the difference, ΔT_{char}, in their characteristic temperatures.

Equation (9.21) can also be expressed in a dimensionless form. It has been shown in Chapter 8 that elution mobilities of all highly interactive solutes approach the same asymptotic level. Therefore, these eluites fall in the category described in Eq. (9.17). Let $\Theta_R = T_R/\theta_{char}$ and $\Theta_{char} = T_{char}/\theta_{char}$ be dimensionless elution temperature and dimensionless characteristic temperature, Eq. (8.65), of a highly interactive solute. It follows from Eq. (8.117) that

$$\Delta\Theta_R = \Delta\Theta_{char} \tag{9.22}$$

where

$$\Delta\Theta_R = \Theta_{R,2} - \Theta_{R,1}, \quad \Delta\Theta_{char} = \Theta_{char,2} - \Theta_{char,1} \tag{9.23}$$

9.3.2
Solutes with Equal Characteristic Temperatures

When Eq. (9.16) (Figure 9.2b) rather than Eq. (9.17) takes place, Eqs. (9.5) and (8.117) yield

$$\Delta T_R = \theta_{char,2} \ln(e^{r_{T2}} - 1) - \theta_{char,1} \ln(e^{r_{T1}} - 1) \tag{9.24}$$

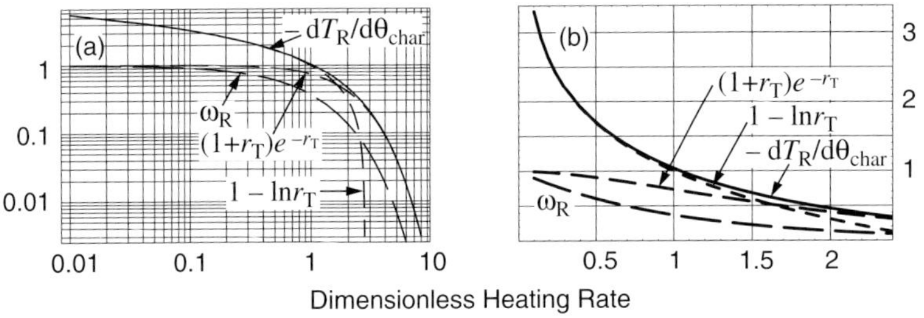

Figure 9.3 Rate, $dT_R/d\theta_{char}$, Eq. (9.25), of change in ΔT_R with change in $\Delta\theta_{char}$; its approximations, Eqs (9.28) and (9.29); and immobility (ω_R, Eq. (8.114)) of a solute eluting during a heating ramp with dimensionless rate r_T. Graphs in (a) provide a broad overview, while those in (b) zoom in the region with moderate heating rates.

The sensitivity, $dT_R/d\theta_{char}$ (Figure 9.3), of ΔT_R to change in the difference $\Delta\theta_{char}$ can be expressed (Appendix 9.A.3) as

$$\frac{dT_R}{d\theta_{char}} = \lim_{\Delta\theta_{char}\to 0}\left(\frac{\Delta T_R}{\Delta\theta_{char}}\right) = \ln(e^{r_T}-1) - \frac{r_T}{1-e^{-r_T}} \tag{9.25}$$

where r_T that can be defined by either of the following two equivalent relations

$$r_T = \frac{r_{T2}\theta_{char,2}}{\theta_{char}} = \frac{r_{T1}\theta_{char,1}}{\theta_{char}} \tag{9.26}$$

is the dimensionless heating rate corresponding to the average characteristic thermal constant,

$$\theta_{char} = (\theta_{char,1} + \theta_{char,2})/2 \tag{9.27}$$

of the two solutes.

According to Eq. (9.25) and Figure 9.3, the rate, $dT_R/d\theta_{char}$, is a negative quantity. For a single solute with a given characteristic temperature, T_{char}, it means that the higher is the solute characteristic thermal constant, θ_{char}, the lower is its elution temperature, T_R. Of course, this is in agreement with the conclusions that follow from previously established relations. Indeed, according to Eq. (8.95), higher θ_{char} means lower dimensionless heating rate, r_T. That, according to Eq. (8.117), implies lower T_R. Also, according to Eq. (8.115), lower r_T implies lower elution mobility, μ, and, therefore, lower elution temperature, Eqs. (8.13) and (8.12).

The magnitude, $|dT_R/d\theta_{char}|$, of quantity $dT_R/d\theta_{char}$ increases when r_T declines. It is important to recognize, however, that even the strongest dependence of $dT_R/d\theta_{char}$ on r_T is not very strong: a three order of magnitude change in r_T, from $r_T = 0.001$ to $r_T = 1$, causes less than an order of magnitude change in $dT_R/d\theta_{char}$ from about -8 to about -1.

The following two approximations simplify the formula for $dT_R/d\theta_{char}$.

When r_T approaches zero, $dT_R/d\theta_{char}$ approaches $\ln r_T - 1$. This remains a good approximation for $dT_R/d\theta_{char}$, Figure 9.3, as long as $r_T < 1$, that is,

$$\frac{dT_R}{d\theta_{char}} \approx \ln r_T - 1 \quad \text{when} \quad r_T < 1 \tag{9.28}$$

When the heating is fast ($r_T > 3$), $|dT_R/d\theta_{char}|$ is small and rapidly (almost exponentially) declines with the increase in r_T. When r_T approaches infinity, $dT_R/d\theta_{char}$ approaches $-(1 + r_T)e^{-r_T}$. This remains a good approximation for $dT_R/d\theta_{char}$, Figure 9.3, as long as $r_T > 2$. In this approximation, quantity e^{-r_T} is the immobility (ω_R, Eq. (8.114)) of a solute at the time of its elution. One can write

$$\frac{dT_R}{d\theta_{char}} \approx -(1 + r_T)e^{-r_T} = -(1 + r_T)\omega_R, \quad \text{when} \quad r_T > 2 \tag{9.29}$$

The graphs of ω_R in Figure 9.3 highlight the fact that the rapid decline in $|dT_R/d\theta_{char}|$ due to the fast heating goes hand-in-hand with the rapid decline in the engagement (ω_R) of a solute with a column stationary phase. The lower is the engagement (the solute–column interaction) at the time of the solute elution, the shorter is the initial fraction of the column where the solute interacts with the stationary phase, and the weaker are the effects of θ_{char} on T_R of a given solute and of $\Delta\theta_{char}$ on ΔT_R, in two different solutes.

A transitional region between the slow and the fast heating is shown in Figure 9.3b. Its portion around $r_T = 0.5$ is close to optimal heating rates (Example 8.13). This also means that, for practical applications, Eq. (9.28) provides suitable approximation to Eq. (9.25).

While $dT_R/d\theta_{char}$ is generally somewhat complicated function of r_T, quantity ΔT_R for a fixed r_T is a nearly linear function (Appendix 9.A.3) of $\Delta\theta_{char}$ for all practically feasible heating rates, and for all experimentally known (Figure 5.6, Eq. (5.48)) departures of θ_{char} from their averages at a fixed T_{char}. As shown in Appendix 9.A.3,

$$\Delta T_R = \frac{dT_R}{d\theta_{char}} \times \Delta\theta_{char} = \left(\ln(e^{r_T} - 1) - \frac{r_T}{1 - e^{-r_T}}\right)\Delta\theta_{char}, \quad \text{when } T_{char,2} = T_{char,1} \tag{9.30}$$

This, in view of Eq. (5.48), can be estimated as

$$|\Delta T_R| \le 0.3\left|\ln(e^{r_T} - 1) - \frac{r_T}{1 - e^{-r_T}}\right|\theta_{char}, \quad \text{when } T_{char,2} = T_{char,1} \tag{9.31}$$

Thus, for the benchmark heating rate ($r_{T0} = 0.7$, Eq. (8.125)), one has

$$|\Delta T_R| \le 0.4\theta_{char}, \quad \text{when} \quad T_{char,2} = T_{char,1} \tag{9.32}$$

where θ_{char} rarely exceeds 40 °C, Figure 5.6, for a typical stationary phase film thickness, and 50 °C, Figure 5.8, for thicker films. This means that

Statement 9.4

In a typical temperature-programmed GC analysis, the spacing (ΔT_R) of elution temperatures of two solutes having equal characteristic temperatures, T_{char}, can rarely be significantly larger than 10 °C.

Therefore,

Statement 9.5

In a typical temperature-programmed GC analysis, large thermal spacing of two solutes can only result from large difference in their characteristic temperatures as described in Eq. (9.21).

It is also useful to notice that because $\theta_{char,1}$ and $\theta_{char,2}$ corresponding to the same T_{char} are not greatly different from each other (Figure 5.6), one might not need to pay significant attention to the fact that, as defined in Eq. (9.26), r_T corresponds to the average θ_{char}. Typically, it is acceptable to assume that, outside the calculation of $\Delta\theta_{char}$ in Eq. (9.12), the quantities $\theta_{char,1}$ and $\theta_{char,2}$ are roughly equal to each other. As a result, one might treat r_T as a quantity described in Eq. (9.19).

9.3.3 Reversal of Elution Order

Change in the sign of elution interval, ΔT_R, of two solutes due to the change in operational conditions of GC analysis indicates reversal of the solute elution order [2, 3, 28, 29, 32, 37–45].

In two previously considered cases described in Eqs. (9.15) and (9.16), two solutes had equal characteristic temperatures or their characteristic temperatures related to characteristic thermal constants as described in Eq. (9.20). In both these cases, elution order of two solutes was a function of their characteristic parameters. The order did not depend on operational conditions such as dimensional heating rate (r_T). The latter had significant effect on the size of the peak spacing, but not on its sign. As a result, r_T did not affect the elution order.

Reversal of elution order due to change in r_T is possible only when the order is caused by a combination of two factors partially compensating each other. A change in r_T can cause a change in the balance between conflicting components of the peak spacing and can cause the change in the peak elution order.

The case where elution order of two solutes, say, "1" and "2," can depend on the heating rate can be treated as a combination, Figure 9.2c, of the two special cases described in Eqs. (9.15) and (9.16). The net difference (ΔT_R) in the solute elution temperatures can be found as (Appendix 9.A.4)

$$\Delta T_R \approx \Delta T_{char} + \frac{dT_R}{d\theta_{char}} \times \Delta\theta_{char} = \Delta T_{char} + \left(\ln(e^{r_T}-1) - \frac{r_T}{1-e^{-r_T}}\right)\Delta\theta_{char} \tag{9.33}$$

Because the derivative, $dT_R/d\theta_{char}$, in this formula is a negative quantity (Figure 9.3), a change in a heating rate can cause the reversal of elution order of two solutes only when both ΔT_{char} and $\Delta\theta_{char}$ have the same sign. Furthermore, the reversal can only take place when $\Delta\theta_{char}$ is comparable with ΔT_{char}.

Figure 5.6 shows that $|\Delta\theta_{char}|$ at a fixed T_{char} seldom exceeds 10 °C, while the range of T_{char} values can cover hundreds of degree celsius. As a result, the solutes in a mixture covering a wide range of characteristic temperatures generally elute in the order of increase in their T_{char} (Figure 8.3), and only the elution order of closely eluting pairs can change with change in the heating rate (Figure 9.1). It follows from Eqs. (9.33) and (5.48) that reversible range of ΔT_{char} values can be estimated as

$$|\Delta T_{char}| \le 0.3\left(\frac{r_T}{1-e^{-r_T}} - \ln(e^{r_T}-1)\right)\theta_{char} \tag{9.34}$$

For the benchmark heating rate ($r_{T0} = 0.7$, Eq. (8.125)), this becomes

$$|\Delta T_{char}| \le 0.4\theta_{char} \tag{9.35}$$

which, due to Eq. (5.46), can be estimated as

$$\left|\frac{\Delta T_{char}}{T_{char}}\right| \le 0.4 \times \frac{22\,°C}{T_{st}} \times (10^3\varphi)^{0.09} \times \left(\frac{T_{char}}{T_{st}}\right)^{-0.3} \approx 0.032 \times (10^3\varphi)^{0.09} \times \left(\frac{T_{char}}{T_{st}}\right)^{-0.3} \tag{9.36}$$

The functions $(10^3\varphi)^{0.09}$ and $(T_{char}/T_{st})^{0.3}$ typically do not significantly depart from unity. As a result, one can conclude that

Statement 9.6

Typically, the change in a heating rate can change the elution order of two solutes only if their characteristic temperatures, T_{char}, differ by no more than a few percentage points.

9.3.4
Sensitivity of the Solute Elution Order to Heating Rate

Equation (9.33) is suitable for evaluation of the differences between solute elution temperatures, and, therefore, for prediction of a solute elution order, as long as linearized model accurately describes migration of each solute. Figure 9.1 sheds some light on this subject. It shows that both the characteristic and the ideal thermodynamic model of a solute–column interaction predict the same elution order in many – probably in most – cases. However, the predicted elution order is not always the same. This means that one or both models and, therefore, Eq. (9.33) might be unsuitable for reliable prediction of the solute elution order.

At the first glance, it appears that Figure 9.1 shows chaotic difference in temperature intervals of closely eluting solutes. This might be interpreted as an implication that one or both models carry a little information regarding the elution order. Fortunately, this is not the case. While predicting different elution temperature intervals in some cases, both models predict similar trends in changes of those intervals due to changes in conditions of GC analyses. This means that both models have sufficient information for the prediction of changes in the elution order once that order is known for a specific set of conditions. This information is useful for a column optimization, and it is worthwhile to try to extract that information from Eq. (9.33).

The sensitivity, γ_r, of ΔT_R to a relative change, dR_T/R_T, in the heating rate can be defined as the rate of change in ΔT_R for a given pair of solutes with a relative change, dR_T/R_T, in R_T. This means that the sensitivity, γ_r, is a factor in the formula

$$d(\Delta T_R) = \gamma_r \times \frac{dR_T}{R_T} \tag{9.37}$$

and can be defined as

$$\gamma_r = \frac{d(\Delta T_R)}{dR_T/R_T} \tag{9.38}$$

According to Eq. (8.95), $dR_T/R_T = dr_T/r_T$. This, due to Eq. (9.33), allows one to express γ_r as, Figure 9.4,

$$\gamma_r = r_T \times \frac{d(\Delta T_R)}{dr_T} = \frac{r_T^2 e^{r_T}}{(e^{r_T}-1)^2}\Delta\theta_{char} \tag{9.39}$$

For moderate heating rates, Eq. (8.128), γ_r can be approximated as

$$\gamma_r \approx \Delta\theta_{char} \tag{9.40}$$

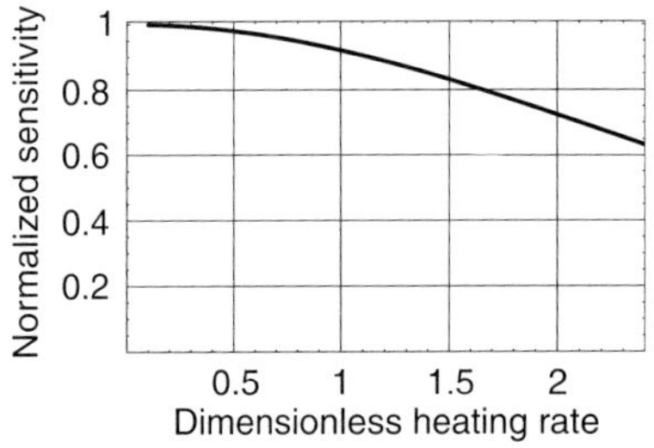

Figure 9.4 Normalized sensitivity, $\gamma_r/\Delta\theta_{char}$, Eq. (9.39), of thermal spacing (ΔT_R) of two solutes to relative change (dR_T/R_T) in the heating rate (R_T) as a function of dimensionless heating rate (r_T). Quantity $\Delta\theta_{char}$ is the difference in characteristic thermal constants of the solutes.

and, therefore,

$$d(\Delta T_R) \approx \Delta\theta_{char} \times \frac{dR_T}{R_T} \tag{9.41}$$

This means that each 1% change in R_T changes ΔT_R for a given solute pair by about $\Delta\theta_{char}/100$ where $\Delta\theta_{char}$ is the difference in the solute characteristic thermal constants.

Example 9.1

If $\Delta\theta_{char} = 10$ K (almost the worst experimentally known difference, Figure 5.6), then 1% increase in R_T causes 0.1 K increase in ΔT_R. According to Eq. (9.30) and Figure 9.3, the sign of ΔT_R is opposite to the sign of $\Delta\theta_{char}$. Therefore, $\Delta T_R < 0$, and the increase in ΔT_R implies reduction in its magnitude $|\Delta T_R|$. □

Due to Eq. (5.48), the magnitude, $|d(\Delta T_R)|$, of $d(\Delta T_R)$ in Eq. (9.41) can be estimated as

$$|d(\Delta T_R)| \leq 0.3\theta_{char}|dR_T/R_T| \tag{9.42}$$

In some cases, it is convenient to deal with dimensionless thermal spacing. Let

$$\Theta_R = \frac{T_R}{\theta_{char}} \tag{9.43}$$

be the dimensionless elution temperature of a solute. A dimensionless thermal spacing, $\Delta\Theta_R$, of two solutes becomes

$$\Delta\Theta_R = \Theta_{R2} - \Theta_{R1} = \frac{T_{R2} - T_{R1}}{\theta_{char}} = \frac{\Delta T_R}{\theta_{char}} \tag{9.44}$$

Due to Eq. (9.42), the magnitude $|d(\Delta\Theta_R)|$ of a change in $\Delta\Theta_R$ due to the relative change, dR_T/R_T, in R_T can be estimated as

$$|d(\Delta\Theta_R)| \leq 0.3|dR_T/R_T| \tag{9.45}$$

This means that the change $d(\Delta\Theta_R)$ in $\Delta\Theta_R$ caused by 1% change in R_T does not typically exceed 0.003.

Example 9.2

Let $\theta_{char} = 30$ K. According to Eq. (9.45), the magnitude, $|d(\Delta\Theta_R)|$, of a change, $d(\Delta\Theta_R)$, in $\Delta\Theta_R$ corresponding to 1% change in R_T does not exceed 0.003. Therefore, $|d(\Delta T_R)| \leq 0.003 \times 30\text{ K} = 0.09$ K. (Compare this with Example 9.1.) □

Generally, the temperature intervals of different solute pairs can have different sensitivity to the heating rate. The sensitivities of the intervals for the consecutive (in order of increase in T_{char}) solute pairs in Figure 9.1 are shown in Figure 9.5. For the conditions in Figure 9.1, the dimensionless heating rates are about 0.4 (at $R_T = 30\,°C$) and 0.5 (at $R_T = 36\,°C$). At these rates, the normalized sensitivity,

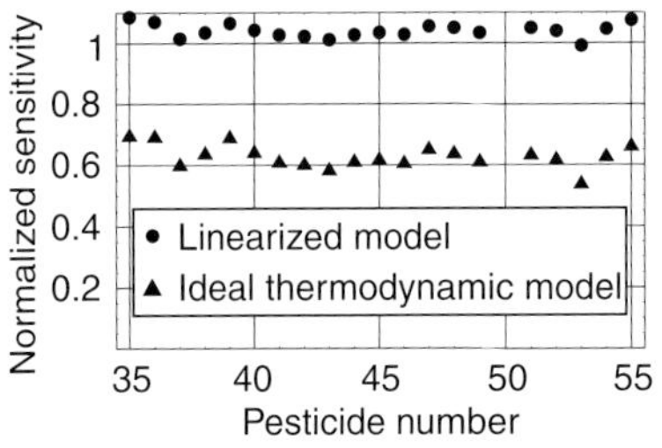

Figure 9.5 Normalized sensitivity ($\gamma_r/\Delta\theta_{char}$, Eqs. (9.37)) of thermal spacing (ΔT_R) of consecutive (in the order of increase in T_{char}) pairs of pesticides in Figure 9.1 to changes in heating rate. To calculate γ_r/θ_{char} corresponding to the solute number i, values $\Delta T_{R,i} = T_{R,i} - T_{R,i-1}$, $\Delta\theta_{char,i} = \theta_{char,i} - \theta_{char,i-1}$, $\Delta\Delta T_{R,i} = \Delta T_{R,i,36} - \Delta T_{R,i,30}$, $\Delta R_{T,i} = R_{T,i,36} - R_{T,i,30}$, $R_{T,i} = (R_{T,i,30} + R_{T,i,36})/2$ for each solute were found from chromatograms in Figure 9.1. The subscripts 30 and 36 indicate heating rates of 30 and 36 °C/min, respectively. According to Eq. (9.37), quantity $\gamma_{r,i}/\Delta\theta_{char,i}$ for each solute is calculated as $\gamma_{r,i}/\Delta\theta_{char,i} = R_T\Delta\Delta T_{R,i}/(\Delta\theta_{char,i}\Delta R_{T,i})$. With $t_M = 0.32$ min at 300 K, $r_{T,30}$ and $r_{T,36}$ are, Eq. (8.98), about 0.4 and 0.5, respectively, suggesting that, shown here simulation results for linearized model are close to predictions by Eq. (9.39), Figure 9.4.

$\gamma_r/\Delta\theta_{char}$, Eqs. (9.37) and (9.39), of ΔT_R to changes in R_T should be close to unity, Figure 9.4. Figure 9.5 confirms this prediction for the linearized model. For ideal thermodynamic model, the sensitivity appears to be about 30% lower, Figure 9.5. Although the difference is significant, it is not decisive. Both results are in the same ballpark. It is also interesting that, as functions of a solute-pair number (i.e., of a combination of characteristic parameters, T_{char} and θ_{char}, of a solute pair) the sensitivities in both model are very similar in all details. Similar results were obtained from testing other data sets.

These observations allow one to conclude that Eq. (9.40) can be treated as an estimate for γ_r, and the formulae that follow from it can be used for the evaluation of the effects of the heating rate on the solute elution order.

9.4
Properties of Generic Solutes

It follows from Eq. (8.15) that generic solutes satisfy Eq. (9.20). Therefore, all highly interactive generic solutes eluting during the same heating ramp elute with equal mobilities and have all properties of the solutes eluting with equal mobilities. Particularly, Eq. (9.21) is valid for highly interactive generic solutes.

Equation (9.21) might not be valid for the generic solutes (and for all other solutes satisfying Eq. (9.20)) if these solutes are not highly interactive during a given heating ramp (not highly retained at the beginning of the ramp). These solutes might also elute with different mobilities. However, as shown in Appendix 9.A.1, the level of the solute–column interaction at the beginning of a heating ramp does not affect the solute elution order. In other words,

Statement 9.7

No change in initial temperature or in heating rate can reverse elution order of two generic solutes no matter how close are their retention times.

In fact, there is hardly any change in conditions of GC analysis that can reverse the elution order of the generic solutes.

Appendix 9.A Elution Temperature Interval

9.A.1 Useful Approximations

Differences of the type $y_2^x - y_1^x$, where x, y_1, and y_2 are positive numbers with the properties $1 \leq y_2/y_1 \leq 2$ and $0 \leq x \leq 2$ appear in expressions for peak spacing. There are several ways to approximate $y_2^x - y_1^x$.

The expression $y_2^x - y_1^x$ can be modified as $y_2^x - y_1^x = (y^x - 1)y_1^x$ where $y = y_2/y_1$ and, therefore, $1 \leq y \leq 2$. This indicates that approximations to $y_2^x - y_1^x$ can be expressed via approximations to $y^x - 1$.

A simple approximation to $y^x - 1$ by its tangent, $x \cdot (y - 1)$, at $y = 1$ is (Figure 9.A.1)

$$y^x - 1 \approx x \times (y - 1) \tag{A1}$$

This approximation is not very accurate when y is significantly different from one. More accurate are approximations (Figure 9.A.1)

$$y^x - 1 \approx \frac{x \times (y-1)}{(y^{1-x} + 1)/2} \tag{A2}$$

$$y^x - 1 \approx \frac{x \times (y-1)}{((y+1)/2)^{1-x}} \tag{A3}$$

The last two approximations might not look simpler than the original formula $y^x - 1$. However, they lead to simple interpretations. All approximations emphasize the key role of the difference $y - 1$ in the value of $y^x - 1$.

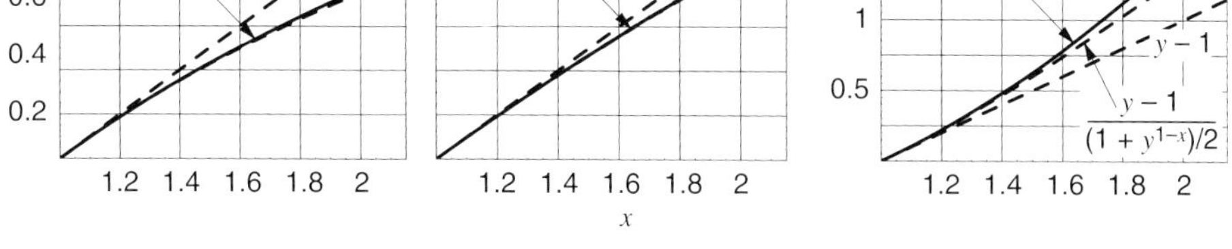

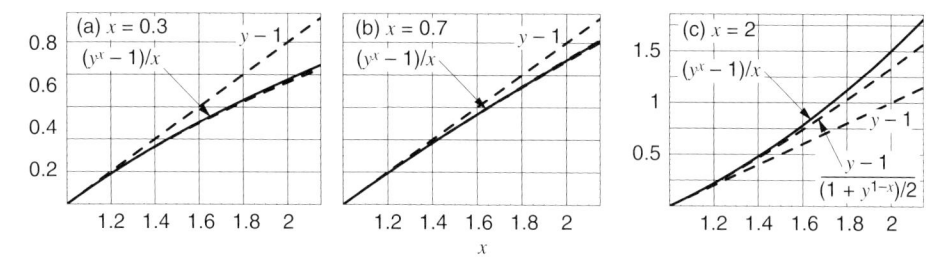

Figure 9.A.1 Approximations (dashed lines) to expression $y^x - 1$ (solid lines) for several values of x. In (a) and (b), approximations by Eqs. (A2) and (A3) are almost indistinguishable from $y^x - 1$. The same is true for Eq. (A3) in (c).

Example 9.3

According to Eq. (8.15), for generic solutes,

$$\begin{aligned}\frac{T_{\text{char},2}}{\theta_{\text{char},2}} - \frac{T_{\text{char},1}}{\theta_{\text{char},1}} &= \frac{T_{\text{st}}}{\theta_{\text{char,st}}}\left(\left(\frac{T_{\text{char},2}}{T_{\text{st}}}\right)^{1-\xi} - \left(\frac{T_{\text{char},1}}{T_{\text{st}}}\right)^{1-\xi}\right) \\ &= \frac{T_{\text{st}}}{\theta_{\text{char,st}}}\left(\frac{T_{\text{char},1}}{T_{\text{st}}}\right)^{1-\xi}\left(\left(\frac{T_{\text{char},2}}{T_{\text{char},1}}\right)^{1-\xi} - 1\right)\end{aligned} \tag{A4}$$

After approximation by Eq. (A1), this becomes

$$\frac{T_{\text{char},2}}{\theta_{\text{char},2}} - \frac{T_{\text{char},1}}{\theta_{\text{char},1}} \approx \frac{(1-\xi)(T_{\text{char},2} - T_{\text{char},1})}{\theta_{\text{char,st}}(T_{\text{char},1}/T_{\text{st}})^{\xi}} \approx \frac{(1-\xi)(T_{\text{char},2} - T_{\text{char},1})}{\theta_{\text{char},1}}$$

$$\begin{aligned}\frac{T_{\text{char},2}}{\theta_{\text{char},2}} - \frac{T_{\text{char},1}}{\theta_{\text{char},1}} &\approx \frac{(1-\xi)\cdot(T_{\text{char},2} - T_{\text{char},1})}{\theta_{\text{char,st}}\left((T_{\text{char},2}/T_{\text{st}})^{\xi} + (T_{\text{char},1}/T_{\text{st}})^{\xi}\right)/2} \\ &= \frac{(1-\xi)\cdot(T_{\text{char},2} - T_{\text{char},1})}{\bar{\theta}_{\text{char}}}\end{aligned}$$

where $\bar{\theta}_{\text{char}} = (\theta_{\text{char},1} + \theta_{\text{char},1})/2$. Equation (A3) is an indication that $\bar{\theta}_{\text{char}}$ can be also found as $\bar{\theta}_{\text{char}} \approx (\bar{T}_{\text{char}}/T_{\text{st}})^{\xi}$ where $\bar{T}_{\text{char}} = (T_{\text{char},1} + T_{\text{char},2})/2$. □

9.A.2 Equal Elution Mobilities

It follows from Eqs. (8.95), (7.96), and (8.47) that r_{T2} can be expressed as

$$r_{\text{T2}} = \frac{R_{\text{T2}} t_{\text{M,char},2}}{\theta_{\text{char},2}} = \left(\frac{T_{\text{char},2}}{T_{\text{char},1}}\right)^{\xi} \frac{\theta_{\text{char},1}}{\theta_{\text{char},2}} \cdot \frac{R_{\text{T2}} t_{\text{M,char},1}}{\theta_{\text{char},1}} = \left(\frac{T_{\text{char},2}}{T_{\text{char},1}}\right)^{\xi} \frac{\theta_{\text{char},1}}{\theta_{\text{char},2}} r_{\text{T1}} \tag{A5}$$

where R_{T} is the absolute heating rate and ξ is a gas parameter listed in Table 6.12. Equation (A5) implies that expression

$$\frac{\theta_{\text{char},2}}{\theta_{\text{char},1}} = (T_{\text{char},2}/T_{\text{char},1})^{\xi} \tag{A6}$$

follows from Eq. (9.19) and vice versa.

Furthermore, Eqs. (A5) and (9.19) imply Eq. (9.20) which allows one to express Eq. (9.12) as

$$\Delta\theta_{\text{char}} = ((T_{\text{char},2}/T_{\text{char},1})^{\xi} - 1)\theta_{\text{char},1} \tag{A7}$$

In practical mixtures analyzed by GC, the ratio $T_{\text{char},2}/T_{\text{char},1}$ is not much larger than two, and $\xi \approx 0.7$ (Table 6.12). This allows one to approximate the last formula as (Appendix 9.A.1)

$$\Delta\theta_{\text{char}} \approx (T_{\text{char},2}/T_{\text{char},1} - 1)\xi\theta_{\text{char},1} = \xi\Delta T_{\text{char}}/\Theta_{\text{char},1} \tag{A8}$$

After the substitution in Eq. (9.18), this yields

$$\Delta T_R \approx \Delta T_{char}(1-\xi \ln k_R/\Theta_{char,1}) \tag{A9}$$

For moderate values, Eq. (8.128), of r_T, Eq. (8.116) yields $\ln k_R < 1.6$. This together with the estimates for Θ_{char} (Eq. (8.72)) and the data for ξ (Table 6.12) allow one to approximate Eq. (A9) as

$$\Delta T_R \approx \Delta T_{char} \tag{A10}$$

The error induced by this approximation is equal to $\xi \ln k_R \Delta T_{char}/\Theta_{char,1}$. In view of Eq. (8.116), the error can be also expressed as

$$-\xi\Delta T_{char}\ln(e^{r_T}-1)/\Theta_{char,1} \tag{A11}$$

which, for the moderate heating rates, Eq. (8.128), is a small fraction of $\Delta T_{char.}$

Because Eq. (8.116) was used in its derivation, the approximation in Eq. (A10) is valid only for the highly interactive solutes. It is important, however, that the sign of ΔT_R is the same as the sign of ΔT_{char} for all solute pairs with characteristic parameters satisfying Eq. (A6).

Indeed, as mentioned earlier, Eq. (A6) implies Eq. (9.19) which further implies Eq. (9.15) allowing to reduce Eq. (9.14) to Eq. (9.18). Combining Eqs. (9.18) and (A6), one has

$$\Delta T_R = \theta_{char,1}\left(\frac{T_{char,1}}{\theta_{char,1}}\left(\frac{T_{char,2}}{T_{char,1}}-1\right)-\left(\left(\frac{T_{char,2}}{T_{char,1}}\right)^{\xi}-1\right)\ln k\right)$$

Suppose that $T_{char,2} > T_{char,1}$. Because ξ is a positive number smaller than one, Table 6.12, the only way for ΔT_R to be a negative number is for $\ln k$ to be significantly larger than $T_{char,1}/\theta_{char,1}$, which is not realistic because $T_{char,1}/\theta_{char,1} > 10$ as follows from the data in Chapter 5.

9.A.3 Equal Characteristic Temperatures

Equation (9.16) allows one to reduce Eq. (A5) to the form

$$R_{T2}\theta_{char,2} = r_{T1}\theta_{char,1} \tag{A12}$$

Introduction of the average (θ_{char}, Eq. (9.27)) of $\theta_{char,1}$ and $\theta_{char,2}$ allows one to define dimensionless heating rate (r_T) corresponding to θ_{char} as

$$r_T\theta_{char} = r_{T2}\theta_{char,2} = r_{T1}\theta_{char,1} \tag{A13}$$

Due to Eqs. (9.27) and (9.12), quantities $\theta_{char,1}$ and $\theta_{char,2}$ can be found as

$$\theta_{char,1} = \theta_{char}-\Delta\theta_{char}/2, \quad \theta_{char,2} = \theta_{char}+\Delta\theta_{char}/2 \tag{A14}$$

The latter together with Eq. (A13) allows one to express Eq. (9.24) as

$$\Delta T_R = \left(\frac{\Delta\theta_{char}}{2} - \theta_{char}\right) \times \ln\left(\exp\left(\frac{-2r_T\theta_{char}}{2\theta_{char} - \Delta\theta_{char}}\right) - 1\right) - \\ - \left(\theta_{char} + \frac{\Delta\theta_{char}}{2}\right) \times \ln\left(\exp\left(\frac{2r_T\theta_{char}}{2\theta_{char} + \Delta\theta_{char}}\right) - 1\right) \quad (A15)$$

The limit of $\Delta T_R/\Delta\theta_{char}$ at $\Delta\theta_{char} \to 0$ can be described as in Eq. (9.25) of the main text.

It is instructive to evaluate dimensionless thermal spacing, $\Delta\Theta_R = \Delta T_R/\theta_{char}$, of two peaks representing two solutes having equal T_{char}. According to Eq. (A15), $\Delta\Theta_R$ depends on r_T and on dimensionless difference, $\Delta\theta_{char}/\theta_{char}$, in characteristic thermal constants of two solutes having the same T_{char}. Typically, $0.1 \le r_T \le 2$. A significantly wider range $0.01 \le r_T \le 4$ of r_T values can be viewed as a range of all practically feasible heating rates. Experimentally known values of $\Delta\theta_{char}/\theta_{char}$ can be observed in Figure 5.6. They are summarized in Eq. (5.48). Figure 9.A.2 shows that for all these values of r_T and $\Delta\theta_{char}/\theta_{char}$, quantity $\Delta\Theta_R$ is proportional to $\Delta\theta_{char}/\theta_{char}$. This implies that Eq. (9.30) of the main text is a good approximation to ΔT_R for all practical values of r_T and $\Delta\theta_{char}$.

9.A.4 Two Arbitrary Solutes

Equation (9.14) can be modified as

$$\Delta T_R = \Delta T_{R,\mu} + \Delta T_{R,T} \quad (A16)$$

where

$$\Delta T_{R,\mu} = \Delta T_{char} - (\theta_{char,0} - \theta_{char,1})\ln k_{R,1} \quad (A17)$$

$$\Delta T_{R,T} = \theta_{char,0}\ln k_{R,0} - \theta_{char,2}\ln k_{R,2} \quad (A18)$$

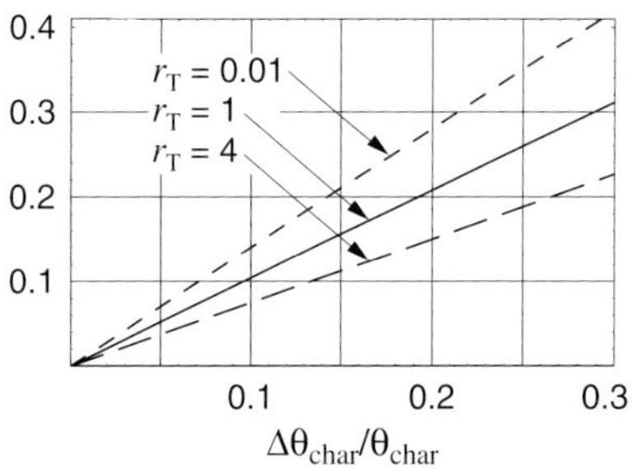

Figure 9.A.2 Dimensionless thermal spacing, $\Delta\Theta_R = \Delta T_R/\theta_{char}$, of two peaks having the same T_{char} vs. dimensionless difference, $\Delta\theta_{char}/\theta_{char}$, in their characteristic thermal constants. Quantity r_T is dimensionless heating rate. Quantity ΔT_R was obtained from Eq. (A15). Typically $|\Delta\theta_{char}/\theta_{char}| < 0.3$ (Eq. (5)). Solid line: $\Delta\Theta_R$, short dashes: $\Delta\Theta_R/4$, long dashes: $8\Delta\Theta_R$.

This is equivalent to bringing an addition solute – solute "0" (Figure 9.2c) – in the picture. The elution mobility, $\mu_{R,0}$, of solute "0" is the same as that of solute "1," and the characteristic temperature, $T_{char,0}$, of solute "0" is the same as that of solute "2," that is,

$$\mu_{R,0} = \mu_{R,1}, \quad T_{char,0} = T_{char,2} \tag{A19}$$

Because $\mu_{R,0} = \mu_{R,1}$ implies $k_{R,0} = k_{R,1}$, Eq. (A17) represents the same case as does Eq. (9.18). Therefore, $\Delta T_{R,\mu}$ in Eq. (A17) can be expressed in the same way as ΔT_R is expressed in Eq. (9.21). Similarly, $\Delta T_{R,T}$ in Eq. (A18) can be expressed like ΔT_R in Eq. (9.30). Equation (A16) becomes

$$\Delta T_R \approx \Delta T_{char} + \frac{dT_R}{d\theta_{char}} \times \Delta\theta_{char.2} \tag{A20}$$

where (Figure 9.2c)

$$\Delta\theta_{char.2} = \theta_{char.2} - \theta_{char.0} = \Delta\theta_{char} - ((T_{char.}/T_{char,1})^{\xi} - 1)\theta_{char.1} \tag{A21}$$

and $\Delta\theta_{char}$ is the difference in $\theta_{char,2}$ and $\theta_{char,1}$ (Eq. (9.12)). The last formula can be approximated as (Appendix 9.A.1)

$$\Delta\theta_{char.2} \approx \Delta\theta_{char} - (T_{char.2}/T_{char.1} - 1)\xi\theta_{char.1} = \Delta\theta_{char} - \xi\Delta T_{char}/\Theta_{char.1} \tag{A22}$$

After substitution of this expression in Eq. (A20), one has

$$\Delta T_R \approx \left(1 - \frac{dT_R}{d\theta_{char}} \times \frac{\xi}{\Theta_{char,1}}\right) \times \Delta T_{char} + \frac{dT_R}{d\theta_{char}} \times \Delta\theta_{char} \tag{A23}$$

Let us approximate this formula as

$$\Delta T_R \approx \Delta T_{char} + \frac{dT_R}{d\theta_{char}} \times \Delta\theta_{char} \tag{A24}$$

The approximation error will be evaluated shortly. Accounting for Eq. (9.25), Eq. (A24) can be expressed as shown in Eq. (9.33).

The error of transition from Eqs. (A23) to (A24) can be expressed as

$$\frac{dT_R}{d\theta_{char}} \times \frac{\xi}{\Theta_{char,1}} \times \Delta T_{char} = \left(\ln(e^{r_T} - 1) - \frac{r_T}{1 - e^{-r_T}}\right) \times \frac{\xi}{\Theta_{char,1}} \times \Delta T_{char} \tag{A25}$$

Another major error in Eq. (A24) was the one introduced when Eq. (A9) was approximated by its simplified version, $\Delta T_R \approx \Delta T_{char}$. That error is described in Eq. (A11). Combining the two errors, one has for major portion, $\Delta T_{R,err}$, of the net error in ΔT_R of Eq. (A24):

$$\Delta T_{R,err} = -\frac{r_T}{1 - e^{-r_T}} \times \frac{\Delta T_{char}}{\Theta_{char,1}} \times \xi = -\frac{r_T}{1 - e^{-r_T}} \times \frac{\Delta T_{char}}{T_{char,1}} \times \xi\theta_{char,1} \tag{A26}$$

When ΔT_{char} is small compared to $T_{\mathrm{char},1}$, quantity $T_{\mathrm{char},1}$ can be estimated as $T_{\mathrm{char},1} \approx T_{\mathrm{char},2} \approx T_{\mathrm{char}}$. In this case, the difference in $\theta_{\mathrm{char},1}$ and $\theta_{\mathrm{char},2}$ cannot be larger than their difference at the same T_{char} (Eqs. (5.41) and (5.48), Figure 5.6). One can write Eq. (A26) as

$$\Delta T_{\mathrm{R,err}} \approx -\frac{r_{\mathrm{T}}}{1-e^{-r_{\mathrm{T}}}} \times \frac{\Delta T_{\mathrm{char}}}{\Theta_{\mathrm{char}}} \times \xi \qquad \text{(A27)}$$

The magnitude of this error monotonically increases with the increase in r_{T}. Accounting for the values of ξ and Θ_{char} (Table 6.12 and Eq. (8.72)), the magnitude of this error for typical conditions can be estimated as

$$|\Delta T_{\mathrm{R,err}}| \approx \left|\frac{r_{\mathrm{T}}}{1-e^{-r_{\mathrm{T}}}} \times \frac{\Delta T_{\mathrm{char}}}{\Theta_{\mathrm{char}}} \times \xi\right| \le 0.14|\Delta T_{\mathrm{char}}| \qquad \text{(A28)}$$

References

1 Giddings, J.C. (1990) in *Multidimensional Chromatography Techniques and Applications* (ed. H.J. Cortes), Marcel Dekker, New York and Basel, pp. 1–27.
2 Bautz, D.E., Dolan, J.W., and Snyder, L.R. (1991) *J. Chromatogr.*, **541**, 1–19.
3 Dolan, J.W., Snyder, L.R., and Bautz, D.E. (1991) *J. Chromatogr.*, **541**, 21–34.
4 Dolan, J.W. (2003) *LC-GC*, **21**, 350–354.
5 Harris, W.E. and Habgood, H.W. (1966) *Programmed Temperature Gas Chromatography*, John Wiley & Sons, Inc., New York.
6 Guiochon, G. (1969) *Advances in Chromatography*, vol. **8** (eds J.C. Giddings and R.A. Keller), Marcel Dekker, New York, pp. 179–270.
7 Cramers, C.A. and Leclercq, P.A. (1988) *Crit. Rev. Anal. Chem.*, **20**, 117–147.
8 Guiochon, G. and Guillemin, C.L. (1988) *Quantitative Gas Chromatography for Laboratory Analysis and On-Line Control*, Elsevier, Amsterdam.
9 Ettre, L.S. and Hinshaw, J.V. (1993) *Basic Relations of Gas Chromatography*, Advanstar, Cleveland, OH.
10 Purnell, J.H. (1960) *J. Chem. Soc.*, 1268–1274.
11 Giddings, J.C. (1991) *Unified Separation Science*, Wiley, New York.
12 Ettre, L.S. (1993) *Pure Appl. Chem.*, **65**, 819–872.
13 Jennings, W., Mittlefehldt, E., and Stremple, P. (1997) *Analytical Gas Chromatography*, 2nd edn, Academic Press, San Diego.
14 Hyver, K.J. (1989) Chapter 1, in *High Resolution Gas Chromatography*, 3rd edn (ed. K.J. Hyver), Hewlett-Packard Co., USA.
15 Maurer, T., Engewald, W., and Steinborn, A. (1990) *J. Chromatogr.*, **517**, 77–86.
16 Repka, D., Krupčík, J., Benická, E., Maurer, T., and Engewald, W. (1990) *J. High Resolut. Chromatogr.*, **13**, 333–337.
17 Akard, M. and Sacks, R. (1994) *Anal. Chem.*, **66**, 3036–3041.
18 Akard, M. and Sacks, R. (1996) *Anal. Chem.*, **68**, 1474–1479.
19 McGuigan, M. and Sacks, R. (2001) *Anal. Chem.*, **73**, 3112–3118.
20 Giddings, J.C. (1965) *J. Chromatogr.*, **18**, 221–225.
21 Blumberg, L.M. (1992) *Anal. Chem.*, **64**, 2459–2460.
22 Blumberg, L.M. (1994) *Chromatographia*, **39**, 719–728.
23 Blumberg, L.M. (1995) *Chromatographia*, **40**, 218.
24 Snow, N.H. and McNair, H.M. (1992) *J. Chromatogr. Sci.*, **30**, 271–275.
25 Eveleigh, L.J., Ducauze, C.J., and Arpino, P.J. (1996) *J. Chromatogr.*, **A725**, 343–350.

26 Snijders, H., Janssen, H.-G., and Cramers, C.A. (1995) *J. Chromatogr.*, **718**, 339–355.

27 Snijders, H., Janssen, H.-G., and Cramers, C.A. (1996) *J. Chromatogr.*, **756**, 175–183.

28 Bautz, D.E., Dolan, J.W., Raddatz, L.R., and Snyder, L.R. (1990) *Anal. Chem.*, **62**, 1560–1567.

29 Abbay, G.N., Barry, E.F., Leepipatpiboon, S., Ramstad, T., Roman, M.C., Siergiej, R.W., Snyder, L.R., and Winniford, W.L. (1991) *LC-GC*, **9**, 100–114.

30 Girard, B. (1996) *J. Chromatogr.*, **721**, 279–288.

31 Martire, D.E. (1989) *J. Chromatogr.*, **471**, 71–80.

32 González, F.R. and Nardillo, A.M. (1997) *J. Chromatogr. A*, **779**, 263–274.

33 Tudor, E. (1997) *J. Chromatogr. A*, **779**, 287–297.

34 Beens, J., Tijssen, R., and Blomberg, J. (1998) *J. Chromatogr. A*, **822**, 233–251.

35 González, F.R. (2002) *J. Chromatogr. A*, **942**, 211–221.

36 Héberger, K., Görgényi, M., and Kowalska, T. (2002) *J. Chromatogr. A*, **973**, 135–142.

37 Hayes, P.C. Jr. and Pitzer, E.W. (1985) *J. High Resolut. Chromatogr.*, **8**, 230–242.

38 Pell, R.J. and Gearhart, H.L. (1987) *J. High Resolut. Chromatogr.*, **10**, 388–391.

39 Bingcheng, L., Bingchang, L., and Koppenhoefer, B. (1988) *Anal. Chem.*, **60**, 2135–2137.

40 Krupčík, J., Repka, D., Hevesi, T., Nolte, J., Paschold, B., and Mayer, H. (1988) *J. Chromatogr.*, **448**, 203–218.

41 White, C.M., Hackett, J., Anderson, R.R., Kail, S., and Spoke, P.S. (1992) *J. High Resolut. Chromatogr.*, **15**, 105–120.

42 Schlauch, M. and Frahm, A.W. (2001) *Anal. Chem.*, **73**, 262–266.

43 Chu, S. and Hong, C.-S. (2004) *Anal. Chem.*, **76**, 5486–5497.

44 Levkin, P.A., Levkina, A., Czesla, H., and Schurig, V. (2007) *Anal. Chem.*, **79**, 4401–4409.

45 Engewald, W. (2007) *The Restek Advantage*, **2**, 1. 23.

10
Formation of Peak Widths

10.1
Overview

Random motions of molecules of a solute and a carrier gas cause the solute *diffusion* within the gas. As a result, the solute zones in a column get gradually wider in a longitudinal (axial) direction. This is a process of solute (longitudinal) dispersion. The dispersion increases the widths of chromatographic peaks and reduces their separation. Prediction of the widths of chromatographic peaks is the task for this chapter.

Solute dispersion is not the only cause for the broadening of solute zones in chromatography. Decompression of a carrier gas in GC can cause a substantial broadening of the zones as they travel toward the column outlet. The gas decompression and similar nondispersive mechanisms can play a dominant role in the zone broadening. However, the increase in the widths (in space) of the solute zones due to the nondispersive mechanisms has typically a minor effect on the widths of the peaks (in time). On the other hand, nondispersive mechanisms substantially complicate peak width evaluation.

Note 10.1

This is a typical case of interference of secondary factors (in this case, nondispersive zone broadening) with the primary ones (solute dispersion). While having a little effect on the net outcome (in this case, peak widths), the secondary factors can (and actually do in this case) significantly complicate the analysis of the effects of the primary ones. □

Gas decompression is a source of many confusions surrounding the issue of peak widths in GC. Another source of the confusions comes from the dual role of diffusion in the process of solute separation in chromatography.

On one hand, the diffusion causes the solute dispersion which is unfavorable for their separation (the subject of this chapter). On the other hand, the solute diffusion

Temperature-Programmed Gas Chromatography. Leonid M. Blumberg

ISBN: 978-3-527-32642-6

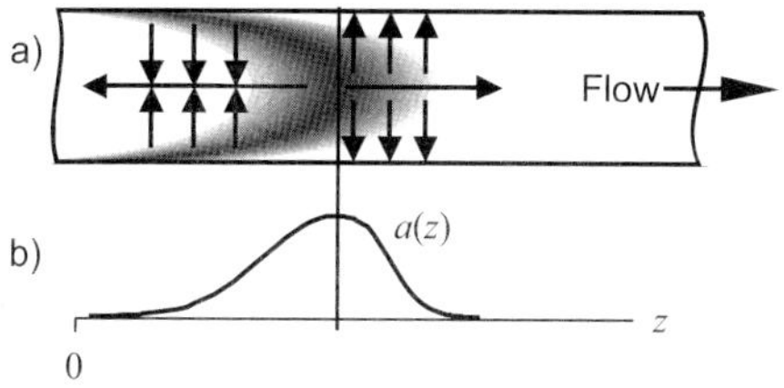

Figure 10.1 Distribution of a solute within a column due to parabolic velocity profile, Figure 7.1, of gas velocities. Darker area in (a) represents higher solute concentration. The arrows with short arrow-heads show directions of a solute diffusion. Graph $a(z)$ in (b) shows net axial distribution of the solute. Full effect of the solute diffusion is not shown in (a). Due to the diffusion, a typical solute distribution looks more like in Figure 1.3a and 8.1.

plays indispensable role in solute separation: it delivers the solute molecules from the gas stream to the stationary phase where the molecules can be retained for some time (the average retention depends on the solute type); it returns the solute molecules desorbed from the stationary phase back into the gas stream; it also mixes the solute molecules within the gas stream causing all molecules of the same solute to migrate along the column with the same apparent velocity in spite of a parabolic profile of the gas flow, Figure 10.1. The net result of the useful aspects of the diffusion – the migration of different solutes with different velocities – has already been analyzed in previous chapters, and is not considered here.

The dual role – favorable and unfavorable for the separation – of a solute diffusion in the separation process is a source of conflicting interpretations of a column *plate height* – a metric of a solute dispersion rate. Sometimes, the plate height is viewed as a measure of the column efficiency or of similar positive characteristics of separation. In this book, the plate height is treated only as a measure of a solute zone dispersion rate unfavorable to the separation.

More damaging than the confusions in its perception are the outright errors in a wide perception of plate height theory. Commercial literature, practical recommendations for method developers, and even majority of peer reviewed publications describe the theory in the form that is in direct contradiction with known rigorous studies of the subject. Widely accepted errors lead to incorrect recommendations for column optimization, and have other negative consequences. The need to address these problems significantly affected the layout of this chapter. In order to make possible to eventually pinpoint actual events that lead to the distortions, a historical perspective plays significant role in the flow of the chapter material.

The chapter addresses three topics: plate height theory (Sections 10.2–10.10), evolution of the plate height concept (Section 10.11), and discussion of the roots of a widely accepted formula that erroneously describes the basics of the plate height theory (Section 10.12).

Standard deviation is the only metric of the width of distribution of a solute material in a column and of the width of a peak in a chromatogram that can be predicted from method and solute parameters (Chapter 3). As before, the standard deviation is treated as the primary metric of the width of any pulse-like function. Unless otherwise is explicitly stated, the terms *width* and *standard deviation* (of a solute zone and/or of a peak) are treated as synonyms.

Before going to description of the zone broadening, a few words should be said about the peak width metrics.

Standard deviations, σ, of each peak can be calculated (or estimated) and reported by a data analysis system. If σ is not available from a report, it can be found from alternative peak width metrics for *a priori* known distinctive peak shapes such as those shown in Figure 3.1. A simple way to calculate σ is through the area-over-height width [1] (effective width [2, 3], equivalent width [4], area/height ratio [5], area-to-height [6]), w_A, Eq. (3.1) – an approach known from James and Martin since 1952 [7, 8]. The quantity w_A can be converted into σ by using known conversion factors, such as those available from Table 10.1. Thus, for a Gaussian peak

$$\sigma = \frac{w_A}{\sqrt{2\pi}} \approx 0.399 w_A = 0.399\frac{\text{peak area}}{\text{peak height}} \tag{10.1}$$

In the early days of chromatography, "a ruler and a pencil" were used for the measurement of peak parameters. The peak width metrics that can be found this way are the half-height width, w_h, and the base width, w_b. These metrics are defined as [9], respectively, the "length of the line parallel to the peak base at 50% of the peak height that terminates at the intersection with the two limbs of the peak," and the "segment of the peak base intercepted by the tangents drawn to the inflection points on either side of the peak." The formulae for conversion of these metrics into σ and vice versa for several peak shapes can be found in Table 10.1.

The earliest known use of the *base width*, w_b, can be traced to famous paper by van Deemter *et al.* [10] who described the metric (page 275, column 1, and page 276, column 2) as "the width . . . of the elution curve [measured] as the distance between the points of intersection of the tangents in the inflection points with the horizontal axis" because, "for the case of an equimolar mixture of solutes, the purity of the eluted material is about 97.7% if a separation is effected at the minimum between the two concentration peaks." In the same year, the metric was recommended [11], and endorsed [12, 13] by Nomenclature Committees appointed by the first three *International Symposia on Gas Chromatography*, in the accompanying papers [14, 15], and by IUPAC [9, 16, 17].

Table 10.1 Conversion of standard deviations, σ, of the peaks shown in Figure 3.6 into the width parameters shown in Figure 3.1.

Peak shape	w_A (area-over-height width)	w_b (base width)	w_h (half-height width)
Gaussian	$\sqrt{2\pi}\sigma \approx 2.507\sigma$	4σ	$\sqrt{8\ln 2}\sigma \approx 2.355\sigma$
Exponential	σ	σ	$\ln 2\sigma \approx 0.693\sigma$
EMG	$\sigma < w_A < 2.507\sigma$	$\sigma < w_b < 4\sigma$	$0.693\sigma < w_h < 2.355\sigma$
Rectangular	$2\sqrt{3}\sigma \approx 3.464\sigma$	$2\sqrt{3}\sigma \approx 3.464\sigma$	$2\sqrt{3}\sigma \approx 3.464\sigma$

According to Ettre [18], the first use of the half-height width, w_h, as a peak width metric in chromatography can be found in 1950 PhD dissertation by Müller [19]. As an alternative to the base width, the half-height width has been discussed (but not accepted) on the 1957 Nomenclature Committee [12]. Later, it has been adopted as the peak width metric in the definition of *separation number* (*Trennzahl*) [20, 21].

Both metrics, w_b and w_h, are currently recommended by the International Union of Pure and Applied Chemistry (IUPAC) [9].

10.2 Local Plate Height

10.2.1 Diffusion in One-Dimensional Stationary Medium

The story of development of a chromatographic theory of broadening of a solute zone can start from 1822 – the year when Fourier published "Théorie Analytique de la Chaleur" [22] ("The Analytical Theory of Heat" [23, 24]). Fourier has shown that (all quotations are taken from Freeman translation [23, 24], pages 45–52, and 102) a *flux*, J, of heat [energy per time per area] along the z-axis in "a solid body formed of some homogeneous substance" and having temperature, T; and that *conservation of heat* at a point (x, y, z) in a three-dimensional body are governed by equations, respectively,

$$J = -K\frac{\mathrm{d}T}{\mathrm{d}z}, \quad \text{and} \quad \frac{\mathrm{d}T}{\mathrm{d}t} = \frac{K}{CD}\left(\frac{\mathrm{d}^2T}{\mathrm{d}x^2} + \frac{\mathrm{d}^2T}{\mathrm{d}y^2} + \frac{\mathrm{d}^2T}{\mathrm{d}z^2}\right) \tag{10.2}$$

with "coefficient K ... to be the measure of the specific conducibility of each substance; this number has very different values for different bodies" while "C denotes the capacity of the substance for heat; D [is] its density", and t denotes time. (The designation of symbols C, D, and K in Eq. (10.2) is the same as it is in the original text [23, 24], but different from the designations elsewhere in this book.)

In 1855, Fick [25, 26] adopted Fourier equations to the description of diffusion of materials. In one-dimensional form, Fick's first and second laws of diffusion in a stationary (not moving), uniform medium can be described [27] as, respectively, $J = -D\partial c/\partial z$ and

$$\frac{\partial c}{\partial t} = D\frac{\partial^2 c}{\partial z^2} \tag{10.3}$$

where c, J, and D are, respectively, the concentration (amount per volume), the flux (amount per area per time), and the diffusivity (Chapter 6) of a material. The latter is measured in units of length^2/time (such as cm^2/s) and is a binary property whose value depends on the diffusing material as well as on the medium where the diffusion takes place. Equation (10.3) describes conservation of matter in a stationary one-dimensional medium.

10.2.2
Dispersion of a Solute Zone in One-Dimensional Stationary Medium

In 1827, Brown observed [28, 29] "a peculiar character in the motions of the particles of pollen [of several plants] in water." He expanded the observations to the "minute" particles of many other organic and inorganic substances in water and in other liquids, and found the same pattern of "rapid oscillatory motion."

In 1905, Einstein has shown [30, 31] that what had become known as *Brownian motion* (*Brownian movement*) could be described by Eq. (10.3). Einstein also solved the equation and found that due to diffusion of tiny "suspended particles" or "solute molecules" in liquid, distribution of a small zone of this material placed in a stationary (not moving), static (not changing with time), uniform (the same at all locations) liquid tends to become Gaussian (Chapter 3). The width of the zone (in units of length, such as centimeter) expressed via the standard deviation, $\tilde{\sigma}$, of spatial distribution of the zone material can be found from *Einstein formula*

$$\tilde{\sigma}^2 = 2Dt \qquad (\text{conditions: stationary, static, uniform, } \tilde{\sigma} = 0 \text{ at } t = 0) \qquad (10.4)$$

Note 10.2

Traditionally, standard deviation is denoted as σ. Following IUPAC recommendations [9] and widely accepted practice, symbol σ is reserved for the (temporal) standard deviation of a peak in a chromatogram. □

Let us pause here, and asses Eq. (10.4) from contemporary point of view of broadening of a solute zone in chromatography.

The quantity $\tilde{\sigma}^2$ is the variance (Chapter 3) of a solute zone. According to Eq. (10.4), quantity $2D$ is the rate of increase in $\tilde{\sigma}^2$. Therefore, the diffusivity, D, is one-half of the rate of increase in the zone variance.

Later, zone-broadening mechanisms other than its diffusion will be explored. However, the broadening due to the diffusion and related mechanisms plays a special role in chromatography. Considering an increase in $\tilde{\sigma}^2$ of a zone due to diffusion of its material as a process of the zone dispersion, one can view quantity $2D$ as the zone dispersion rate. In chromatography and in other fields [32, 33], the emphasis on $2D$ as a zone dispersion rate is more important than the emphasis on D as a diffusivity of the zone material. To avoid the nuisance of dealing with one-half (D) of the dispersion rate, it is convenient to introduce the dispersion rate or dispersivity, $\mathcal{D}$, of a solute zone as

$$\mathcal{D} = 2D \qquad (10.5)$$

and express Einstein formula in Eq. (10.4) as

$$\tilde{\sigma}^2 = \mathcal{D}t \qquad (\text{conditions: stationary, static, uniform, } \tilde{\sigma} = 0 \text{ at } t = 0) \qquad (10.6)$$

It is also worth mentioning that in mathematical literature, Eq. (10.3) is sometimes expressed in the form [32]

$$\frac{\partial c}{\partial t} = \frac{\mathscr{D}}{2}\frac{\partial^2 c}{\partial z^2} \tag{10.7}$$

emphasizing dispersion rate ($\mathscr{D}$) rather than diffusivity (D) as in Eq. (10.3).

10.2.3
Diffusion in a Uniform Flow in a Capillary Column

A mass-conserving migration of a solute along a uniform flow of a fluid (liquid or gas) in a (round) capillary column with internal diameter, d, Figure 2.1 (Fick's second law of diffusion in that medium) can be described as [27, 34–36]

$$\frac{\partial c}{\partial t} = D\left(\frac{\partial^2 c}{\partial r^2} + \frac{1}{r}\frac{\partial c}{\partial r} + \frac{\partial^2 c}{\partial z^2}\right) - 2v\left(1-4\frac{r^2}{d^2}\right)\frac{\partial c}{\partial z} \quad \text{(uniform conditions)} \tag{10.8}$$

where r is the radial distance from the column axis, Figure 7.1, t is time, and z is longitudinal (axial) distance along the column – all three are mutually independent variables. The variables $c = c(z, t, r)$, D, and v are, respectively, the solute concentration (amount/volume), its diffusivity, and velocity defined as

$$v = \mu u = \frac{u}{1+k} \tag{10.9}$$

where u is the fluid velocity, and μ and k are solute mobility and retention factor, respectively. The term $2v(1 - 4r^2/d^2)$ reflects the parabolic flow profile, Figure 7.1, of the fluid.

10.2.4
Dispersion of a Solute Zone in a Uniform Flow in an Inert Tube

In the case of an inert tube, $k = 0$. Equation (10.9) yields $v = u$ indicating that the solute velocity at any point within the tube is the same as the flow velocity of the fluid.

A systematic study of a longitudinal dispersion of a solute zone migrating with flow of liquid in an open, inert, round tube is known from Taylor [34, 37]. Analyzing Eq. (10.8) in a paper [34] published in 1953, Taylor came to conclusion that a "soluble substance" migrating in a tube "is dispersed relative to a plane which moves with [the fluid] velocity... exactly as though it were being diffused by a process which obeys the same law as molecular diffusion but with a diffusion coefficient" increased, compared to the "molecular diffusion" (coefficient of pure diffusion [38]), D, by the amount

$$\Delta D = \frac{d^2 u^2}{192 D} \tag{10.10}$$

known as *Taylor diffusion coefficient* [39, 40] (the virtual coefficient of diffusion [41], the dynamic diffusion constant [42]). In other words, Taylor has shown that Eq. (10.8) – a three-dimensional equation of a mass-conserving migration of a solute in a uniform flow in inert round tube (Fick's three-dimensional second law of diffusion) – can be reduced to one-dimensional equation of a mass-conserving migration (one-dimensional Fick's second law of diffusion)

$$\frac{\partial a}{\partial t} = D_{\text{eff}}\frac{\partial^2 a}{\partial z^2} - v\frac{\partial a}{\partial z} \quad \text{(uniform conditions)} \tag{10.11}$$

where $a = a(z,t)$ is the specific amount of a solute (in units of amount per length, such as g/cm, mol/mm, and so forth), and [38, 39, 42, 43]

$$D_{\text{eff}} = D + \frac{d^2 u^2}{192 D} \tag{10.12}$$

is the solute effective diffusivity [44, 45] (effective coefficient of remixing [38], effective diffusion constant [39], total diffusion constant [42], dynamic diffusion constant [43], effective diffusion coefficient [46]).

Note 10.3

In his 1947 paper, Westhaver [38] did not explicitly mention Eq. (10.12). However, it was transparently implied by Eqs. (12) and (24) in Westhaver paper so that the first proof of Eq. (10.12) can be credited to Westhaver. □

Let us pause again and asses the new result from chromatographic point of view. Taylor derived Eq. (10.10) for the flow of a liquid. However, the nature of a fluid in a tube is irrelevant as long as no decompression of the fluid along the tube occurs. This means that Eqs. (10.10) and (10.12) are also valid for nondecompressing flow of a gas (i.e., for all cases of gas flow where pressure drop along a tube is much smaller than outlet pressure).

Applying Taylor's treatment of Eq. (10.8) to Eq. (10.11), one can conclude that the term $(\partial a/\partial z)v$ in Eq. (10.11) affects location, z, of a solute zone at a given time, t, but it does not affect the zone dispersion. As a result, the variance, $\tilde{\sigma}_z^2$, of the zone can be found from the Einstein formula as

$$\tilde{\sigma}_z^2 = \mathcal{D}t \quad \text{(conditions : static, uniform, } \tilde{\sigma} = 0 \text{ at } t = 0\text{)} \tag{10.13}$$

where dispersion rate,

$$\mathcal{D} = 2D_{\text{eff}} \tag{10.14}$$

is a generalization of $\mathcal{D}$ in Eq. (10.5). Due to Eq. (10.12), one has

$$\mathcal{D} = 2D + \frac{d^2 u^2}{96} \tag{10.15}$$

Expansion of the scope of the definition of dispersion rate ($\mathcal{D}$) from stationary conditions in Eq. (10.5) to flow conditions in Eq. (10.15) underwrites similar expansion in the scope of Einstein formula from Eq. (10.6) to Eq. (10.13). The requirement for the stationary condition is no longer present in Eq. (10.13).

10.2.5 Plate Height as a Spatial Dispersion Rate of a Moving Solute

Continuing with the evaluation of Taylor's results [34] from chromatographic point of view, let us recall that, instead of time (t), the distance (z) traveled by a solute migrating in a tube can be used as an independent variable. In chromatography, this approach can be convenient because the net distance (L) – the length of a column – traveled by each solute before its elution is the same for all solutes injected in the column.

Let z be the distance traveled by a solute from the inlet of a tube during time t. As $t = z/u$, where u is the solute velocity, one can rewrite Eq. (10.13) as

$$\tilde{\sigma}_z^2 = \mathscr{H} z \quad (\text{conditions: static, uniform, } \tilde{\sigma} = 0 \text{ at } t = 0) \tag{10.16}$$

where

$$\mathscr{H} = \frac{\mathscr{D}}{u} \tag{10.17}$$

For historical reasons (a subject of forthcoming discussion), quantity $\mathscr{H}$ in the last two formulae is called as plate height. (The reason for the departure at this point from traditional notation, H, for the plate height will become evident later.) Due to Eqs. (10.17) and (10.12), the plate height in inert tube can be expressed as

$$\mathscr{H} = \frac{2D}{u} + \frac{d^2 u}{96D} \tag{10.18}$$

Comparing Eqs. (10.13) and (10.16), one can notice that, similarly to quantity $\mathscr{D}$, quantity $\mathscr{H}$ can be also viewed as a solute dispersion rate (the rate of increase in the variance, $\tilde{\sigma}_z^2$, of the zone) [35, 47–52]. However, while $\mathscr{D}$ is temporal dispersion rate – the rate of increase in $\tilde{\sigma}_z^2$ with time, t, – $\mathscr{H}$ is spatial dispersion rate – the rate of increase in $\tilde{\sigma}_z^2$ with distance, z. Similarly to the unit – length^2/time (such as cm^2/s) – of measure of $\mathscr{D}$, the unit of measure of $\mathscr{H}$ can be length^2/length (such as cm^2/m).

Example 10.1

Suppose that $\mathscr{H} = 1\,\text{cm}^2/\text{m}$ and $L = 10\,\text{m}$, where L is the length of the tube. Equation (10.16) implies that if $\tilde{\sigma}_z^2 = 0$ at $z = 0$, then after the solute zone has traveled a distance $z = 1\,\text{m}$, its variance ($\tilde{\sigma}_z^2$) became $1\,\text{cm}^2/\text{m} \times 1\,\text{m} = 1\,\text{cm}^2$. When the zone continued to migrate along the tube, $\tilde{\sigma}_z^2$ becomes $2\,\text{cm}^2$ at $z = 2\,\text{m}$ and, finally, $10\,\text{cm}^2$ at the end of the tube. □

Consider again the unit (length^2/length) of measurement of $\mathscr{H}$. Cancellation of similar terms in the numerator and the denominator of the unit leads to the length (in microns, millimeters, and so forth) as a unit of measure of H.

Example 10.2

$1\,\text{cm}^2/\text{m} = 1\,\text{cm}^2/(100\,\text{cm}) = 0.01\,\text{cm} = 0.1\,\text{mm} = 100\,\mu$. □

Its historically established name – the plate height – and the length as its unit of measure obscure to some degree the fact that quantity $\mathscr{H}$ in Eq. (10.16) is a rate. However, these factors do not contradict to the interpretation of $\mathscr{H}$ as a rate. They only make that interpretation less intuitive.

Since $\mathscr{H}$ is measured in units of length, there should be an interpretation of $\mathscr{H}$ as the length of a column segment playing a certain role in a solute dispersion. What is special about the $\mathscr{H}$-long column segment? A simple answer to this question is that the length, $\mathscr{H}$, of that segment is numerically equal to the spatial rate, $\mathscr{H}$, of a solute dispersion expressed in units of length. To get more insightful answer, let us write Eq. (10.16) as

$$\tilde{\sigma}_z^2 = \mathscr{H}^2 \frac{z}{\mathscr{H}} \tag{10.19}$$

This expression shows that every time when a zone advances along the z-axis by a distance, Δz, equal to $\mathscr{H}$, quantity $z/\mathscr{H}$ increases by one unit, and, therefore, $\tilde{\sigma}_z^2$ increases by $\Delta\tilde{\sigma}_z^2 = \mathscr{H}^2$. This allows to interpret a plate as such an $\mathscr{H}$-long distance along a tube that when a zone travels that distance, the variance ($\tilde{\sigma}_z^2$) of the zone increases by the square ($\mathscr{H}^2$) of the distance itself.

Example 10.3

With $\mathscr{H} = 1\,\text{cm}^2/\text{m} = 0.1\,\text{mm}$, Example 10.1 can be rephrased as follows. Every time when a solute zone advances along the tube by 0.1 mm, its variance ($\tilde{\sigma}_z^2$) increases by $(0.1\,\text{mm})^2 = 0.01\,\text{mm}^2$. It takes 10^4 of 0.1-mm-long segments to travel 1 m, it takes 2×10^4 of those segments to travel 2 m, and 10^5 segments to reach the end of the tube. Hence, 1 m away from the inlet of the tube, $\tilde{\sigma}_z^2$ becomes $0.01\,\text{mm}^2 \times 10^4 = 100\,\text{mm}^2 = 1\,\text{cm}^2$. Further along, $\tilde{\sigma}_z^2 = 0.01\,\text{mm}^2 \times (2 \times 10^4) = 2\,\text{cm}^2$ at $z = 2\,\text{m}$, and $\tilde{\sigma}_z^2 = 0.01\,\text{mm}^2 \times 10^5 = 10\,\text{cm}^2$ at $z = L = 10\,\text{m}$. □

It further follows from Eq. (10.19) that if displacement Δz of a zone along a column is smaller than $\mathscr{H}$ then the increment $\Delta\tilde{\sigma}_z^2$ in $\tilde{\sigma}_z^2$ is smaller than $\mathscr{H}^2$. On the other hand, if $\Delta z > \mathscr{H}$ then $\Delta\tilde{\sigma}_z^2 > \mathscr{H}^2$.

Example 10.4

If, in Eq. (10.19), $\Delta z = \mathscr{H}/2$ then $\Delta\tilde{\sigma}_z^2 = (\Delta z/\mathscr{H})\mathscr{H}^2 = \mathscr{H}^2/2 < \mathscr{H}^2$. On the other hand, if $\Delta z = 2\mathscr{H}$ then $\Delta\tilde{\sigma}_z^2 = (\Delta z/\mathscr{H})\mathscr{H}^2 = 2\mathscr{H}^2 > \mathscr{H}^2$. □

It also follows from Eq. (10.19) that the larger is the plate height the proportionally smaller is the number, $\Delta z/\mathscr{H}$, of plates in a Δz-long segment of the tube. On the other hand, the increment $\mathscr{H}^2$ in $\tilde{\sigma}_z^2$ per plate is proportional to the square of the plate height. The net result, according to Eqs. (10.16) and (10.19), is that the larger is the plate height ($\mathscr{H}$) the larger is the increase in $\tilde{\sigma}_z^2$ per a given increase in z, thus confirming once more that $\mathscr{H}$ can be viewed as the rate of increase in $\tilde{\sigma}_z^2$ with the increase in z.

As mentioned earlier, being spatial dispersion rate defined in Eq. (10.16), the concept of the plate height is similar to the concept of the temporal dispersion rate defined in Eq. (10.13) (the latter has no special name). Although these observations are very important for the understanding of the plate height concept, even more important are the following facts.

So far, we considered only the inert tubes where, unlike in chromatographic columns, there is no stationary phase and, hence, no interaction of a solute with internal walls of the tube. Therefore, while the concept of a plate height, $\mathcal{H}$, has its historic roots in chromatography, the very existence of Eq. (10.18) for $\mathcal{H}$ in an inert tube implies that

Statement 10.1

There is nothing exclusively chromatographic in the plate height concept.

Equation (10.17) indicates that

Statement 10.2

The only requirement for $\mathcal{H}$ to be meaningful as a dispersion rate is $u \neq 0$. Therefore, $\mathcal{H}$ is a suitable measure of a rate of dispersion of any statistically large collection of objects moving as a group (a zone) along the same path.

Example 10.5

Consider dispersion of a group of participants of 2001 New Your City Marathon. Based on statistics of its 24 000 finishers, the 42-km marathon route had $\mathcal{H} = 0.6$ km $= 0.6\,\text{km}^2/\text{km}$. This means that the variance, $\tilde{\sigma}_z^2$, of the group of the marathon participants grew by $0.6\,\text{km}^2$ per every 1 km of advancement of the mostly populated region of the group along the route. An observer standing at a certain location along the route could measure the distribution of participants along the route by counting the number of participants that are passing by in equal time intervals. An observer standing at the finishing line could have found that, at the time when the most populated region of the group crossed the finishing line, the group standard deviation was about $(0.6 \times 42)^{1/2}\,\text{km} \approx 5\,\text{km}$. From another perspective, there were about $42/0.6 = 70$ plates along the marathon route. The group advancement by each plate increased the value of $\tilde{\sigma}_z^2$ by $(0.6\,\text{km})^2 = 0.36\,\text{km}^2$ up to its maximum of about $\tilde{\sigma}_z^2 = 70 \times 0.36\,\text{km}^2 \approx 25\,\text{km}^2$, or $\tilde{\sigma}_z \approx 5\,\text{km}$. □

Introduction of the plate height, $\mathcal{H}$, as a part of discussion of Taylor's 1953 study [34] of a solute dispersion in inert tubes and comparison of $\mathcal{H}$ with temporal dispersion rate ($\mathcal{D}$) highlights Statement 10.1. Historically, however, the fact that quantities $\mathcal{D}$ and $\mathcal{H}$ describe different aspects of the same phenomenon – dispersion rate of a migrating solute – has been recognized and highlighted by Golay [35] several years later.

10.2.6 Plate Height in a Capillary Column

In 1941 Martin and Synge [53–55] described *partition chromatography* utilizing solvation of solutes in liquid stationary phase as a retention mechanism. Only packed columns were known at that time.

Golay came up with the idea of a capillary (open tubular) column for partition GC in 1956 [56, 57]. At the international meeting on GC in August of 1957, Golay reported his experimental results and theoretical reasoning behind the expected performance of a capillary column [58]. During the meeting, Martin, remarking that he had been "thinking on exactly the same lines," asked Golay if he had compared his theoretical predictions with the effects of diffusion in "a parabolic distribution of velocities" known from Taylor. "I confess I did not know of that work" was the answer. At the next symposium on GC in May of 1958, Golay presented his "Theory of chromatography in open and coated tubular columns with round and rectangular cross-section" [35]. Taylor's theory [34] played a prominent role in Golay's theory.

As Taylor [34] before him, Golay used Eq. (10.8) as the starting point for his study. However, in Golay's treatment, k in Eq. (10.9) for a solute velocity, v, in Eq. (10.8) could be any non-negative number. In addition to that, Golay evaluated not only temporal dispersion rate ($\mathscr{D}$), but also spatial dispersion rate ($\mathscr{H}$). Both rates were defined as derivatives

$$\mathscr{D} = \frac{d\tilde{\sigma}_z^2}{dt} \quad \text{(uniform conditions)} \tag{10.20}$$

$$\mathscr{H} = \frac{d\tilde{\sigma}_z^2}{dz} \quad \text{(uniform conditions)} \tag{10.21}$$

where $\mathscr{H}$ is described in *Golay formula*

$$\mathscr{H} = \frac{2D}{u} + \frac{(1+6k+11k^2)d^2u}{96(1+k)^2 D} + \frac{2d_f^2 ku}{3(1+k)^2 D_S} \tag{10.22}$$

and, for a solute migrating with velocity, v,

$$\mathscr{D} = \mathscr{H} v \tag{10.23}$$

(Actually, Golay evaluated the derivatives in Eqs. (10.20) and (10.21) without assigning any notations to them.)

Shortly after it became known from Golay [35, 47, 48], Eq. (10.21) was adopted by other workers [46, 49–52, 59] as a definition of $\mathscr{H}$. Soon we will see that this definition is suitable only for uniform medium [44, 45].

Golay's results expand previously known ones in many ways.

Equation (10.20) is differential and, therefore, a more general form of Eq. (10.13). Being concerned only with the rates of increase in $\tilde{\sigma}_z^2$ (rather than with the value of $\tilde{\sigma}_z^2$ itself, as in Eq. (10.13)), Eq. (10.20) is valid for any initial value of $\tilde{\sigma}_z^2$ at $t=0$.

Furthermore, Eq. (10.8) – the basis for Golay's results – is a valid description [27] of a mass-conserving migration of a solute in dynamic medium. This, together with Golay's treatment of the subject [35] and the differential form of Eq. (10.20), made static conditions unnecessary for Eq. (10.20). As a result, the conditions for Eq. (10.20) are less restrictive than those for Eq. (10.13).

More important for chromatography is Eq. (10.21), which Golay treated as an alternative to Eq. (10.20). This unconventional treatment of a solute dispersion not only allowed Golay to derive Eq. (10.22) for the plate height ($\mathcal{H}$) in chromatography, but also to highlight the similarity of $\mathcal{H}$ and $\mathcal{D}$, and to stress the fact that $\mathcal{H}$ is just another form (spatial rather than temporal) of the dispersion rate.

There are several reasons for Eq. (10.21) to be more important for chromatography than Eq. (10.20).

First, as it has already been mentioned earlier, all eluting solutes travel the same distance – the column length, L. Therefore, any solute when it arrived to location z along the column went through the same fraction of its dispersion as did any other solute arriving to the same z. On the other hand, elution time can be substantially different for different solutes. For example, 10 min after the injection can be long after its elution for one solute, half-way through its journey for another one, and a barely noticeable departure from the inlet for yet another solute. One can conclude that distance traveled by solutes is a more consistent measure of progress of their migration than their migration time is. For that reason, it is preferable to use parameter z rather than t as an independent variable in the studies of a solute migration. Along with that, it becomes preferable to deal with spatial rather than temporal dispersion rates.

Another reason for the preference of $\mathcal{H}$ over $\mathcal{D}$ becomes apparent from comparison of Eqs. (10.22) and (10.23). Although the former describes $\mathcal{H}$ as a function of solute parameters (its diffusivity and retention), close examination allows one to conclude that under typical conditions, the dependence does not cause large difference in $\mathcal{H}$ for the majority of solutes. As a result, $\mathcal{H}$ can be treated as a column parameter. This is especially true for the solutes that are significantly retained in isothermal analysis or at the beginning of a heating ramp in temperature-programmed analysis. For these solutes, the plate height is roughly the same [60] as the column internal diameter (d) allowing one to estimate the variance ($\tilde{\sigma}_z^2$) of all eluting solute zones as $\tilde{\sigma}^2 = Ld$. On the other hand, the value of $\mathcal{D}$ is, according to Eqs. (10.23) and (8.2), proportional to solute mobility, μ. There can be several orders of magnitude difference between the values of μ and, therefore, the values of $\mathcal{D}$ for different solutes in the same analysis. As a result, there could be as many values of $\mathcal{D}$ in analysis of the same sample as the number of components in the sample. Furthermore, during temperature-programmed GC analysis, μ of a given solute could change within several orders of magnitude. And so would $\mathcal{D}$. Therefore, $\mathcal{D}$, being a solute- and the run-time-dependent parameter, cannot be treated as a column operational parameter.

Why then almost the same attention has been given so far to $\mathcal{D}$ as it has been to $\mathcal{H}$?

A physical concept of molecular diffusion is widely known and well understood in chromatography. So is the concept of diffusivity, D, as a measure of the rate of the diffusion in stationary medium. Natural extensions of the concept of molecular diffusivity are the concepts of effective diffusivity (D_{eff}) in a moving fluid and of the temporal dispersion rate ($\mathcal{D} = 2D_{\text{eff}}$). All these quantities are measured in the same familiar units of length^2/time.

On the other hand, some imagination is needed to accept the spatial dispersion rate that is unknown outside of chromatography. In addition to that, the widely accepted units of measure of $\mathcal{H}$ are the units of length (rather than length^2/length that would be more appropriate for the spatial dispersion rate) which, at the first glance, do not look like the units of a rate. And it also does not help that the historically evolved term *plate height* for $\mathcal{H}$ has no intuitive association with a concept of a rate.

The concept of temporal dispersion rate ($\mathcal{D}$) was used here as a convenient bridge to the concept of spacial dispersion rate ($\mathcal{H}$). Once the bridge from the familiar term of (molecular) diffusivity through the temporal dispersion rate ($\mathcal{D}$) to the plate height ($\mathcal{H}$) has been established, there is no need to treat quantity $\mathcal{D}$ on a par with quantity $\mathcal{H}$.

The fact that Eq. (10.21) is valid only for uniform conditions raises an interesting question. If $\mathcal{H}$ is a uniform quantity ($\partial\mathcal{H}/\partial z = 0$), then what does the differential form of Eq. (10.21) represent? As mentioned earlier, its differential form makes Eq. (10.21) suitable for any $\tilde{\sigma}_z$ at $z = 0$. But there is much more than that. Although $\partial\mathcal{H}/\partial z = 0$, quantity $\mathcal{H}$ in Eq. (10.21) can be a function of a solute location, z, if the medium is dynamic ($\partial\mathcal{H}/\partial t \neq 0$) like it is in temperature-programmed analysis. Here is why. When conditions of a solute migration change with time, the value of $\mathcal{H}$ at any particular z is an instant quantity (Chapter 4). For a given solute, quantity $\mathcal{H}$ at any given instant, t_1, can be different from $\mathcal{H}$ at another instant, t_2. The change in $\mathcal{H}$ with the change in z occurs because it takes time for a solute to travel along the column, and, because it changes with time, the quantity $\mathcal{H}$ has different instant values at different locations z. As a result, $\mathcal{H}$, corresponding to a given solute, can change with its location (i.e., $d\mathcal{H}/dz \neq 0$) even if $\partial\mathcal{H}/\partial z = 0$.

Golay differential equations (Eqs. (10.20) and (10.21)) for the variance ($\tilde{\sigma}_z^2$) of a solute zone can be viewed as a generalization of the Einstein formula in Eq. (10.6). From that perspective, the understanding of the mechanisms of a solute dispersion as it evolved from Einstein to Taylor and to Golay can be viewed as a process of removing the restrictions from the original Einstein formula. Thus, Einstein formula was valid for stationary, static, and uniform medium, and for ideal initial conditions ($\tilde{\sigma} = 0$ at $t = 0$). Taylor formula in Eq. (10.13) does not require stationary medium. Golay formula only requires uniform medium. Not only the Golay equations remove the requirements for static medium and ideal initial conditions in Taylor formula, but they also remove implicit requirement for no retention in Taylor formula.

Maybe even more important was the introduction by Golay of the spatial dispersion rate, $\mathcal{H}$, of a solute, and the discovery that (with the clarifications that he described) the mysterious at that time concept of height equivalent to one theoretical plate (H.E.T.P., discussed later) that had been known in chromatography for more than 15 previous years [53] can be identified with the newly introduced and quantified spatial derivative $d\tilde{\sigma}_z^2/dz$. Thus Golay reintroduced H.E.T.P. as a by-product of his theory. For the first time, H.E.T.P. had a clear actionable definition (discussed later).

While its development was stimulated by the needs of chromatography, and while it was developed for chromatography, Eq. (10.22) provides another illustration that, as has been mentioned earlier, plate height is not an exclusively chromatographic concept. Understanding of this fact is important for better understanding of chromatography.

The plate number, $\mathcal{N} = L/\mathcal{H}$ (discussed later), is generally viewed as a measure of a column efficiency or its separation efficiency. This means that the smaller is $\mathcal{H}$ of an L-long column, the more efficient is the column. It follows from Eq. (10.22) that $\mathcal{H}$ is the smallest when $k = 0$, meaning that the most efficient column is an inert tube having no stationary phase and no separation power.

The ratio t_R/σ of a peak retention time (t_R) to its width (σ) can be viewed as a measure of sharpness of a peak. The larger is the ratio, the sharper is the peak. In a simple case of static analysis using a column with no gas decompression, quantity t_R/σ can be found as $t_R/\sigma = \mathcal{N}^{1/2}$. This suggests that $\mathcal{N}$ can be viewed as a measure of efficiency of delivery of the peaks to the column outlet – the larger is $\mathcal{N}$, the sharper are the peaks. However, it does not seem to be proper to view $\mathcal{N}$ as a measure of separation efficiency because an inert tube incapable of separation has the largest $\mathcal{N}$.

In closing, it is worth emphasizing again that plate height ($\mathcal{H}$) is no more and no less than a dispersion rate of a solute migrating in a column. As follows from its definition in Eq. (10.21), $\mathcal{H}$ is the spatial dispersion rate (dispersion per unit of distance) similar to temporal dispersion rate (dispersion per unit of time) defined in Eq. (10.20).

10.2.7
Structure and Parameters of Golay Formula for Plate Height

Golay formula in Eq. (10.22) can be expressed as

$$\mathcal{H} = \frac{2D}{u} + \frac{\vartheta_{G1}^2 d^2 u}{96D} + \frac{2\vartheta_{G2}^2 d_f^2 u}{3D_S} \qquad (10.24)$$

where, due to Eq. (8.8), quantities ϑ_{G1} and ϑ_{G2} can be described as, Figure 10.2,

$$\vartheta_{G1} = \frac{\sqrt{1+6k+11k^2}}{1+k} = \sqrt{11-16\mu+6\mu^2} = \sqrt{1+4\omega+6\omega^2} \qquad (10.25)$$

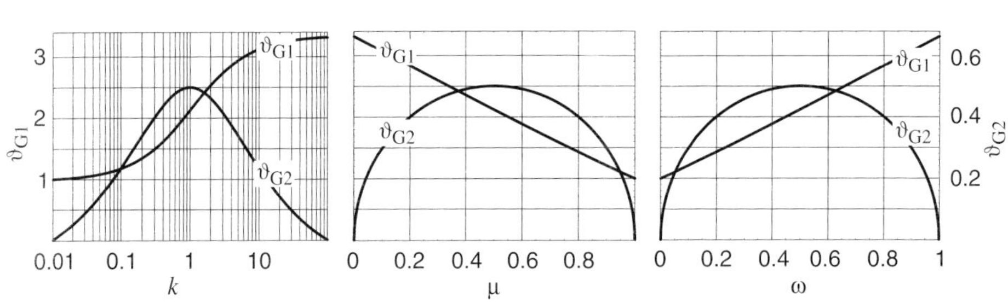

Figure 10.2 Quantities ϑ_{G1} (left scale) and ϑ_{G2} (right scale) in Eqs. (10.25) and (10.26).

$$\vartheta_{G2} = \sqrt{k}/(1+k) = \sqrt{(1-\mu)\mu} = \sqrt{(1-\omega)\omega} \tag{10.26}$$

It appears from Figure 10.2 that ϑ_{G1} is almost a linear function of μ and ω. Indeed, for any value of ω, the error of linear approximation $\vartheta_{G1} \approx 1 + 2.25\omega$ does not exceed 2%. Due to Eq. (8.7), this implies that, for any value of ω, μ, and k within their proper ranges ($0 \leq \omega \leq 1$, $0 \leq \mu \leq 1$, $k \geq 0$), the errors of approximations

$$\vartheta_{G1} \approx 1 + 2.25\omega = 3.25 - 2.25\mu = (1 + 3.25k)/(1+k) \tag{10.27}$$

are also limited to 2%. When it becomes necessary to emphasize a particular variable – k, μ or ω – in formulae for ϑ_{G1}, the following notations will be used:

$$\vartheta_{G1k} = \frac{\sqrt{1+6k+11k^2}}{1+k} \approx \frac{1+3.25k}{1+k} \tag{10.28}$$

$$\vartheta_{G1\mu} = \sqrt{11-16\mu+6\mu^2} \approx 3.25 - 2.25\mu \tag{10.29}$$

$$\vartheta_{G1\omega} = \sqrt{1+4\omega+6\omega^2} \approx 1 + 2.25\omega \tag{10.30}$$

Let us go back to Eq. (10.24). The last term in it determines an increase in $\mathscr{H}$ due to the time that it takes for a solute to diffuse in and out of stationary phase film. That time depends on the amount of stationary phase and on a solute diffusivity, D_S, in it. Equation (10.24) can be expressed as the sum

$$\mathscr{H} = \mathscr{H}_{\text{thin}} + \mathscr{H}_{\text{f}} \tag{10.31}$$

of the terms

$$\mathscr{H}_{\text{thin}} = \frac{2D}{u} + \frac{\vartheta_{G1}^2 d^2 u}{96D} \tag{10.32}$$

$$\mathscr{H}_{\text{f}} = \frac{2\vartheta_{G2}^2 d_f^2 u}{3D_S} \tag{10.33}$$

where $\mathscr{H}_{\text{thin}}$ is the plate height in a thin film column and $\mathscr{H}_{\text{f}}$ is an increase in $\mathscr{H}$ at low diffusivity (D_S) of solutes in a stationary phase.

The first term of Eq. (10.32) represents molecular diffusion of a solute in a carrier gas. The second term of Eq. (10.32) represents the interference of the parabolic profile of the gas velocity with molecular diffusion of a solute in the gas and with the solute retention if it exists (if $k \neq 0$). This term can be interpreted as resistance to mass transfer in a carrier gas. The quantity $\mathscr{H}_{\text{f}}$ represents the interference of the time of a solute residence in stationary phase with parabolic gas velocity profile. The term can be interpreted as resistance *to* mass transfer in a stationary phase.

A wall-coated capillary column can be viewed as a thin film one when the effect of resistance to mass transfer in the stationary phase is negligible compared to that resistance in the carrier gas, that is, when $\mathscr{H}_{\text{f}}$ can be ignored compared to the second term in Eq. (10.32) This condition exists when $64(\vartheta_{G2}/\vartheta_{G1})^2\varphi^2 \ll D_S/D$ where $\varphi = d_f/d$ (Eq. (2.7)). It follows from Eqs. (10.25) and (10.26) that [61] the quantity

$(\vartheta_{G2}/\vartheta_{G1})^2$ is the highest and equal to $(11^{1/2} - 3)/4 \approx 0.08$ at $k = 1/11^{1/2}$. This allows one to conclude that in a thin film column,

$$\varphi^2 \ll \frac{0.2 D_S}{D} \quad \text{(thin film)} \tag{10.34}$$

Larger plate height causes broader peaks, which is detrimental to a column performance. Therefore, it is desirable to avoid thick film columns in which component $\mathcal{H}_f$ significantly affects the net value of $\mathcal{H}$. Thicker film increases the column loadability (sample capacity), which helps to reduce distortion of large peaks [10, 62–65]. The same or better result can be obtained by proportional increase in all dimensions – length, diameter and film thickness – of a thin film column without transforming it into a thick film one. In addition to their less efficient performance, analysis of thick film columns is more complex than the analysis of thin film ones [61]. Being simpler to analyze and no less efficient than thick film columns, thin film columns enjoy primary attention in this chapter. A study of the columns with arbitrary film thickness will continue, but only to some point.

As mentioned earlier, Eq. (10.21) defining plate height ($\mathcal{H}$) is valid only for uniform medium. Nevertheless, as shown later, Eqs. (10.22) and (10.24) quantifying dependence of $\mathcal{H}$ on column and solute parameters can be used outside that restriction where parameters D and u in Eqs. (10.22) and (10.24) can be a function of distance (z) from the column inlet.

In all applications where the column outlet is at vacuum and in majority of analyses of complex samples, gas velocity (u) and pressure (p) can significantly change along the column (Figure 7.3). The same is true for D, which is inversely proportional to p (Eq. (6.36)).

The local velocity (u) of a gas and the local diffusivity (D) of a solute can be different at different locations (z) along the column. However, in a uniformly heated uniform column, products Dp and up (p is the local pressure at coordinate z) do not change with z (Eqs. (6.26) and (7.35)). As a result, the ratio D/u and, therefore, plate height ($\mathcal{H}_{\text{thin}}$, Eq. (10.32)) in a thin film column do not change with z. However, the term $\mathcal{H}_f$ in Eq. (10.33) and, therefore, plate height as a whole are coordinate-dependent local quantities.

Furthermore, although quantity $\mathcal{H}_{\text{thin}}$ in Eq. (10.32) is independent of z, the dependence of u on z in this formula is a source of significant complications. Thus, the fact that the ratio D/u is independent of z while u depends on z implies that D and u are mutually dependent quantities. This mutual dependence prevents independent treatment of gas flow variable (u) and parameter D in Eq. (10.32).

Note 10.4

It might be tempting to express Eq. (10.32) in the form

$$\mathcal{H} = B/u + Cu \quad \text{where} \quad B = 2D, \quad C = (\vartheta_{G1} d)^2/(96D) \tag{10.35}$$

implying that quantities B and C are independent of u and Eq. (10.35) is, therefore, the equation of hyperbola [66] in (u, $\mathcal{H}$)-plane.

Unfortunately, quantities B and C in Eq. (10.35) can be treated as being practically independent of u only when carrier gas decompression along the column is negligible. Otherwise, D and u – both functions of local pressure (p) – are mutually dependent quantities, B and C in Eq. (10.35) are functions of u, Eq. (10.35) is not a hyperbola in (u, $\mathscr{H}$)-plane, and implications of Eq. (10.35) are in contradiction with reality. (Further discussion on this subject can be found in Section 10.12.) □

Mutual dependence of the gas flow variable u and parameter D in Eq. (10.32) can be avoided by using flow variable other than u. To that end, Giddings and others [67–70] used energy flux [71] $J = pu$ (Eqs. (7.42) and (7.126)) as the gas flow variable in Eq. (10.32). With J as the flow variable, Eq. (10.32) can be expressed as

$$\mathscr{H}_{\text{thin}} = \frac{B}{J} + CJ, \quad B = 2Dp, \quad C = \frac{\vartheta_{G1}^2 d^2}{96Dp} \tag{10.36}$$

where, for each solute, product Dp is independent of p (Eq. (6.26)) and parameters B and C are independent of J.

Further analysis suggests that even more suitable as a gas flow variable in Eq. (10.32) is carrier gas specific flow rate (Eq. (7.53))

$$f = \frac{T}{T_{\text{ref}}}\frac{F}{d}$$

where F is gas flow rate (Eq. (7.25)), T is column temperature, and T_{ref} is the temperature (25 °C in this book) at which F is measured. The choice of T_{ref} affects numerical value of F, but not numerical value of f. Measured in units of volume/time/length (such as mL/min/mm), quantity f is gas flow rate per unit of column diameter (d). Advantages of using f as gas flow variable will be discussed shortly.

If f is known, then F can be found as

$$F = \frac{T_{\text{ref}}}{T} df \tag{10.37}$$

which, for $T = T_{\text{ref}}$, becomes $F = df$.

Example 10.6

Suppose that $f = 10$ mL/min/mm and $T = T_{\text{ref}}$. Flow rate (F) of a column with $d = 0.1$ mm is 1 mL/min. In a column with $d = 0.25$ mm, $F = 2.5$ mL/min. □

Since the product Dp is a pressure-independent quantity, it can be replaced with the product $D_{\text{pst}}p_{\text{st}}$ of standard pressure ($p_{\text{st}} = 1$ atm) and diffusivity (D_{pst}) at that pressure. This, together with Eqs. (7.58) and (2.7), allows one to express Eq. (10.24) as

$$\mathscr{H} = \frac{b}{f} + c_1 f + c_{2u} u \tag{10.38}$$

where

$$b = \frac{\pi D_{\text{pst}} d}{2} \tag{10.39}$$

$$c_1 = \frac{\vartheta_{G1}^2 d^2}{48b} \tag{10.40}$$

$$c_{2u} = \frac{2\vartheta_{G2}^2 d_f^2}{3D_S} = \frac{2d^2\varphi^2\vartheta_{G2}^2}{3D_S} \tag{10.41}$$

The parameters b, c_1, and c_{u2} in Eq. (10.38) depend on many factors (carrier gas type, solute, column dimensions and temperature, and so forth). It is important, however, that

Statement 10.3

In a uniformly heated uniform column, parameters b, c_1, and c_{u2} in Eq. (10.38) are uniform (independent of z) even in the presence of gas decompression and do not depend on f or u.

As mentioned earlier, although parameters b, c_1, and c_{u2} in Eq. (10.38) are uniform, plate height $\mathcal{H}$ can be nonuniform (can be different at different z) because, due to the gas decompression along a column, variable u can be nonuniform. Things are different in a thin film column.

Substitution of Eqs. (10.39), (10.40) and (10.41) in Eqs. (10.32) and (10.33) yields

$$\mathcal{H}_{\text{thin}} = \frac{b}{f} + c_1 f \tag{10.42}$$

$$\mathcal{H}_{\text{f}} = c_{u2} u \tag{10.43}$$

Equation (10.38) becomes Eq. (10.42) when

$$c_{u2} = 0 \quad \text{(thin film)} \tag{10.44}$$

which is a valid approximation when condition in Eq. (10.34) is satisfied.

As mentioned earlier, f does not change with z. This, in view of uniform column heating and Statement 10.3, allows one to conclude that

Statement 10.4

Even in the presence of gas decompression, plate height, ($\mathcal{H}_{\text{thin}}$) in a thin film column is uniform (independent of z).

As a mathematical object, Eq. (10.42) is a hyperbola [66] (Figure 10.3). It can be described in a symmetric form as

$$\mathcal{H}_{\text{thin}} = \frac{\mathcal{H}_{\text{min,thin}}}{2}\left(\frac{f_{\text{opt,thin}}}{f} + \frac{f}{f_{\text{opt,thin}}}\right) \tag{10.45}$$

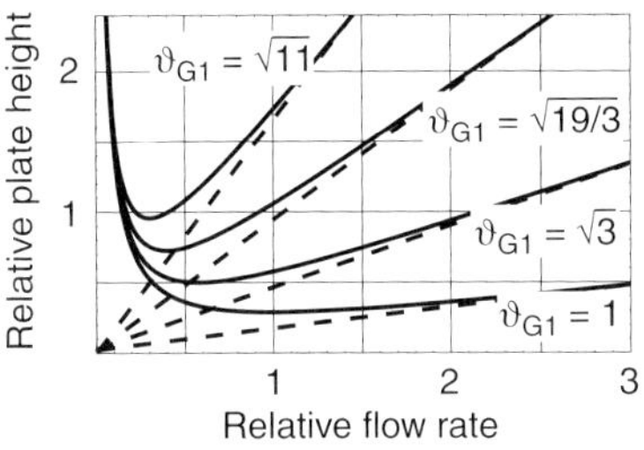

Figure 10.3 Hyperbolas (solid lines) in Eqs. (10.42) and (10.45) for $\vartheta_{G1} = 1$ ($k = \omega = 0$, $\mu = 1$), $\vartheta_{G1} = 3^{1/2}$ ($k = 1/2$, $\omega = 1/3$, $\mu = 2/3$), $\vartheta_{G1} = (19/3)^{1/2}$ ($k = 2$, $\omega = 2/3$, $\mu = 1/3$), and $\vartheta_{G1} = 11^{1/2}$ ($k = \infty$, $\omega = 1$, $\mu = 0$). Relative flow rate and relative plate height are, respectively, $F/F_{\text{opt,thin,1}} = f/f_{\text{opt,thin,1}}$ and $\mathscr{H}_{\text{thin}}/\mathscr{H}_{\text{min,thin,1}}$ where $F_{\text{opt,thin,1}}$, $f_{\text{opt,thin,1}}$ and $\mathscr{H}_{\text{min,thin,1}}$ are parameters at $\vartheta_{G1} = 1$. Each dashed line is a tangent at $f = \infty$ to a respective hyperbola. All tangents have zero intercepts.

where

$$\mathscr{H}_{\text{min,thin}} = 2\sqrt{bc_1} = \frac{d\vartheta_{G1}}{2\sqrt{3}} \tag{10.46}$$

$$f_{\text{opt,thin}} = \sqrt{\frac{b}{c_1}} = \frac{2\sqrt{3}\pi D_{\text{pst}}}{\vartheta_{G1}} \tag{10.47}$$

are, respectively, the lowest plate height in a thin film column and the corresponding optimal specific flow rate.

Equation (10.45) shows that plate height in a thin film column can be expressed via more meaningful parameters – minimal plate height ($\mathscr{H}_{\text{min,thin}}$) and optimal specific flow rate ($f_{\text{opt,thin}}$) – rather than through somewhat faceless parameters b and c_1 of Eq. (10.42). Equations (10.46) and (10.47) indicate that thin film columns have the following unique properties.

Statement 10.5

Optimal specific flow rate ($f_{\text{opt,thin}}$) in a thin film column is independent of the column dimensions (internal diameter and length).

This is the key advantage of specific flow rate (f) as gas flow variable in Eq. (10.32). In this regard, f is similar to Giddings' reduced (dimensionless) gas velocity [46, 72, 73]. However, related to practically important flow rate (F) via simple formula in Eq. (10.37), f is more practically oriented than reduced gas velocity is. Thus F_{opt} can be found from F_{opt} as

$$F_{\text{opt}} = \frac{T_{\text{ref}}}{T} df_{\text{opt}} \tag{10.48}$$

which, for $T = T_{\text{ref}}$, becomes $F_{\text{opt}} = d\,f_{\text{opt}}$.

Another important fact following from Eq. (10.45) is that in the vicinity of optimal f, quantity $\mathcal{H}_{\text{thin}}$ is a weak function of f. Thus, departure of f from $f_{\text{opt,thin}}$ by a factor of 2 in either direction (and the same departure of gas flow rate from F_{opt}) increases $\mathcal{H}_{\text{thin}}$ by no more than 25%. As for the $\mathcal{H}_{\text{thin}}$ itself, for significantly retained solutes ($k \geq 1$), Figure 5.4, quantity ϑ_{G1} in Eq. (10.46) changes from 2.1 (at $k = 1$) to 3.3 (at $k = \infty$). As a result, $\mathcal{H}_{\text{min,thin}}$ changes from about $0.6d$ to about d. One can conclude that for significantly retained solutes, $\mathcal{H}_{\text{thin}}$ can be estimated as

$$\mathcal{H}_{\text{thin}} \approx d, \quad (k \geq 1, \text{ near optimal flow}) \tag{10.49}$$

According to Eq. (10.46), ϑ_{G1} is the only solute-dependent quantity that affects $\mathcal{H}_{\text{min,thin}}$. On the other hand, according to Eq. (10.47), not only ϑ_{G1} affects $f_{\text{opt,thin}}$. The latter also depends on the solute diffusivity, which can be different for different solutes even in the same column with the same carrier gas.

In order to theoretically predict $f_{\text{opt,thin}}$ for a given method of chromatographic analysis, there must be a way to find solute diffusivity from method parameters. Unfortunately, it is not known how to predict the diffusivity of majority of the solutes. Empirical formulae for diffusivity of n-alkanes are described in Eqs. (6.40) and (6.41). Both were derived from previously published experimental data and empirical formulae [74]. Several experimental facts justify adoption of Eqs. (6.40) and (6.41) for other solutes.

It is known that diffusivity of all solutes strongly correlates with the solute retention. Thus, at a fixed temperature (T), solutes with higher molecular weight have generally lower diffusivities [74–77] and higher retention factors (k). An increase in T increases diffusivity of a given solute and reduces its k. Moreover, it is known from numerous observations of experimental data that,

Statement 10.6

If there are n-alkanes in a sample, then the width of an n-alkane peak is roughly the same as the widths of its neighbors of almost any chemical structure.

This suggests that relationship between retention of any solute and its diffusivity is roughly the same as that relationship for n-alkanes. In other words, Eqs. (6.41) and (6.40) reasonably well represent general trend of dependence of solute diffusivities on their properties and on parameters of GC analysis.

While Eqs. (6.40) and (6.41) definitely represent general trend and accepted for the study of column performance in this book, it is hard to quantify their accuracy. One reason for that is unknown accuracy of experimental diffusivity data and empirical formulae from which Eqs. (6.40) and (6.41) were derived [76]. Insufficient accuracy of reported peak widths is another reason. As a result, the approximations in Eqs. (6.40) and (6.41) were accepted merely because they were the simplest that could have been deduced (Chapters 5 and 6) from available sources. The following factors were also considered. Diffusivity of a particular solute affects carrier gas flow rate that is optimal for that solute. A difference in diffusivities of simultaneously migrating solutes makes it impossible to set up the flow rate that is the best for all solutes. Therefore only the general trends can be used for the evaluation of general optimal conditions. Fine tuning can always be used if it becomes necessary for optimization of conditions for separation of particular critical pairs. Substitution of Eqs. (6.40) and (6.41) in Eqs. (10.39) and (10.40) yields

$$b = \frac{\pi\gamma_D^2 D_{gst} d(10^3\varphi)^{0.09}}{2}\left(\frac{T_{char}}{T_{st}}\right)^{-1.25}\left(\frac{T}{T_{st}}\right)^{1+\xi}, \quad c_1 = \frac{\vartheta_{G1}^2 d^2}{48b} \tag{10.50}$$

$$b = \frac{\pi\gamma_D^2 D_{gst} d(10^3\varphi)^{0.09}}{2k^{0.1}}\left(\frac{T}{T_{st}}\right)^{\xi-0.25}, \quad c_1 = \frac{\vartheta_{G1}^2 d^2}{48b} \quad (k \geq 0.1) \tag{10.51}$$

where γ_D and ξ are empirical parameters of carrier gas and $D_{g,st}$ is the gas self-diffusivity – all three are listed for several gases in Table 6.12. Equations (6.40) and (6.41) describe solute diffusivity as functions of different parameters. The difference is transmitted to parameter b in Eqs. (10.50) and (10.51).

Equation (10.50) describes b as a function of a column temperature (T) and of a solute characteristic temperature (T_{char}) – a solute parameter that depends on the solute itself and on the stationary phase type and its thickness. This makes Eq. (10.50) suitable for the evaluation of peak widths corresponding to the solutes with known thermodynamic parameters (such as characteristic parameters T_{char} and θ_{char}). Thus, peak widths in computer-generated chromatograms in Figures 10.4 and 10.5 were found from plate height formula based on Eq. (10.50).

Equation (10.50) makes it possible to find parameter b for each solute with known characteristic temperature (T_{char}). This means that Eq. (10.50) can be used when composition of a test mixture together with characteristic parameters of each component of the mixture are known. This is rarely the case in practical analyses. Equation (10.51) does not impose these requirements. It expresses b as a function of *a priori* known column temperature and retention factor (k), which is a known function, Eq. (8.22), of time in isothermal analysis, and a function, Eq. (8.116), of a heating rate for the solutes eluting during a heating ramp. Transition from Eq. (10.50) to Eq. (10.51) was based on the approximation in Eq. (5.52), which added additional error to Eq. (10.51). Dependence of parameters b and c_1 in Eq. (10.51) on k is shown in Figure 10.6. Substitution of Eq. (10.51) in Eq. (10.47) yields

$$f_{opt,thin} = \frac{2\sqrt{3}\pi\gamma_D^2 D_{g,st}(10^3\varphi)^{0.09}}{k^{0.1}\vartheta_{G1}}\left(\frac{T}{T_{st}}\right)^{\xi-0.25}, \quad k \geq 0.1 \tag{10.52}$$

Presence of the dimensionless film thickness (φ) in Eqs. (10.50), (10.51) and (10.52) is a reminder that although the property of a column to be a thin film one implies that φ does not affect $\mathcal{H}_{min}$, the film thickness does affect $\mathcal{H}$ at a given f. This effect comes through the influence of φ on optimal specific flow rate in a thin film column.

Equation (10.52) shows that the effect of carrier gas on $f_{opt,thin}$ in a thin film column comes mostly through the product $\gamma_D^2 D_{g,st}$. A minor effect of a small difference in quantities ξ for different gases can be ignored. Relations of $f_{opt,thin}$ for several gases to

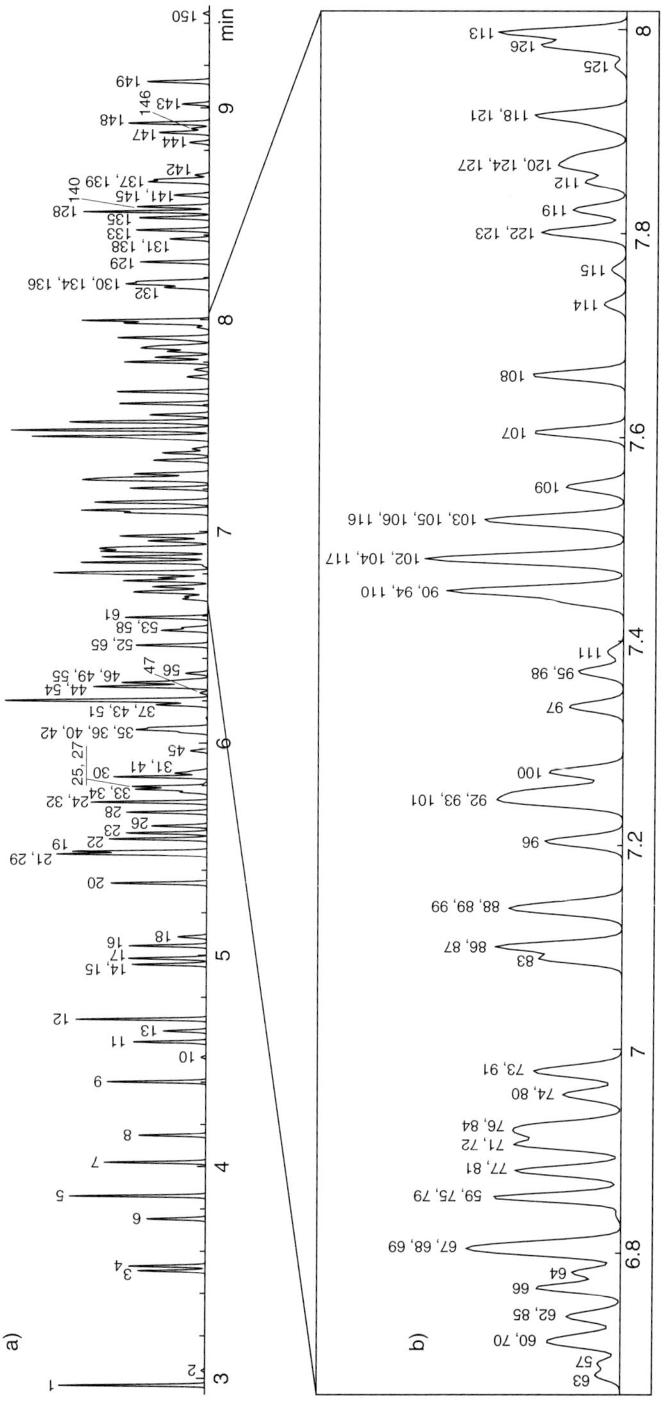

Figure 10.4 Computer-generated chromatogram of the pesticide mixture in Figure 8.3. To compute the chromatogram, numerical integration of Eq. (10.62) was included in the algorithm described in Appendix 8.A.1. All conditions are the same as in Figure 8.3. Conditions for Eq. (10.62): thin film column, $\varphi = 0.001$, $\mathscr{H} = \mathscr{H}_{\text{thin}}$ with D for Eq. (10.32) calculated from Eq. (6.40) for all solutes.

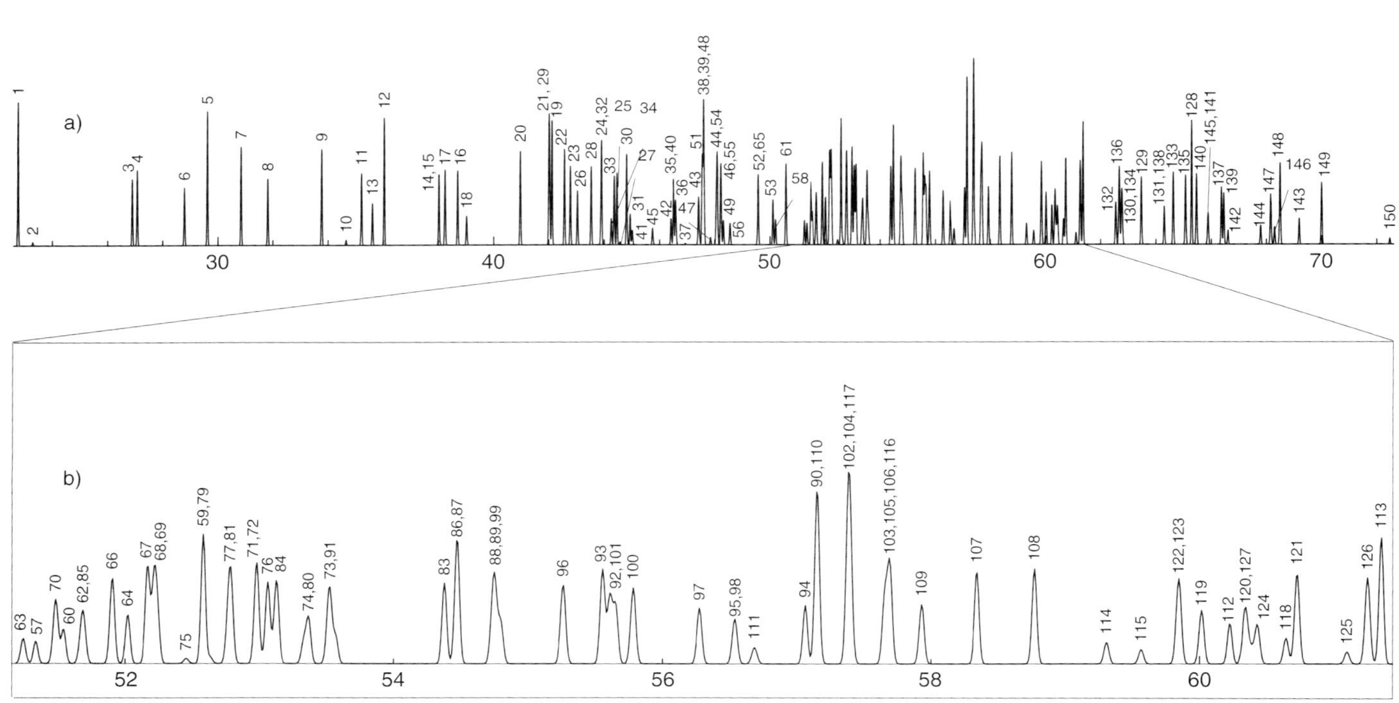

Figure 10.5 Computer-generated chromatogram of the pesticide mixture in Figure 8.3. To compute the chromatogram, numerical integration of Eq. (10.62) was included in the algorithm described in Appendix 8.A.1. Column length: 40 m (initial hold-up time 2.45 min), heating rate 3.91 °C/min. All other conditions are the same as in Figure 10.4. This chromatogram is about 7.67 times longer than the chromatogram in Figure 10.4, but it has fewer overlapped peaks. It also has the same retention pattern as in chromatograms of Figures 8.3a and 10.4.

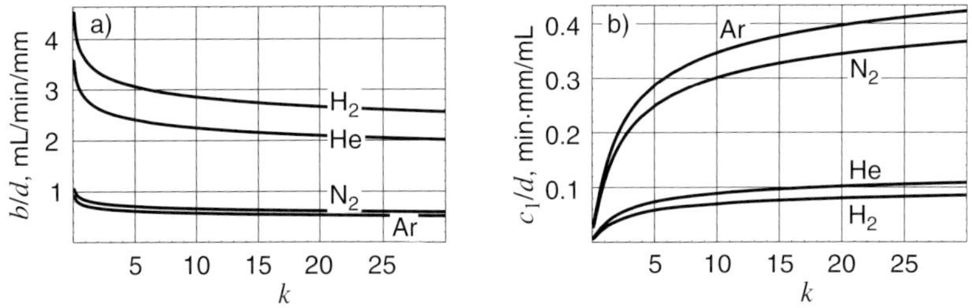

Figure 10.6 Parameters b and c_1 (Eq. (10.51)) at $T = 100\,^\circ C$ and $\varphi = 0.001$. Parameters $D_{g,st}$, ξ and γ_D were taken from Table 6.12.

Table 10.2 Relative values, $f_{opt,thin}/f_{opt,thin,hydrogen}$, of optimal specific flow rates ($f_{opt,thin}$) of several gases in relation to their counterpart for hydrogen[a].

Gas	He	H_2	N_2	Ar
$f_{opt,thin}/f_{opt,thin,hydrogen} = \gamma_D^2 D_{g,st}/(\gamma_D^2 D_{g,st})_{hydrogen}$	0.79	1	0.24	0.2

a) Sources: Eq. (10.52) and Table 6.12.

their counterpart for hydrogen are shown in Table 10.2. As an example, the actual values of $f_{opt,thin}$ at one set of conditions are listed in Table 10.3.

It might be convenient if a GC instrument allowed a direct setting of a specific flow rate. However, this is not currently the case. Therefore, eventually one needs to transform optimal specific flow rate (f_{opt}) into optimal flow rate (F_{opt}) using Eq. (10.48). Both f_{opt} and F_{opt} are temperature-dependent quantities. It follows from Eqs. (10.48) and (10.52) that for two solutes migrating with the same retention factors at temperatures T_1 and T_2, quantities $F_{opt,thin,1}$ and $F_{opt,thin,2}$ relate as

$$F_{opt,thin,2} = F_{opt,thin,1}\left(\frac{T_2}{T_1}\right)^{\xi-1.25} \tag{10.53}$$

where ξ is gas-dependent empirical parameter listed in Table 6.12. For helium and hydrogen, $\xi \approx 0.7$. The last formula becomes

$$F_{opt,thin,2} = F_{opt,thin,1}\left(\frac{T_2}{T_1}\right)^{-0.55} \tag{10.54}$$

Similar results were experimentally obtained elsewhere [60].

Table 10.3 Optimal specific flow rates, $f_{opt,thin}$, of several gases calculated from Eqs. (10.52) with data for $\gamma_D^2 D_{g,st}$ from Table 6.12 (conditions: $k = 1$, $T = 100\,^\circ C$, $\varphi = 0.001$).

Gas	He	H_2	N_2	Ar
$f_{opt,thin}$, mL/min/mm	9.2	11.7	2.8	2.4

10.3
Solute Zone in Nonuniform Medium

10.3.1
Spatial Width of a Zone

The formulae for $\mathscr{H}$ that have been considered so far have several limitations. For one thing, the formulae were derived only for uniform conditions and, therefore, could not yet be applied when strong gas decompression along the column exists. Furthermore, although they are valid for dynamic uniform conditions (such as temperature and pressure programming in a column with weak gas decompression), the formulae do not provide a simple way of predicting the widths of the peaks in chromatograms resulted from dynamic analyses.

Golay proposed workarounds for several special cases of nonuniform static chromatography [35]. Giddings and coworkers found a partial solution [67, 68] for the effect of a gas decompression in static GC (to be described later). Nevertheless, basic questions for peak width in nonuniform dynamic chromatography in general and in temperature- and pressure-programmed GC with decompressing carrier gas in particular remained unanswered [49, 50, 78, 79].

In the late 1980s, several experimental problems exposed the lack of theoretical understanding of nonuniform dynamic chromatography.

Temperature programming by uniform heating of a GC column – the prototype of what was soon to become the mainstream technique for the separation of complex mixtures in GC – was first described by Griffiths, James, and Phillips in 1952 [80–83]. A year earlier, Zhukhovitskii and coworkers described a chromathermography apparatus [84, 85] that, in GC, became the source of the idea of *in-column focusing* of solute *zones* by negative thermal gradients moving from the column inlet to its outlet. By the time when it was given a serious consideration [86], it was already doubtful [87] that the focusing could outperform much simpler and rapidly developing Griffiths' uniform programmable heating. The focusing was forgotten for some time. However, by the late 1980s, the interest to the in-column focusing flared up again [78, 88–90] with very promising expectations. Attempts of theoretical evaluations to avoid costly experiments led to the conclusion that the known theories were inadequate for the task. At about the same time, experiments in another field of chromatography – the supercritical fluid chromatography (SFC) with packed columns – demonstrated a performance significantly exceeding [91] previous predictions based on expected effects of decompression of non-ideal carrier gas in SFC. The known theories could not explain the mystery. Both events were indicative of the need to go back to the drawing board.

Two papers published in 1992 and 1993 addressed the issues of chromatography in nonuniform medium. The first one [44] offered a general solution for a static nonuniform chromatography, while the second [45] extended the solution to dynamic nonuniform conditions. The results helped to resolve both immediate theoretical problems (unfavorably for the in-column focusing [92–95] and favorably for the SFC [44, 96]) and provided a general framework for the treatment of zone broadening in nonuniform dynamic chromatography in general, and in pressure- and/or temperature-programmed GC with decompressing carrier gas in particular.

The theory of nonuniform dynamic chromatography allows one to answer the following two questions important for this book. Do the known formulae for $\mathcal{H}$ (such as Eq. (10.22) and its modifications) remain valid for nonuniform dynamic chromatography? If the answer is "yes," then is Eq. (10.21) a correct equation for $\tilde{\sigma}_z^2$ in a nonuniform dynamic medium?

The first question has a simple and positive answer [44, 45] that only requires some clarification. A nonuniform nature of a medium is irrelevant outside of a solute zone and, therefore, the whole issue of uniformity becomes irrelevant if the zone is infinitesimally narrow. This, according to Chebyshev's inequality (Chapter 3), takes place when $\tilde{\sigma}_z^2$ approaches zero. This, in view of Eqs. (10.21) and (10.20), implies that if the local plate height ($\mathcal{H}$) and local dispersivity ($\mathcal{D}$) are understood as local values (Chapter 4) of coordinate-dependent quantities defined as [44, 45]

$$\mathcal{H} = \lim_{\tilde{\sigma}\to 0} \frac{\mathrm{d}\tilde{\sigma}_z^2}{\mathrm{d}z}, \qquad \mathcal{H} = \lim_{\tilde{\sigma}\to 0} \frac{\mathrm{d}\tilde{\sigma}_z^2}{\mathrm{d}t} \tag{10.55}$$

then Eqs. (10.22) and (10.23) remain valid for a capillary column operating under nonuniform conditions. The definitions further imply that in a column of any type,

Statement 10.7

The values of quantities $\mathcal{H}$ and $\mathcal{D}$ defined in Eq. (10.55) do not depend on the uniformness of conditions of a solute migration.

Note 10.5

Equation (10.55) defines $\mathcal{H}$ and $\mathcal{D}$ as quantities that can be "measured" in a thorough experiment allowing to insert an infinitely narrow solute zone at an arbitrary location (z) along a column.

It is also worth mentioning that Eq. (10.55) defines $\mathcal{H}$ and $\mathcal{D}$ as local and instant quantities (Chapter 4). Equations (10.21) and (10.20) (without the limits at $\sigma \to 0$) providing adequate definitions for $\mathcal{H}$ and $\mathcal{D}$ in uniform medium (static or dynamic) can be viewed as special cases of the definitions in Eq. (10.55). □

Formulae in Eq. (10.55) define parameters $\mathcal{H}$ and $\mathcal{D}$ in nonuniform medium. However, they do not describe the evolution of the width ($\tilde{\sigma}_z$) of solute zone and do not suggest how this quantity can be found. For that, one needs to go back to the basics.

Equations (10.21) and (10.20) for $\tilde{\sigma}_z$ in uniform medium were found from Eq. (10.11) for conservation of mass of a solute migrating in a uniform medium ($\partial D_{\text{eff}}/\partial z = \partial v/\partial z = 0$). It has been shown elsewhere [44] that when the medium is not uniform ($\partial D_{\text{eff}}/\partial z \neq 0$ and/or $\partial v/\partial z \neq 0$), the conservation of mass is described by equation [97]

$$\frac{\partial a}{\partial t} = \frac{\partial}{\partial z}\left(D_{\text{eff}}\frac{\partial a}{\partial z}\right) - \frac{\partial}{\partial z}(va) \tag{10.56}$$

The latter converges to Eq. (10.11) when $\partial D_{\text{eff}}/\partial z = \partial v/\partial z = 0$.

There are substantial technical problems with solving Eq. (10.56). They can be removed if independent variable v defined in Eq. (10.9) as $v = \mu u$ is replaced with independent variable $v_{\text{app}} = v + \partial D_{\text{eff}}/\partial z$. As shown elsewhere [44], it is quantity v_{app} rather than quantity $v = \mu u$ that satisfies general definition of velocity (Chapter 4) as the parameter that can be found as $\mathrm{d}z/\mathrm{d}t$. In other words,

$$v_{\text{app}} = \frac{\mathrm{d}z}{\mathrm{d}t} = v + \frac{\partial D_{\text{eff}}}{\partial z} \tag{10.57}$$

After substitution of $v = v_{\text{app}} - \partial D_{\text{eff}}/\partial z$ in Eq. (10.56), the latter can be transformed into *Fokker–Planck equation* [32, 33]

$$\frac{\partial a}{\partial t} = \frac{\partial^2}{\partial z^2}(D_{\text{eff}}a) - \frac{\partial}{\partial z}(v_{\text{app}}a) \tag{10.58}$$

Equation (10.58) can be also arranged as [32]

$$\frac{\partial a}{\partial t} = \frac{1}{2}\frac{\partial^2}{\partial z^2}(\mathscr{D}a) - \frac{\partial}{\partial z}(v_{\text{app}}a) \tag{10.59}$$

where $\mathscr{D}$ is the solute dispersivity defined in Eq. (10.14).

While Eqs. (10.58) and (10.59) are easier to solve than Eq. (10.56) [45, 98, 99], parameter v_{app} in Eqs. (10.58) and (10.59) is more complex than parameter v in Eq. (10.56). Luckily, the term $\partial D_{\text{eff}}/\partial z$ in Eq. (10.57) is, for typical GC conditions, negligible compared to term $v = \mu u$ (Appendix 10.A.1) so that

$$v_{\text{app}} \approx v = \mu u \tag{10.60}$$

Generally, the coordinate-dependent changes in conditions of a solute migration could be very sharp. For example, a column can be assembled of several segments of different diameters and/or different stationary phases, and so forth. In the vicinity of the sharp transitions, solutions of Eqs. (10.58) and (10.59) can be too complex [45, 98, 99] to be useful for addressing practical problems. More realistically, a medium where a solute migrates – the column itself, the mobile phase, and so forth – can be nonuniform, but smooth [45] so that $\mathscr{H}$ and the spatial gradient ($\partial v/\partial z$) of a solute velocity (v) are nearly uniform within the zone.

It follows from Eq. (10.59) that variance ($\tilde{\sigma}_z^2$) of a solute zone migrating in a smooth dynamic medium can be described by ordinary linear differential equation [45]

$$\frac{d\tilde{\sigma}_z^2}{dz} = \mathcal{H} + \left.\frac{2\tilde{\sigma}_z^2}{v}\frac{\partial v(z,t)}{\partial z}\right|_{t=t(z)} \quad \text{(smooth medium)} \qquad (10.61)$$

where $\mathcal{H}$ is the quantity in Eq. (10.23) and $t(z)$ is the time of migration of the solute to z. (Additional comments for partial and ordinary derivatives in this equation can be found in Chapter 4.)

As mentioned earlier, the evolution of the theory of a solute dispersion in chromatography can be seen in the light of relaxation of restrictions to the scope of the theory from the restrictions in Eqs. (10.6), (10.13), (10.20) and (10.21) to the requirement of smooth medium for Eq. (10.61) – the requirement that is tolerant to all regular conditions in GC (Chapter 2). Less demanding but more complex solutions can be found elsewhere [45, 98, 99]. However, those solutions are not necessary for the evaluation of nonuniform conditions caused by the gas decompression in otherwise uniform column.

Two components of the rate $d\tilde{\sigma}_z^2/dz$ of increase can be recognized in Eq. (10.61).

The first component ($\mathcal{H}$) describes dispersion of the solute zone – the increase in its variance ($\tilde{\sigma}_z^2$) due to the solute diffusion and its side-effects. As a dispersion rate, $\mathcal{H}$ is always a positive quantity. This reflects the fact that the dispersion can *only* cause a zone expansion, but not its contraction. In that sense, the dispersion represented by $\mathcal{H}$ is the zone's irreversible expansion.

The dispersive expansion of a solute zone can be compensated by other mechanisms represented by the second term in Eq. (10.61). This term is positive when the gradient ($\partial v/\partial z$) of solute velocity is positive as, for example, in the case of a decompressing flow in a uniformly heated GC column. Generally, however, $\partial v/\partial z$ can be a negative quantity as shown in Figure 10.7. This means that the second term in Eq. (10.61) can be either positive and cause a zone expansion, or negative and cause a zone contraction. In that sense, the second term in Eq. (10.61) reflects reversible (elastic) changes in the width of a solute zone.

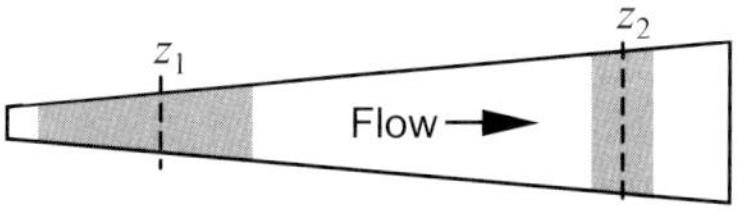

Figure 10.7 Migration of a fixed-volume plug (dark area) along a tube with gradually increasing diameter and mass-conserving flow. The plug's linear velocity and its length are inversely proportion to cross-sectional area of the tube. During its migration from location z_1 to location z_2, the plug gradually becomes narrower in a longitudinal direction so that $\tilde{\sigma}_{z2} < \tilde{\sigma}_{z1}$. (This illustration has been inspired by similar Giddings' illustration [100].)

Example 10.7

Figure 10.7 might be viewed as a depiction of a mass-conserving migration of unretained ($v = u$) nondiffusive ($\mathcal{H} = 0$) solute in static medium where, according to Statement 4.1, $\partial v/\partial z = dv/dz$. This, together with $\mathcal{H} = 0$, allows one to express Eq. (10.61) as

$$\frac{1}{2\tilde{\sigma}_z^2}\frac{d\tilde{\sigma}_z^2}{dz} = \frac{1}{v}\frac{dv}{dz}$$

This can be further rearranged as $d\tilde{\sigma}_z/\tilde{\sigma}_z = dv/v$ indicating that (Figure 10.7) relative nondispersive change, $d\tilde{\sigma}_z/\tilde{\sigma}_z$, in the width ($\tilde{\sigma}_z$) of a zone (its expansion or contraction) is proportional to the relative change, dv/v, in the zone velocity (v). This result represents the zone behavior described by Eq. (10.61), but that cannot be described by Eq. (10.21). □

As mentioned earlier, the second term in Eq. (10.61) diminishes and Eq. (10.61) becomes Eq. (10.21) when solute velocity (v) is uniform ($\partial v/\partial z = 0$). According to Eq. (10.9), $\partial v/\partial z = \mu\partial u/\partial z + u\partial\mu/\partial z$ where u and μ are, respectively, the gas velocity and solute mobility. Because only uniform column heating is considered in this book (Chapter 2), μ is a uniform quantity, Eq. (8.35). As a result, μ can be excluded from Eq. (10.61) by replacing the solute velocity (v) with the gas velocity (u). Equation (10.61) becomes (Appendix 10.A.1)

$$\frac{d\tilde{\sigma}_z^2}{dz} = \mathcal{H} + \left.\frac{2\tilde{\sigma}_z^2}{u}\frac{\partial u(z,t)}{\partial z}\right|_{t=t(z)} \quad (\text{when } \partial\mu/\partial z = 0) \tag{10.62}$$

All numerical evaluations of the peak widths in this book including the ones in Figures 10.4 and 10.5 were based on this formula incorporated in the numerical integration described in Appendix 8.A.1.

In GC, the second term in Eq. (10.62) diminishes and the equation converges to Eq. (10.21) when decompression of a carrier gas is weak, that is, when it can be assumed that $\partial u/\partial z = 0$. In the case of any significant decompression, the second term can be quite large and its contribution to the net value of $\tilde{\sigma}_z$ can be much larger than the contribution from the first term.

Example 10.8

Consider two cases of a solute migration assuming that both cases are static and in both cases, column lengths (L), gas inlet velocities (u_i), and the plate heights ($\mathcal{H}$) are the same fixed quantities.

Case 1. Let the gas velocity be uniform ($\partial u/\partial z = 0$). As a result, Eq. (10.62) becomes Eq. (10.21). Its solution ($\tilde{\sigma}_1^2$) for $\tilde{\sigma}_z^2$ at the column outlet is $\tilde{\sigma}_1^2 = \mathcal{H}L$.

Case 2. Suppose that there is a 10-fold decompression of a carrier gas ($p_o/p_i = 0.1$, therefore, $u_o/u_i = 10$). In view of Eq. (7.67), the gas velocity (u) can be expressed as $u = u_i/(1 - 0.99z/L)^{1/2}$ where z is the distance from the column inlet. Equation (10.62) becomes

$$\frac{d\tilde{\sigma}_z^2}{dz} = \mathscr{H} + \frac{2 \cdot 0.99\tilde{\sigma}_z^2}{L - 0.99z}$$

Its solution ($\tilde{\sigma}_2^2$) for $\tilde{\sigma}_z^2$ at $z = L$ is $\tilde{\sigma}_2^2 = 50.5\mathscr{H}L = 50.5\tilde{\sigma}_1^2$. Therefore, $\tilde{\sigma}_2 = 50.5^{1/2} \approx 7\tilde{\sigma}_1$ showing that 10-fold decompression causes a solute zone at the outlet to be more than seven times wider than it would have been in a similar case without the decompression. □

An important property of the variance of a pulse-like function is its additivity (Chapter 3) in the case where the variance results from the contribution of independent factors. In uniform chromatography, the variance ($\tilde{\sigma}_z^2$) of a solute zone located at z is the accumulation of elementary variances ($d\tilde{\sigma}_z^2$) caused by independent elementary events of dispersion of the zone resulted from traversing all elementary segments (dz) along the path of the zone. Equation (10.21) reflects the additivity of $\tilde{\sigma}_z^2$ in a uniform medium.

Emphasizing the importance of additivity in the evolution of the width of a solute zone in chromatography [73, 100], Giddings also pointed out that in the presence of a significant gas decompression along GC column, $\tilde{\sigma}_z^2$ is a nonadditive function of z. This is because, in the presence of the gas decompression, $\tilde{\sigma}_z^2$ at a given z results not only from independent elementary dispersion events preceding arrival of the zone to point z, but also from the expansion due to the decompression of the width that has been accumulated prior to the zone arrival to z.

Nonadditive nature of $\tilde{\sigma}_z^2$ is represented by the second term in Eq. (10.62), which makes the rate ($d\tilde{\sigma}_z^2/dz$) of change in $\tilde{\sigma}_z^2$ to be a function of $\tilde{\sigma}_z^2$ itself. Only when the medium is uniform, Eq. (10.62) converges to Eq. (10.21). From this point of view, more complex nature of Eq. (10.62) compared to Eq. (10.21) is a result of the absence of additivity of $\tilde{\sigma}_z^2$ in a nonuniform medium.

10.3.2
Temporal Width of a Zone

Effect of the gas decompression on temporal widths of peaks in a chromatogram is much smaller than its effect on spatial widths of the zones in a column.

Speaking of nonadditive nature of $\tilde{\sigma}_z^2$ in nonuniform medium and the associated mathematical complications, Giddings pointed out that carrier gas decompression in GC not only makes the solute zones wider along the z-axis, but it also increases the zone velocities (v). This reduces the net effect of the decompression on the ratios $\tilde{\sigma}_z/v$ and, eventually, on temporal widths of peaks in a chromatogram making the effect

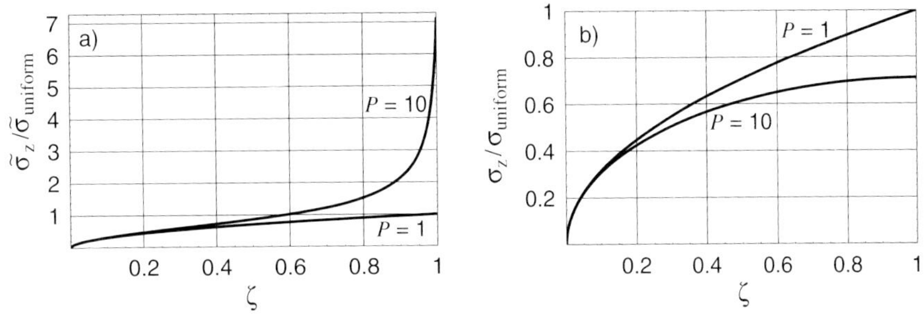

Figure 10.8 Evolution of width-parameters of a solute zone. Spatial width ($\tilde{\sigma}_z$) of a zone in (a) and the ratio ($\sigma_z = \tilde{\sigma}_z/v$) in (b) as functions of dimensionless distance ($\varsigma = z/L$) from inlet at no decompression ($P = p_i/p_o = 1$) and at 10-fold decompression ($P = 10$) of carrier gas. The normalizing quantities $\tilde{\sigma}_{uniform}$ and $\sigma_{uniform}$ are, respectively, $\tilde{\sigma}_z$ and σ_z at $P = 1$ (no decompression), and at $\varsigma = 1$.

much smaller (Figure 10.8b) than it might appear from the dominant effect (Example 10.8, Figure 10.8a) of the decompression on the spatial widths ($\tilde{\sigma}_z$) of solute zones.

Example 10.9

Let us assume that both cases of Example 10.8 describe static migration of a solute in an inert tube ($\mu_1 = \mu_2 = 1$). Therefore, in both cases, the solute velocities (v_1 and v_2) are the same as the gas velocities ($v_1 = u_1$, $v_2 = u_2$). In the case 1 (uniform), $u_1 = u_i$ and, therefore, $v_1 = u_i$. As a result, $\tilde{\sigma}_1/v_1 = \tilde{\sigma}_1/v_i$. In the case 2 of the 10-fold decompression, $u_{o2} = 10u_i$. Hence, outlet solute velocity (v_{o2}) can be found as $v_{o2} = u_{o2} = 10u_i$. As shown in Example 10.8, $\tilde{\sigma}_2 \approx 7\tilde{\sigma}_1$. Therefore, $\tilde{\sigma}_2/v_{o2} \approx 0.7\tilde{\sigma}_1/v_i \approx 0.7\tilde{\sigma}_1/v_1$. The 30% difference in the values of $\tilde{\sigma}_2/v_{o2}$ and $\tilde{\sigma}_1/v_1$ is not as dramatic as the sevenfold difference in $\tilde{\sigma}_2$ and $\tilde{\sigma}_1$ in Figure 10.8. Interestingly, because it has greater effect on u than on $\tilde{\sigma}_z$, the gas decompression slightly reduces the ratio $\tilde{\sigma}/v$ for a solute zone while increasing (dramatically, in some cases) its spatial width $\tilde{\sigma}$. □

The ratio $\tilde{\sigma}_z/v$ is measured in units of time and, as follows from the forthcoming analysis, more directly relates to (temporal) standard deviation (σ) of a peak than to (spatial) standard deviation ($\tilde{\sigma}_z$) of corresponding solute zone. What is more, Giddings demonstrated [73, 100] that quantity $(\tilde{\sigma}_z/v)^2$ could be an additive function of z even in the presence of the carrier gas decompression. This suggests that, even in the presence of a strong gas decompression, an equation describing evolution of quantities like $\tilde{\sigma}_z/v$ should be simpler than Eq. (10.62), and it should provide the results that more directly relate to the widths of the peaks in a chromatogram.

Let $\tilde{\sigma}$ be the spatial standard deviation of eluting solute, that is,

$$\tilde{\sigma} = \tilde{\sigma}_z \text{ at } z = L \tag{10.63}$$

where, as before, $\tilde{\sigma}_z$ is a coordinate-dependent spatial standard deviation of a solute migrating in a column. The temporal standard deviation (in units of time) (σ) of a corresponding peak can be found as

$$\sigma = \frac{\tilde{\sigma}}{v_{oR}} \tag{10.64}$$

where v_{oR} is the solute elution velocity – its outlet velocity (v_o) at the time (t_R) of its elution. Due to Eq. (8.2), v_{oR} can be expressed as

$$v_{oR} = \mu_R u_{oR} \tag{10.65}$$

where u_{oR} is the outlet velocity of a carrier gas at t_R and μ_R is the solute elution mobility – its mobility at t_R.

Note 10.6

During its elution, a portion of a solute still remaining in a column continues to disperse [101–104]. As a result, even under the uniform static conditions, a symmetrically distributed solute yields an asymmetric peak whose standard deviation (σ) is different from the quantity $\tilde{\sigma}/v_{oR}$. This is another side of elution-related aberrations discussed in Notes 8.3 and 8.4. Fortunately [101–104], the relative difference $(\sigma - \tilde{\sigma}/v_{oR})/(\tilde{\sigma}/v_{oR})$ is in the order of d/L where d and L are, respectively, column internal diameter and length. This, due to Eq. (2.9) allows one to ignore the difference between σ and $\tilde{\sigma}/v_{oR}$ as it is done in Eq. (10.64). It is important to recognize, however, that Eq. (10.64) is a convenient and, typically, sufficiently accurate approximation for σ, but not its definition. This fact is not always recognized in the literature and is a source of confusion. □

Equation (10.64) suggests that standard deviation (σ) of a peak corresponding to a solute can be treated as a result of evolution of quantity

$$\sigma_z = \frac{\tilde{\sigma}_z}{v} \tag{10.66}$$

– the temporal standard deviation of what would be a peak if the solute elution took place at location z. This means that σ_z can be viewed as the standard deviation of the evolving in-column peak corresponding to the solute located at z. The quantity σ_z can be also viewed as temporal equivalent of spatial standard deviation ($\tilde{\sigma}_z$) of a solute zone located at z, or simply as (temporal) width of evolving peak.

As shown in Appendix 10.A.1, the evolution of quantity σ_z can be described by an ordinary differential equation

$$\frac{d\sigma_z^2}{dz} = \frac{\mathcal{H}}{v^2} - \frac{2\sigma_z^2}{v^2}\left.\frac{\partial v(z,t)}{\partial t}\right|_{t=t(z)} \tag{10.67}$$

resulting from the transformation of Eq. (10.61) and valid under the same conditions (of smooth nonuniformity). A solution of this equation for σ_z at $z = L$ is peak width (σ) in a chromatogram, that is,

$$\sigma = \sigma_z \text{ at } z = L \tag{10.68}$$

Equation (10.67) is an alternative to Eq. (10.61). Both equations describe the same phenomenon – evolution of the width of a solute zone in a nonuniform dynamic medium – and both are equally valid. However, because they describe different sides of the phenomenon, the equations have essential technical differences and different utility. Thus, Eq. (10.61) is more suitable for uniform (static and dynamic) medium where, as a result of $\partial v/\partial z = 0$, spatial variance ($\tilde{\sigma}_z^2$) of a solute zone is an additive function (Eq. (10.21)) of z. On the other hand, Eq. (10.67) is more suitable for static (uniform and nonuniform) medium where, due to the fact that $\partial v/\partial t = 0$, Eq. (10.67) becomes

$$\frac{d\sigma_z^2}{dz} = \frac{\mathscr{H}}{v^2} \quad \text{(static analysis)} \tag{10.69}$$

indicating that temporal variance (σ_z^2) of the evolving peak is an additive function of z [73, 100].

Note 10.7

Pointing out that quantity σ_z^2 was an additive function of z even in the presence of significant gas decompression [73, 100], Giddings did not mention that this was true only for static conditions (as it is required in order for Eq. (10.67) to become Eq. (10.69)). This created an impression [98] that the additivity was an unconditional property of σ_z^2. □

Equations (10.61) and (10.67) as well as their simpler versions in Eqs. (10.62), (10.21) and (10.69) allow one to find peak widths in a chromatogram by solving ordinary differential equations. This approach might be unsuitable for practical evaluations of a column performance. However, the equations provide a solid basis for practically useful simplifications.

10.4 Apparent Plate Number and Height

10.4.1 Overview

According to Eqs. (10.68) and (10.69), the width (σ) of a peak representing a given solute in the simplest case of static uniform chromatographic analysis with ideal sample introduction ($\sigma = 0$ at $z = 0$) can be found as $\sigma^2 = \mathscr{H} L/v^2$, where $\mathscr{H}$

is the plate height, L is the column length, and v is the solute velocity. The ratio L/v in the last formula is the peak retention time (t_R). Therefore, the formula can be expressed as

$$\sigma = t_R/\sqrt{\mathcal{N}} \quad \text{(static uniform conditions)} \tag{10.70}$$

where

$$\mathcal{N} = L/\mathcal{H} \quad \text{(static uniform conditions)} \tag{10.71}$$

is the column plate number corresponding to the solute.

Even in the same analysis, the values of $\mathcal{N}$ could be different for different solutes. However, in a thin film column, $\mathcal{N}$ is roughly the same for all solutes and, therefore, it can be treated as a column parameter.

Equation (10.70) with $\mathcal{N}$ defined in Eq. (10.71) may not be valid when a medium is nonuniform and/or dynamic. To find σ, one might need to solve Eqs. (10.67) or (10.62) for σ or for its equivalent $\tilde{\sigma}$. Could those solutions for dynamic and/or nonuniform conditions be reduced to the form that is as simple as Eq. (10.70)? Finding answers to this question is the subject of this section. It is always assumed that unless otherwise is specifically stated, the sample introduction is ideal, that is, $\sigma_z = \tilde{\sigma}_z = 0$ at $z = 0$.

Equation (10.70) allows one to find peak width (σ) if its retention time (t_R) and the column plate number ($\mathcal{N}$) are known. The quantities t_R and σ could be measured from experimental data. In that case, Eq. (10.70) could be reversed to become *Glueckauf formula* [105],

$$\mathcal{N} = \frac{t_R^2}{\sigma^2} \quad \text{(static uniform conditions)} \tag{10.72}$$

for finding plate number defined in Eq. (10.71).

In 1956, the scientific committee chaired by Martin recommended Eq. (10.72) as the definition of plate number [11]. Since then, Eq. (10.72) remains to be the prevailing definition of the plate number [8, 9, 46, 47, 59, 106–114]. Recently, the definition was reconfirmed [9] by IUPAC.

Strictly speaking, Eq. (10.72) interpreted as the definition of $\mathcal{N}$ is not equivalent to Eq. (10.71). Equation (10.71) is based on the assumption that σ can be found from Eq. (10.64), while σ in Eq. (10.72) is assumed to be the measured standard deviation of a peak. Due to elution-related aberrations, the latter is slightly different from the former [101–104]. However, the difference is practically insignificant justifying widely accepted and adopted convention [46, 110] to treat quantity σ in Eq. (10.64) and the measured standard deviation (σ) of a peak as the same quantity. With that, the definitions in Eqs. (10.71) and (10.72) become equivalent to each other.

Note 10.8

Several aspects of the elution-related aberrations were described in Notes 8.3, 8.4, and 10.6. □

When conditions of chromatographic analysis are dynamic and/or nonuniform, the number of plates traversed by a solute during its migration along an L-long column can be found as [44]

$$\mathcal{N} = \int_0^L \frac{dz}{\mathcal{H}} \tag{10.73}$$

The quantity $1/\mathcal{H}$ in this formula – the number of plates per unit of a column length at a given location – can be interpreted as a specific plate number [44] at the time when a given solute is at z.

Equation (10.73) is an extension of Eq. (10.71) to general conditions of varying $\mathcal{H}$. When $\mathcal{H}$ is a fixed quantity, Eq. (10.73) converges to Eq. (10.71). Let us call the plate number ($\mathcal{N}$) defined in Eqs. (10.71) and (10.73) as a directly counted plate number. Although the direct counting is a natural way to count the number ($\mathcal{N}$) of plates traversed by a migrating solute, it does not offer a simple formula for finding σ when conditions are dynamic and/or nonuniform. This is where alternative definition of the plate number can be useful.

The alternative plate number should address the following issues. It should (i) lead to a simple formula for the peak width calculation, (ii) be predictable from conditions of GC analysis, and (iii) converge to the intuitive notion of plate number under some conditions such as the uniform static ones.

Foreseeing the need for the alternative plate number concept for dynamic and/or nonuniform conditions, it is proper for the purpose of continuity to expand the concept to static uniform conditions (where there is no practical need for the alternative concept). Using familiar symbol N to denote the alternative plate number, one can write Glueckauf formula in Eq. (10.72) as [8, 9, 11, 46, 47, 59, 105–114]

$$N = \frac{t_R^2}{\sigma^2} \quad \text{(static uniform conditions)} \tag{10.74}$$

Under the static uniform conditions, this formula gives the same results as Eq. (10.71) does. In other words,

$$N = \mathcal{N} \quad \text{(static uniform conditions)} \tag{10.75}$$

Although Eqs. (10.71) and (10.74) describe numerically equal quantities, they define the two plate number concepts from different perspectives. As mentioned earlier, Eq. (10.71) defines quantity $\mathcal{N}$ as the number of $\mathcal{H}$-long segments in the column. On the other hand, the definition in Eq. (10.74) defines quantity N as a measure of quality of delivery of the peaks to their place in a chromatogram – the larger is N, the sharper is the peak. Their numerical equality suggests that while $\mathcal{N}$ is the actual number of plates ($\mathcal{H}$-long segments) in the column, N is what appears from the external measurements of the peak parameters (t_R and σ) to be the number of plates. In view of that, N in Eq. (10.74) can be called as *apparent plate number* [68, 73, 115]. Under nonuniform conditions, apparent plate number (N) is lower [44] than the directly counted plate number ($\mathcal{N}$). The ratio $N/\mathcal{N}$ can be viewed as a measure of losses in a column performance due to nonuniform conditions [65].

As will be shown shortly, Eq. (10.74) is suitable as a definition of apparent plate number (*N*) for all static conditions (uniform and nonuniform). In that broader scope, quantity $N = (t_R/\sigma)^2$ is known in current terminology [9] as the plate number (without the qualifier apparent). To comply with the existing conventions, the terms *plate number* and *apparent plate number* are treated from now on as synonyms with the qualifier *apparent* used only when it is necessary to stress the apparent (based on the appearance from external measurements) nature of quantity *N*.

Note 10.9

Frequently, *N* is defined as

$$N = 16(t_R/w_b)^2 = 5.55(t_R/w_h)^2 \tag{10.76}$$

where w_b and w_h are the base and the half-height widths of a peak, respectively, Figure 3.1. In the early days of chromatography, peak width parameters w_b and w_h were useful because a pencil and a ruler were the primary tools for the peak widths measurement in chromatograms. Currently, one can also obtain the area-over-height width (w_A, Eq. (3.1)) that is either directly available from computer-generated chromatographic reports or can be calculated from the reported peak areas and heights. This makes it possible to calculate *N* from experimental data as

$$N = 2\pi\left(\frac{t_R}{w_A}\right)^2 = 2\pi t_R^2\left(\frac{\text{peak height}}{\text{peak area}}\right)^2 \tag{10.77}$$

It is important to keep in mind, however, that all formulae mentioned in this note are valid only for the Gaussian peaks, while Eq. (10.74) is valid for any peak shape. □

The possibility to experimentally measure parameter *N* is not the only useful feature. The parameter *N* becomes theoretically useful when its value can be predicted from conditions of a chromatographic analysis. A theoretically predicted *N* can be used for theoretical prediction of peak widths in a chromatogram. A concept of apparent [68, 115] (measured [67], observed [46, 52, 100, 116, 117]) plate height (*H*) defined as

$$H = \frac{L}{N} \tag{10.78}$$

provides a bridge for the prediction of *N*.

In contemporary literature, quantity *H* in Eq. (10.78) is typically designated as the plate height (without the qualifier apparent) [110, 111, 113]. The same terminology is recommended by IUPAC [9]. To comply with these conventions, the terms plate height and apparent plate height are treated here as synonyms. As with the terms *plate number* and *apparent plate number*, the qualifier apparent will be used in this text when it is necessary or desirable to stress the apparent nature of *H*.

Recall that actual (local) plate height ($\mathcal{H}$, Eq. (10.55)) is the spatial dispersion rate of a solute. From that point of view, *H* in Eq. (10.78) is the apparent spatial dispersion rate. In static uniform medium, $H = \mathscr{H}$. However, as will be shown shortly, gas decompression makes H to be different from $\mathscr{H}$ even if $\mathscr{H}$ is uniform.

To a large extent, the forthcoming search for the simple formulae for σ is the search for the suitable definitions of the plate number (N) that, through Eq. (10.78), allows one to define H that can be found from conditions of GC analysis. Once that H is found, Eq. (10.78) can be reversed to become a formula,

$$N = \frac{L}{H} \tag{10.79}$$

for N that can be used to find σ from the formulae similar to Eq. (10.70).

Unretained width

$$\sigma_{\mathrm{m}} = \mu_{\mathrm{R}}\sigma \tag{10.80}$$

of a peak where μ_{R} is the elution mobility of the corresponding solute can be useful for predicting H. Frequently, it is easier to predict quantities μ_{R} and σ_{m} than to directly predict quantity σ. In that case, Eq. (10.80) allows one to find σ as

$$\sigma = \frac{\sigma_{\mathrm{m}}}{\mu_{\mathrm{R}}} \tag{10.81}$$

Due to Eqs. (10.64) and (10.65), σ_{m} can be found as

$$\sigma_m = \frac{\tilde{\sigma}}{u_{\mathrm{oR}}} \tag{10.82}$$

where u_{oR} is the carrier gas outlet velocity at time t_{R}.

Note 10.10

In many ways, the concept of an unretained width, σ_{m}, of an arbitrary peak is similar to the concept of the width, σ_{M}, of an unretained peak (if one actually exists in a particular analysis). In the case of σ_{M}, Eq. (10.82) can be expressed as $\sigma_M = \tilde{\sigma}_M / u_{\mathrm{oR}}$ where $\tilde{\sigma}_M$ and u_{oR} are, respectively, the spatial width of unretained solute zone and outlet gas velocity at the solute elution time.

There is an important difference between quantities σ_{m} and σ_{M}. The retention factor (k) in the formula for $\mathscr{H}$ corresponding to σ_{M} is zero. On the other hand, as σ_{m} corresponds to the solute that was retained during its entire migration along the column, quantity k in the formula for $\mathscr{H}$ corresponding to σ_{m} is not zero. □

Example 10.10

Consider the migration of methane and n-decane in a thin film, $10\,\mathrm{m} \times 0.53\,\mathrm{mm}$ column, at $50\,^{\circ}\mathrm{C}$ with hydrogen at $u = 50\,\mathrm{cm/s}$ ($\Delta p \approx 5\,\mathrm{kPa}$). Under these conditions, gas decompression is weak and has negligible effect on gas flow and peak width parameters. Methane would be practically unretained ($k_{\mathrm{M}} = 0$), while retention factor, k_{10}, of decane – its characteristic temperature (T_{char}, Table 5.7) is about $60\,^{\circ}\mathrm{C}$ above the column temperature – could be approximately 10.

Let us assume that $k_{10} = 10$. Diffusivities, D_{M} and D_{10}, of methane and decane can be estimated from Eq. (6.36) as $D_{\mathrm{M}} = 0.8\,\mathrm{cm^2/s}$ and $D_{10} = 0.3\,\mathrm{cm^2/s}$. From

Eq. (10.25), one has for ϑ_{G1} of methane and decane, respectively, $\vartheta_{G1,M} = 1$ and $\vartheta_{G1,10} = 3.1$. Equations (10.32) and (10.82) yield $\mathscr{H}_M = 0.34$ mm, $\mathscr{H}_{10} = 0.59$ mm, $\sigma_M = (\mathscr{H}_M L)^{1/2}/u = 0.12$ s, $\sigma_{m,10} = (\mathscr{H}_{10} L)^{1/2}/u = 0.15$ s. □

In static uniform medium, $\mu_R = \mu$ and, according to Eq. (8.20), $t_R = t_M/\mu_R$ where t_M is the hold-up time. This, together with Eq. (10.81), allows one to express Eq. (10.74) as

$$N = \frac{t_M^2}{\sigma_m^2} \quad \text{(static uniform conditions)} \tag{10.83}$$

This formula is equivalent to Eq. (10.74), but sometimes it is more convenient because it does not directly involve the solute retention parameters in the definition of N. Substitution of the last formula in Eq. (10.78) allows one to describe H as

$$H = \frac{\sigma_m^2}{t_M^2} L \quad \text{(static uniform conditions)} \tag{10.84}$$

10.4.2
Static Conditions

Due to gas decompression along a GC column, conditions of a solute migration can be nonuniform. As shown in Appendix 10.A.3, Eq. (10.62) yields for spatial width ($\tilde{\sigma}$) of an eluite (the width of a solute zone at the column outlet):

$$\tilde{\sigma}^2 = \frac{L}{j^2}(j_G \mathscr{H}_{\text{thin}} + c_{u2}\bar{u}) + P^2\tilde{\sigma}_i^2 \tag{10.85}$$

where L is the column length, j the James–Martin compressibility factor (Eq. (7.114)), the $\mathscr{H}_{\text{thin}}$ local plate height (Eqs. (10.42) and (10.45)) in a thin film column, $P = p_i/p_o$, c_{u2} is the coefficient described in Eq. (10.41), $\bar{u}$ the average carrier gas velocity, and

$$\tilde{\sigma}_i = \tilde{\sigma}_z|_{z=0} \tag{10.86}$$

is the initial width of a zone (spatial standard deviation of the zone concentration immediately after its injection in the column inlet), and

$$j_G = j_G(P) = \frac{9(P^4-1)(P^2-1)}{8(P^3-1)^2} = \frac{9(1+P^2)(1+P)^2}{8(1+P+P^2)^2}, \quad P = \frac{p_i}{p_o} \tag{10.87}$$

is the *Giddings compressibility factor* [44, 67, 68, 100, 118] – a week function (Figure 7.5) of P that changes from $j_G = 1.0$ at $P = 1$ to $j_G = 1.125$ at $P = \infty$.

Note 10.11

Component $\tilde{\sigma}_i$ in Eq. (10.85) represents nonideal sample introduction – the fact that the initial width of injected sample is not zero. Peak broadening due to the factors other than the column itself is known as *extracolumn peak broadening* [119].

In addition to nonideal sample introduction, the extracolumn peak broadening can be caused by a detector, data analysis system, column connections, and so forth. In some studies, extracolumn peak broadening is treated as a component of plate height [120–126]. In this book, plate height accounts only for the column contribution to the peak broadening. Positive and negative effects of several components of extracolumn peak broadening on overall performance of GC system are considered separately. □

Substitution of Eq. (10.85) in Eq. (10.82) and accounting for Eqs. (7.121) and (7.118) yields for σ_m at ideal sample introduction ($\tilde{\sigma}_i = 0$):

$$\sigma_m^2 = (j_G \mathscr{H}_{thin} + c_{u2}\bar{u}) \frac{t_M^2}{L} \tag{10.88}$$

Comparison of Eqs. (10.84) and (10.88) allows one to conclude that the former can be extended to static (possibly nonuniform) conditions, that is,

$$H = \frac{\sigma_m^2}{t_M^2} L \quad \text{(static conditions)} \tag{10.89}$$

where H is described by *Giddings formula* [67, 68, 100],

$$H = j_G \mathscr{H}_{thin} + c_{u2}\bar{u} = \left(\frac{b}{f} + c_1 f\right) \times j_G\left(\frac{p_i}{p_o}\right) + c_{u2}\bar{u} \tag{10.90}$$

Note 10.12

In the original formulae [67, 68], product $p_o u_o$ was used instead of variable f in Eq. (10.90). However, according to Eq. (7.58), f is proportional to $p_o u_o$. Therefore, the difference of Eq. (10.90) from the original formulae is only in the scale of parameters b and c_1. More importantly, the original version of Eq. (10.90) was derived directly from Eq. (10.22) for ideal gas decompression and static conditions. On the other hand, Eq. (10.90) was derived here from Eq. (10.62) (Appendix 10.A.3), which is a special case of Eq. (10.61) suitable not only for the decompression of ideal gas, but for any smooth nonuniform static and dynamic conditions. Therefore, Eq. (10.61) (and its equivalent modifications and subsets in Eqs. (10.62), (10.67) and (10.69)) is broader than Eq. (10.90). □

In static analysis, $\mu_R = \mu$. This, in view of Eqs. (8.20) and (10.80), allows one to express Eq. (10.89) in a familiar form [46, 52, 67, 68, 73, 100, 115–117]

$$H = \frac{\sigma^2}{t_R^2} L \quad \text{(static conditions)} \tag{10.91}$$

Substitution of the last two formulae in Eq. (10.79) yields the formulae

$$N = \frac{t_M^2}{\sigma_m^2} \quad \text{(static conditions)} \tag{10.92}$$

$$N = \frac{t_R^2}{\sigma^2} \quad \text{(static conditions)} \tag{10.93}$$

extending Eqs. (10.83) and (10.74) to static (possibly nonuniform) conditions.

As mentioned earlier, Eq. (10.93) is a familiar Glueckauf formula [105] widely accepted [8, 9, 11, 46, 47, 59, 106–114] as the definition of plate number N without mentioning its apparent (based on the appearance from measurement) nature. To comply with the existing conventions, the terms plate number and apparent plate number are treated here as synonyms with the qualifier apparent used when it is necessary to stress the apparent nature of quantity N.

Equation (10.93) allows one to find σ as

$$\sigma = \frac{t_R}{\sqrt{N}} \quad \text{(static conditions)} \tag{10.94}$$

In view of Eqs. (8.20) and (10.81), this also implies that

$$\sigma = \frac{t_M}{\mu\sqrt{N}} \quad \text{(static conditions)} \tag{10.95}$$

An interesting observation regarding the relationship between local and apparent plate heights ($\mathcal{H}$ and H) in static analysis follows from Eq. (10.90).

In a thin film column,

$$H_{\text{thin}} = j_G \mathcal{H}_{\text{thin}} \tag{10.96}$$

indicating that although, according to Statement 10.4, $\mathcal{H}_{\text{thin}}$ is a fixed quantity and, therefore, is independent of the distance (z) along the column, gas decompression increases apparent plate height (H_{thin}) compared to its local counterpart ($\mathcal{H}_{\text{thin}}$). The fact that $\mathcal{H}_{\text{thin}}$ is a uniform quantity means that H is not an average of $\mathcal{H}$, as it is sometimes suggested, and the term apparent plate height [68] better describes the essence of H.

In a thin film column, it is also easy to find the directly counted plate number ($\mathcal{N}_{\text{thin}}$). Because, according to Statement 10.4, $\mathcal{H}_{\text{thin}}$ is independent of z even in the presence of a carrier gas decompression. It follows from Eq. (10.73) that $\mathcal{N}_{\text{thin}} = L/\mathcal{H}_{\text{thin}}$. On the other hand, it follows from Eq. (10.96) that apparent plate number (N_{thin}) in a thin film column is

$$N_{\text{thin}} = \frac{L}{H_{\text{thin}}} = \frac{L}{j_G \mathcal{H}_{\text{thin}}} = \frac{\mathcal{N}_{\text{thin}}}{j_G} \tag{10.97}$$

One can conclude that, although gas decompression reduces the plate number, the damage is rather minor because, according to Eq. (10.87), j_G cannot be higher than 1.125.

10.4.3
Plate Height and Pneumatic Variables

Pneumatic state of a column (gas flow rate, pressure and gas velocity profiles, and so forth) depends on two pneumatic parameters (Chapter 7) such as the pairs (p_o, $\bar{u}$),

(p_o, F), and so forth. On the other hand, H in Eq. (10.90) is described as a function of four pneumatic variables, f, p_i, p_o, and $\bar{u}$. Because only two variables can be mutually independent, Giddings formula in Eq. (10.90) is incomplete. It offers only *partial* description of H. Without providing additional information regarding relations between f, p_i, p_o, and $\bar{u}$, Eq. (10.90) cannot be used for calculating H, for plotting H as a function of one of its pneumatic variables, and so forth.

To choose the two pneumatic variables in a formula for H, one needs to take into account several practical considerations.

It is desirable to express H as a function of those pneumatic parameters (setpoints) that can be controlled by a GC instrument. The carrier gas pneumatic parameters, such as average velocity ($\bar{u}$), flow rate (F), specific flow rate (f), several pressure parameters, and so forth, fall in this category. On the other hand, the gas local velocity (u) that, due to the gas decompression, can change along a column cannot be used as a setpoint. In vacuum outlet operations, the outlet gas velocity (u_o) approaches infinity and also cannot be used as a setpoint [68].

There is another practical consideration. The outlet pressure (p_o) is frequently predetermined by the choice of a detector (vacuum for mass spectrometers, a fraction of atmosphere above ambient pressure for atomic emission detectors, and so forth) and, therefore, must be treated as one of the two independent pneumatic variables. On the other hand, p_o cannot be arbitrarily changed. This reduces the choice of the controllable independent pneumatic variables to only one.

It appears that the best choice for the controllable independent parameter for the study of a column performance is the specific flow rate (f). Two factors lead to this conclusion.

First, according to Statement 10.5, optimal specific flow rate ($f_{opt,thin}$) in a thin film column representing majority of capillary columns is independent of the column dimensions. According to Eq. (10.52), $f_{opt,thin}$ depends only on the carrier gas type, its temperature, and on a solute retention. This substantially simplifies prediction of $f_{opt,thin}$. On the other hand, there is a simple relationship between f and (actual) flow rate (F) of a carrier gas. According to Eq. (10.48), F in a capillary column is proportional to the product $d \times f$ where d is a column internal diameter.

Example 10.11

It can be found from Eq. (10.52) and Table 6.12 that, at $T = 100\,^\circ$C and $k = 2$, f_{opt} for helium and hydrogen in a thin film column are, respectively, 7.26 and 9.23 mL/min/mm. Therefore, assuming that F is measured at normal temperature ($T_{ref} = 25\,^\circ$C), it can be found from Eq. (10.48) that in a 0.1-mm column ($d = 0.1$ mm) of any length, $F_{opt,helium} = 0.58$ mL/min and $F_{opt,hydrogen} = 0.74$ mL/min. □

Frequently, average velocity ($\bar{u}$) of a carrier gas is treated as the independent variable in the formulae for H. For that reason, in addition to function $H(f,p_o)$, function $H(\bar{u},p_o)$ is also evaluated below for comparison.

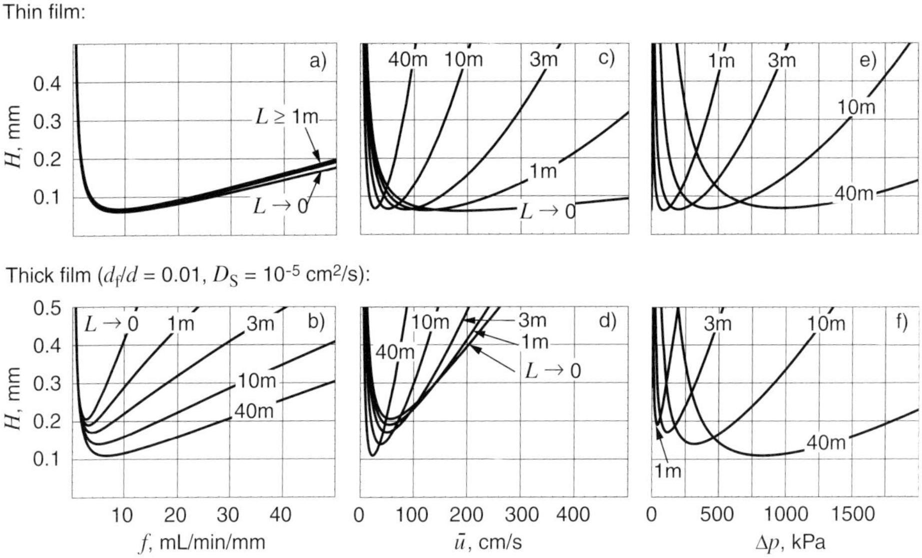

Figure 10.9 Apparent plate height (H, Eqs. (10.98)–(10.100) for normal decane in L-long capillary column vs. gas-specific flow (f), average velocity ($\bar{u}$), and pressure drop (Δp). Conditions: $d = 0.1$ mm, carrier gas helium, $D_S = 10^{-5}$ cm^2/s, $k = 1$, $p_o = p_{st} = 1$ atm, $T = 100\,^\circ$C. Diffusivity, D_{pst}, of C_{10} at p_{st} and T for the parameters in Eqs. (10.39) and (10.40), and gas viscosity, η, at T were calculated from Eqs. (6.39) and (6.20). Curves at $L \to 0$ represent negligible gas decompression. Column length (L) has no effect on optimal flow (f_{opt}) corresponding to the lowest plate height (H_{min}) in a thin film column [127] (a). On the other hand, an increase in L increases Δp_{opt}, (e, f), but reduces [128, 129] $\bar{u}_{opt}$, (c, d). There is also a general trend [61, 127–129] of reduction in f_{opt}, Δp_{opt}, and $\bar{u}_{opt}$ in thick film columns (second row) compared to their thin film counterparts (first row) with the same L.

It is convenient to first find a function $H(p_i, p_o)$ as transitional step to functions $H(f,p_o)$ and $H(\bar{u},p_o)$. Substitution of Eqs. (7.60) and (7.120) in Eq. (10.90) yields a complete function (Figures 10.9 and 10.10),

$$H = H(p_i, p_o) = \left(\frac{8b\Omega p_{st}}{\pi d(p_i^2 - p_o^2)} + \frac{\pi c_1 d(p_i^2 - p_o^2)}{8p_{st}\Omega}\right) \times j_G\left(\frac{p_i}{p_o}\right) + \frac{3c_{u2}(p_i^2 - p_o^2)^2}{4(p_i^3 - p_o^3)\Omega} \tag{10.98}$$

of p_i and p_o. The latter, after the substitutions of Eqs. (7.52) and (7.146), yields complete functions [127–129] (Figures 10.9–10.11),

$$H = H(f\bar{u}, p_o) =$$

$$\left(\frac{b}{f} + c_1 f\right) \times j_G\left(\sqrt{1 + \frac{8fp_{st}\Omega}{\pi d p_o^2}}\right) + \frac{48c_{u2}f^2\Omega}{\pi^2 d^2 p_{st}}\left(\left(\frac{p_o^2}{p_{st}^2} + \frac{8f\Omega}{\pi d p_{st}}\right)^{\frac{3}{2}} - \frac{p_o^3}{p_{st}^3}\right)^{-1} \tag{10.99}$$

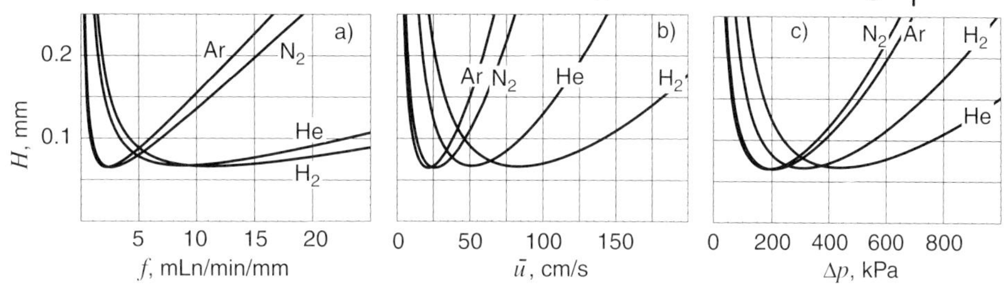

Figure 10.10 Apparent plate height, H, in a 10 m-long, thin film column with different carrier gases. All conditions other than the column length and the gas types are the same as in Figure 10.9 for a thin film column. Hydrogen has the highest f_{opt} and $\bar{u}_{opt}$, and helium has the highest Δp_{opt}. All three optimal parameters (f_{opt}, $\bar{u}_{opt}$, and Δp_{opt}) for nitrogen and argon are almost the same, and are lower than their helium and hydrogen counterparts.

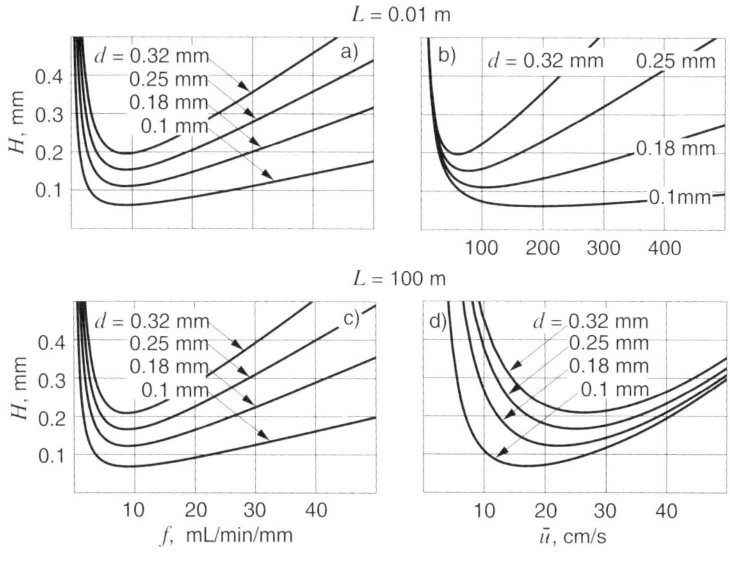

Figure 10.11 Apparent plate height (H) in a thin film column as a function of a gas-specific flow rate (f) and average velocity ($\bar{u}$). All conditions except for the column internal diameter (d) and length (L) are the same as in Figure 10.9. These graphs show that d and L have no effect on f_{opt}, (a, c). The only difference between the curves in (a) and (c) is that, due to the effect of j_G on H at any f, all curves in (c) are about 12.5% higher compared to their counterparts in (a). On the other hand, $\bar{u}_{opt}$ – the average gas velocity at H_{min} – depends on d and L. In a short columns (b), $\bar{u}_{opt}$ declines when d increases (Eq. (10.121)). In a long columns (d), $\bar{u}_{opt}$ increases when d gets smaller (Eq. (10.124)). The difference in the scales of $\bar{u}$ in (b) and (d) reflects the fact that $\bar{u}_{opt}$ also depends on L.

$$H = H(\bar{u}, p_o) =$$

$$\left(\frac{8b\Omega p_{st}}{\pi d p_o^2(\Phi^2(\bar{u}\Omega/p_o)-1)} + \frac{\pi c_1 d p_o^2(\Phi^2(\bar{u}\Omega/p_o)-1)}{8p_{st}\Omega}\right) j_G(\Phi(\bar{u}\Omega/p_o)) + c_{u2}\bar{u} \tag{10.100}$$

of the pairs of variables (f, p_o) and $(\bar{u}, p_o)$. The parameters b, c_1, and c_{u2} in the last three formulae are described in Eqs. (10.39)–(10.41) and (10.51). A large collection of computer-generated graphs of H as a function of $\bar{u}$ can be found in literature [113, 130].

Each of the last three formulae is valid for an arbitrary decompression of a carrier gas along a column with arbitrary film thickness in a static analysis. Unfortunately, the formulae are too complex. Two extremes in the decompression of a carrier gas allow one to substantially simplify the formulae.

When gas decompression is weak, Eqs. (10.99) and (10.100) become

$$H = \mathscr{H} = \frac{b}{f} + c_1 f + c_2 f = \frac{b}{f} + cf \quad \text{at} \quad \Delta p \ll p_o \tag{10.101}$$

$$H = \mathscr{H} = \frac{b_u}{u} + c_{u1}u + c_{u2}u = \frac{b_u}{u} + c_u u \quad \text{at} \quad \Delta p \ll p_o \tag{10.102}$$

where the parameters b, c_1, and c_{u2} are described in Eqs. (10.39), (10.40) and (10.41). These formulae allow one to find other parameters as

$$c_2 = \frac{8d_f^2 p_{st}\vartheta_{G2}^2}{3\pi d D_S p_o}, \qquad c = c_1 + c_2 \tag{10.103}$$

$$b_u = \frac{4bp_{st}}{\pi d p_o}, \qquad c_{u1} = \frac{\pi d^3 p_o \vartheta_{G1}^2}{192 b p_{st}}, \qquad c_u = c_{u1} + c_{u2} \tag{10.104}$$

In many cases, $p_o = p_{st}$. As a result, $b_u = 2D_{pst}$. However, this is not always the case even if the column outlet is not at vacuum. It might be necessary, for example, to account for the pressure drop across a detector. In that case, each pressure, p_{st} and p_o, in Eq. (10.104) plays its role.

Equations (10.102) and (10.101) are none other than alternative forms of Golay formula previously expressed as Eqs. (10.22), (10.24) and (10.38). Because the gas decompression along the column vanishes when the column length approaches zero, the graphs of Eqs. (10.102) and (10.101) are those in Figure 10.9 that correspond to $L = 0$. For a thin film column, the graphs of Eqs. (10.102) and (10.101) are shown in Figures 10.11a and b.

A strong decompression of a carrier gas takes place when $\Delta p \gg p_o$. In the vacuum outlet operations, this is always the case. Typically, a strong gas decompression also exists in packed columns, in small bore or very long capillary columns, and so forth. Equations (10.90), (10.99), (10.100) and (10.98) can be used for deriving simplified forms corresponding to $\Delta p \gg p_o$. Equation (10.90) combined with Eqs. (7.200) and (7.38) yields

$$H = \frac{9}{8}\left(\frac{b}{f} + c_1 f\right) + C_2\sqrt{f} \quad \text{at} \quad \Delta p \gg p_o \tag{10.105}$$

$$H = \frac{B\bar{u}}{\bar{u}^2} + C_{\bar{u}1}\,\bar{u}^2 + C_{\bar{u}2}\,\bar{u} \quad \text{at} \quad \Delta p \gg p_o \tag{10.106}$$

where, as follows from Eqs. (10.39), (10.40) and (10.41),

$$C_2 = \frac{d_f^2 \vartheta_{G2}^2}{4D_S}\sqrt{\frac{dp_{st}}{\pi L \eta}} \tag{10.107}$$

$$B_{\bar{u}} = \frac{81 d p_{st} b}{512\pi L \eta}, \quad C_{\bar{u}1} = \frac{\pi d L \eta \vartheta_{G1}^2}{6 b p_{st}}, \quad C_{\bar{u}2} = c_{u2} \tag{10.108}$$

Equation (10.106) is known since 1985 from Leclercq and Cramers [131]. A detailed derivation of Eqs. (10.106) and (10.105) as well as discussion of their parameters can be found elsewhere [127–129]. Of particular interest for this study is the independence of parameters b and c_1 of the column length (L) and dependence of parameters $B_{\bar{u}}$ and $C_{\bar{u}1}$ on L. As a consequence, f_{opt} in a thin film column is independent of L, while $\bar{u}_{opt}$ strongly depends on L (Figures 10.9 and 10.11).

Ideally, Eqs. (10.102) and (10.101) are valid when $|\Delta p| \ll p_o$, while Eqs. (10.106) and (10.105) are valid when $|\Delta p| \gg p_o$. However, practical separation of the two regions is less demanding and, to some degree, can be arbitrary. It is desirable for practical reasons to have an explicit description of a borderline between the regions. One way to identify such a borderline is to notice that, according to Eq. (10.98), H in a thin film column can be expressed as essentially a function of pneumatic variable $p_i^2 - p_o^2$. Inequality $|\Delta p| \ll p_o$ is equivalent to the approximating $p_i^2 - p_o^2$ as $p_i^2 - p_o^2 = 2p_o\Delta p$ where $\Delta p = p_i - p_o$. The error of this approximation is $2p_o\Delta p - (p_i^2 - p_o^2) = -\Delta p^2$. On the other hand, inequality $|\Delta p| \gg p_o$ is equivalent to the approximation of $p_i^2 - p_o^2$ as $p_i^2 - p_o^2 = p_i^2$ with the error being p_o^2. The magnitudes of both errors are equal when $\Delta p = p_o$, that is, when $\Delta p = \Delta p_B$ where Δp_B is the borderline pressure drop described in Eq. (7.74). When $\Delta p < \Delta p_B$, Eq. (10.102) and its equivalents such as Eq. (10.101) cause smaller error in calculation of H in a thin film column than Eq. (10.106) and its equivalents such as Eq. (10.105) do. The opposite is true when $\Delta p > \Delta p_B$.

10.4.4
Dimensionless Plate Height

The plate height, H, is a function of column dimensions and pneumatic conditions. As a result, both components (L and H) of the plate number in Eqs. (10.79) depend on column dimensions. Use of dimensionless plate height [72] (reduced plate height [9, 46, 72, 73, 104, 109–111, 114, 132–134]),

$$h = \frac{H}{d_{cx}} = \begin{cases} H/d, & \text{capillary columns} \\ H/d_p, & \text{packed columns} \end{cases} \tag{10.109}$$

where d is internal diameter of capillary column and d_p is the particle size of packing material of a packed column, allows one to express Eq. (10.79) in dimensionless form

$$N = \frac{\ell}{h} \tag{10.110}$$

where ℓ is the column dimensionless length defined in Eq. (7.96).

10.5 Thin Film Columns

Considering only the weak or only the strong decompression of a carrier gas in a column is a way to simplify the plate height formulae. Another way to simplify the formulae is to consider only the thin film columns – the subject of this section – or only the thick film columns considered later. It will be shown later that the thin film columns offer the simplest solutions to many known practical problems of column optimization.

A column behaves as a thin film one when contribution to the peak broadening from resistance to mass transfer in the stationary phase is much smaller than that contribution from resistance in the carrier, that is, when Eq. (10.34) is satisfied and assumption $c_{u2} = 0$ (Eq. (10.44)) for parameter c_{u2} defined in Eq. (10.41) becomes a valid approximation. The quantity D in Eq. (10.34) is a solute diffusivity at any location along the column. When gas decompression in the column is strong, the change in D along the column can be very large. It can be shown that Eq. (10.34) is always satisfied and approximation in Eq. (10.44) remains valid when

$$\varphi^2 \ll 0.15 D_S / D_i \quad \text{(thin film)} \tag{10.111}$$

where D_i is a solute diffusivity at the column inlet.

10.5.1 Plate Height and Flow Rate

At $c_{u2} = 0$, Eq. (10.99) yield for the plate height (H_{thin}) in a thin film column:

$$H_{thin} = \left(\frac{b}{f} + c_1 f\right) j_G \tag{10.112}$$

or, in symmetric form,

$$H_{thin} = \frac{H_{min,thin}}{2}\left(\frac{f_{opt,thin}}{f} + \frac{f}{f_{opt,thin}}\right) \tag{10.113}$$

where $f_{opt,thin}$ is described in Eq. (10.47), and

$$H_{min,thin} = 2 j_G \sqrt{b c_1} = j_G \mathscr{H}_{min,thin} = \frac{j_G d \vartheta_{G1}}{2\sqrt{3}} \tag{10.114}$$

where $\mathscr{H}_{min,thin}$ is described in Eq. (10.46). One can verify that substitution of Eqs. (10.47) and (10.114) in Eq. (10.113) yields Eq. (10.112). Equation (10.113) expresses the plate height via more meaningful parameters – minimal plate height ($H_{min,thin}$) and optimal specific flow rate ($f_{opt,thin}$) – than somewhat faceless parameters b and c_1 in Eq. (10.112).

The quantity j_G in Eqs. (10.112) and (10.114) is a function of f. Therefore, strictly speaking, $H_{min,thin}$ in Eq. (10.114) is a function of f and not a fixed quantity representing the lowest value of H_{thin}. Accordingly, $f_{opt,thin}$ in Eq. (10.47) is not the true value of optimal f. This also means that (unlike function $\mathscr{H}_{thin}(f)$ in Eq. (10.42) –

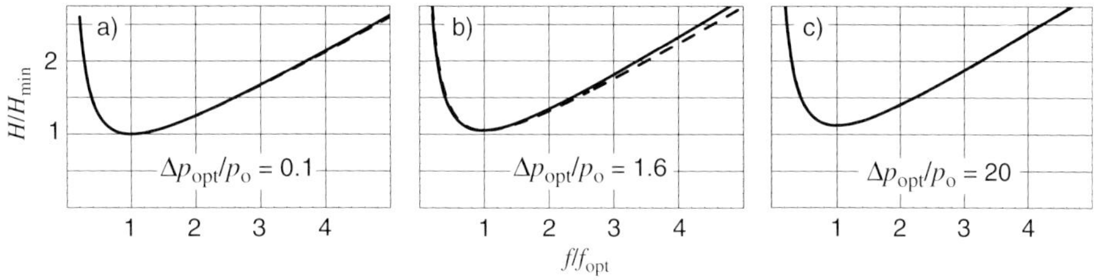

Figure 10.12 Normalized plate height, $H/H_{min} = H_{thin}/H_{min,thin}$, as a function of normalized flow rate, $f/f_{opt} = f/f_{opt,thin}$, in a thin film column for several levels of decompression of a carrier gas along the column at $f = f_{opt}$. Quantities Δp_{opt} and p_o are pressure drop at $f = f_{opt}$ and outlet pressure, respectively (p_o is fixed and the same in all cases). Solid lines represent exact value of H_{thin} calculated from Eq. (10.112) where j_G is a function of f. Dashed lines represent hyperbolas calculated from Eq. (10.112) with $j_G = j_{G,opt}$ where $j_{G,opt}$ is j_G at $\Delta p = \Delta p_{opt}$ ($j_{G,opt}$ is independent of variable f, but can be different for different curve). Only in the case (b) of moderate decompression, the solid line is visibly different from hyperbola (dashed line). There is no visible difference when the decompression is weak (a) or strong (c).

a hyperbola in the (f, $\mathcal{H}_{thin}$)-plane) function $H_{thin}(f)$ is, strictly speaking, not a hyperbola in the (f, H_{thin})-plane (Figure 10.12).

Due to the presence of flow-dependent parameter j_G in Eq. (10.112), the exact formulae for the lowest plate height and for optimal specific flow rate are complex. Fortunately, dependence of j_G on pressure and, therefore, on flow rate is weak (Figure 7.5a). Very small error is added to quantities H_{min} and f_{opt} if, in any formula for apparent plate height, j_G is approximated by its value, $j_{G,opt}$, at $f = f_{opt}$, that is

$$j_G = j_{G,opt} = j_G|_{f=f_{opt}} \qquad (10.115)$$

For thin film columns, this means that quantity j_G in Eqs. (10.112) and (10.114) is approximated by its value at $f = f_{opt,thin}$ where $f_{opt,thin}$ is described in Eq. (10.47). The errors of this approximation are illustrated in Figures 10.12 and10.13.

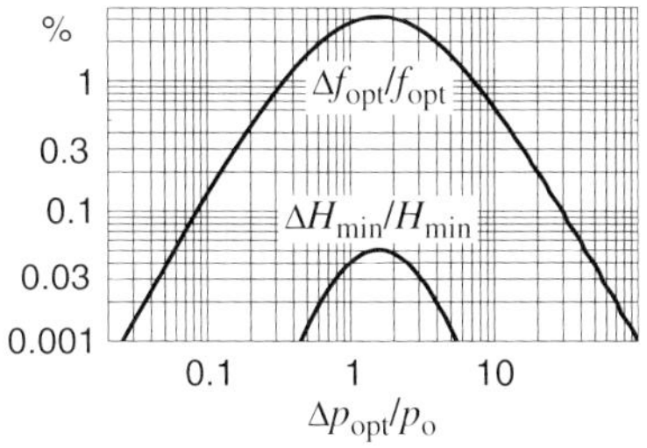

Figure 10.13 Relative %-departures, $\Delta f_{opt}/f_{opt}$ and $\Delta H_{min}/H_{min}$, of f_{opt} and H_{min} in a thin film column from their actual values as functions of normalized optimal pressure drop, $\Delta p_{opt}/p_o$. Quantities $\Delta f_{opt}/f_{opt}$ and $\Delta H_{min}/H_{min}$ are the largest (about 3% and 0.05%, respectively) at moderate gas decompression along a column (at $\Delta p_{opt} \approx 1.6 p_o$) and diminish when, under optimal conditions, the gas decompression is either small or large.

Symmetric forms such as the one in Eq. (10.113) are useful in many studies of column performance. One of them is evaluation of the effect of departure of column pneumatic conditions from their optimal settings on the analysis time.

Example 10.12

Let us find how the values $f_{1.4} = \sqrt{2}f_{opt}$, $f_2 = 2f_{opt}$, and $f_4 = 4f_{opt}$ affect plate number (N) and the analysis time in a thin film column with weak gas decompression.

It follows directly from Eq. (10.113) that the corresponding values of H are: $H_{1.4} = (1.5/2^{1/2})\ H_{min} \approx 1.06\ H_{min}$, $H_2 = 1.25\ H_{min}$, and $H_4 = 2.125\ H_{min}$.

Due to scalability of chromatograms (Section 8.5), a change in the analysis time is proportional to the change in hold-up time (t_M). At weak decompression, t_M is inversely proportional to f. Therefore, $t_{M,1.4} \approx 0.71 t_{M,opt}$, $t_{M,2} = 0.5 t_{M,opt}$, $t_{M,4} = 0.25\ t_{M,opt}$ where $t_{M,opt}$ is t_M at $f = f_{opt}$.

An increase in the plate height leads to a proportional decrease in the plate number, N. This might be an unacceptable price for the reduction in analysis time. Scott and Hazeldean noticed [135] that the loss in N can be recovered by the column length increase equal to the increase in H. This will also raise the analysis time. However, some net gain in the analysis time will remain. At weak decompression, t_M at a fixed f is proportional to L. Therefore, the values of t_M corresponding to no net loss in N are $t_{M,1.4} \approx 1.06 \times 0.71 t_{M,opt} = 0.75 t_{M,opt}$, $t_{M,2} = 1.25 \times 0.5 t_{M,opt} = 0.625 t_{M,opt}$, $t_{M,4} = 2.125 \times 0.25 t_{M,opt} \approx 0.53 t_{M,opt}$. One can conclude that at weak gas decompression, significant (almost 50%) reduction in the analysis time compared to $t_{M,opt}$ can be made without reduction in N and without changing the column internal diameter. □

Example 10.13

At strong gas decompression in a thin film column, flow rates of the previous example have the same effect on the column plate height, but different effect on the hold-up time and, therefore, on the analysis time. At strong decompression, average gas velocity ($\bar{u}$) at a fixed column length (L) is proportional to $f^{1/2}$ (Eq. (7.200)). Therefore, using notation of Example 10.12, one can write: $t_{M,1.4} \approx 0.84 t_{M,opt}$, $t_{M,2} \approx 0.71 t_{M,opt}$, $t_{M,4} = 0.5 t_{M,opt}$.

As in Example 10.12, the loss in the plate number due to the departure of f from f_{opt} can be compensated by an increase in L. At strong decompression, t_M at a fixed f is proportional to $L^{3/2}$, Eqs. (7.118) and (7.200). Therefore, the values of t_M corresponding to no loss in N are $t_{M,1.4} \approx 1.09 \times 0.84 t_{M,opt} \approx 0.92 t_{M,opt}$, $t_{M,2} \approx 1.4 \times 0.71 t_{M,opt} \approx t_{M,opt}$, $t_{M,4} = 3.1 \times 0.5 t_{M,opt} \approx 1.55 t_{M,opt}$. The largest reduction in the analysis time in these examples is about 8% reduction in $t_{M,1.4}$. Quadrupling the flow rate compared to its optimal value accompanied by the increase in the column length in order to preserve N leads to a substantial increase (55%) in the analysis time. □

Due to Eqs. (10.109), (10.113) and (10.114), dimensionless plate height (h_{thin}) in a thin film column can be found as

$$h_{\text{thin}} = \frac{H_{\text{thin}}}{d} = \frac{h_{\text{min,thin}}}{2}\left(\frac{f_{\text{opt,thin}}}{f} + \frac{f}{f_{\text{opt,thin}}}\right) \tag{10.116}$$

where

$$h_{\text{min,thin}} = \frac{H_{\text{min,thin}}}{d} = \frac{j_G \vartheta_{G1}}{2\sqrt{3}} \tag{10.117}$$

is the minimal dimensionless plate height (Figure 10.14) in a thin film column. Specific flow rate ($f_{\text{opt,thin}}$) in that column is described in Eqs. (10.47) and (10.52).

Of all parameters affecting h_{thin} and $h_{\text{min,thin}}$ (Eqs. (10.116) and (10.117)), only the compressibility factor (j_G) can change with the column dimensions. However, all possible values of j_G are confined within a relatively narrow range. In a bigger picture of large changes in column dimensions and pneumatic condition, the minor changes in j_G can be ignored. All parameters in the right-hand side of Eq. (10.116) become independent of column dimensions. One can conclude that

Statement 10.8

At any value of specific flow rate (f), dimensionless plate height (h_{thin}) in a thin film capillary column is independent of a column diameter and length.

Due to Eqs. (10.116) and (7.97), plate height (H_{thin}) and plate number (N_{thin})$= L/H_{\text{thin}}$, in a thin film column can be expressed as

$$H_{\text{thin}} = h_{\text{thin}} d, \quad N_{\text{thin}} = \frac{L}{H_{\text{thin}}} = \frac{\ell}{h_{\text{thin}}} \tag{10.118}$$

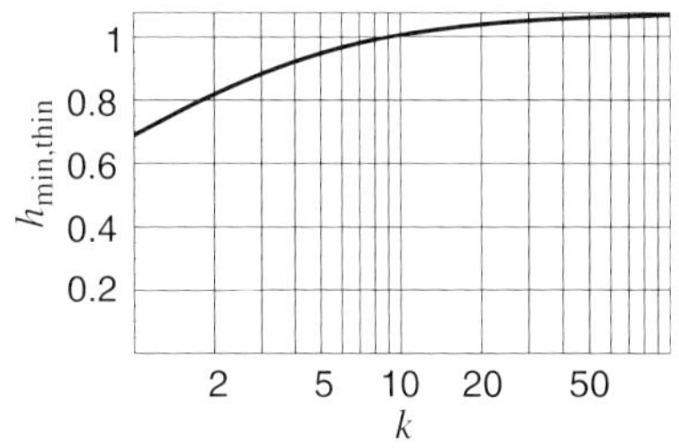

Figure 10.14 Minimal dimensionless plate height, $h_{\text{min,thin}} = H_{\text{min,thin}}/d$, Eq. (10.117), in a thin film column as a function of retention factor, k, for significantly retained solutes ($k \geq 1$) in isothermal analysis. The graph shows $h_{\text{min,thin}}$ under the strong decompression of carrier gas ($j_G = 9/8$). When the decompression is weak ($j_G = 1$), $h_{\text{min,thin}}$ is about 10% lower at any k.

where $\ell = L/d$ is a column dimensionless length. The formulae indicate that H_{thin} and N_{thin} can be expressed as the product of two independent factors. One factor, d or ℓ, represents a column dimension. Another factor, h_{thin}, Eqs. (10.116) and (10.117), represents the column pneumatic conditions and solute retention.

10.5.2
Plate Height and Average Gas Velocity

At $c_{u2} = 0$, Eq. (10.100) yield for the plate height (H_{thin}) in a thin film column:

$$H_{\text{thin}} = \left(\frac{8b\Omega p_{\text{st}}}{\pi d p_{\text{o}}^2(\Phi^2(\bar{u}\Omega/p_{\text{o}})-1)} + \frac{\pi c_1 d p_{\text{o}}^2(\Phi^2(\bar{u}\Omega/p_{\text{o}})-1)}{8p_{\text{st}}\Omega}\right) j_G \tag{10.119}$$

This formula expressing H_{thin} as a function of carrier gas average velocity ($\bar{u}$) is significantly more complex than Eq. (10.112) expressing H_{thin} as a function of the gas-specific flow rate.

The optimal average gas velocity, $\bar{u}_{\text{opt,thin}}$, in a thin film column at an arbitrary gas decompression can be found directly from Eq. (10.119). However, it is convenient to express $\bar{u}_{\text{opt,thin}}$ via already known $f_{\text{opt,thin}}$, Eq. (10.47).

It follows from Eqs. (7.178) and (7.180) that $\bar{u}_{\text{opt,thin}}$ could be found as [127]

$$\bar{u}_{\text{opt,thin}} = \frac{48\Omega p_{\text{st}}^2 f_{\text{opt,thin}}^2}{\pi^2 d_{\text{cx}}^2 p_{\text{o}}^3\left(\left(1+\dfrac{8\Omega p_{\text{st}} f_{\text{opt,thin}}}{\pi d_{\text{cx}} p_{\text{o}}^2}\right)^{3/2} - 1\right)} = \frac{9(L/L_{\text{crit,opt}})u_{\text{o,opt,thin}}}{2((1+3L/L_{\text{crit,opt}})^{3/2}-1)} \tag{10.120}$$

where

$$u_{\text{o,opt,thin}} = \sqrt{\frac{b_u}{c_{1u}}} = \frac{8\sqrt{3}D_{\text{pst}}p_{\text{st}}}{dp_{\text{o}}\vartheta_{\text{G1}}} = \frac{4p_{\text{st}}}{\pi d p_{\text{o}}} f_{\text{opt,thin}} \tag{10.121}$$

is the optimal outlet velocity corresponding to minimal H in a thin film column and $L_{\text{crit,opt}}$ is the column critical length (Eq. (7.76)) corresponding to $f = f_{\text{opt,thin}}$ (and, therefore, to $u_{\text{o}} = u_{\text{o,opt,thin}}$). Diffusivity, D_{pst}, of a solute at standard pressure, p_{st}, depends on the gas and the solute. Generally, D_{pst} is not known for most of the solutes. However, as follows from the discussion leading to Eq. (10.52), approximation in Eq. (6.41) allows one to eliminate the uncertainty. After the substitution of Eq. (10.52), the last formula becomes

$$u_{\text{o,opt,thin}} \approx \left(\frac{T}{T_{\text{st}}}\right)^{\xi-0.25} \frac{8\sqrt{3}\gamma_D^2 D_{\text{g,st}} p_{\text{st}}}{dk^{0.1} p_{\text{o}} \vartheta_{\text{G1}}}, \quad k \geq 0.1 \tag{10.122}$$

It follows from the last two formulae that $u_{\text{o,opt,thin}}$ does not depend on the degree of the gas decompression. On the other hand, average velocity, $\bar{u}_{\text{opt,thin}}$, in Eq. (10.120) does, and does it in a rather complex way.

Equation (10.120) becomes significantly simpler at two opposite extremes – when the decompression is weak and when it is strong. As shown in Chapter 7, the decompression is weak when the column is pneumatically short ($L \ll L_{crit,opt}$). The decompression is strong when the column is pneumatically long ($L \gg L_{crit,opt}$).

At $L \ll L_{crit,opt}$ (weak decompression), there is no significant difference between average gas velocity and its local velocity at any location along the column (Figure 7.3a). Equation (10.120) and (10.121) yields

$$\bar{u}_{opt.thin} = u_{opt.thin} = u_{o.opt.thin} = \frac{8\sqrt{3}D_{pst}p_{st}}{dp_o\vartheta_{G1}} = \frac{4f_{opt.thin}p_{st}}{\pi dp_o} \quad \text{at} \quad \Delta p \ll p_o \tag{10.123}$$

At $L \gg L_{crit,opt}$, Eqs. (10.120), Eqs (7.76) and (10.47) yield [3, 128, 129]

$$\bar{u}_{opt.thin} = \frac{6f_{opt.thin}p_{st}}{\pi dp_o\sqrt{3L/L_{crit.opt}}} = \sqrt{\frac{9\sqrt{3}dD_{pst}p_{st}}{32L\vartheta_{G1}\eta}} \quad \text{at} \quad \Delta p \gg p_o \tag{10.124}$$

After substitution of Eq. (6.19), this becomes

$$\bar{u}_{opt.thin} = \sqrt{\frac{81\pi\sqrt{3}}{640}\frac{(10^3\varphi)^{0.04}}{\sqrt{k^{0.1}\vartheta_{G1k}}}\frac{\gamma_D\upsilon_{st}}{\sqrt{\ell}}\left(\frac{T}{T_{st}}\right)^{-0.13}} \quad \text{at} \quad \Delta p \gg p_o \tag{10.125}$$

The last formula expresses $\bar{u}_{opt.thin}$ via the carrier gas molecular properties. In that formula, $\ell = L/d$ is the column dimensionless length, υ_{st} the gas average molecular speed at standard temperature, T_{st}, and γ_D is an empirical gas parameter. The quantities υ_{st} and γ_D as well as their product are listed in Table 6.12.

Equations (10.123) and (10.124) provide an additional insight into relations between column dimensions and its optimal pneumatic conditions (Figure 10.11). In a pneumatically short column (weak decompression under optimal conditions), $\bar{u}_{opt.thin}$ is independent of L and decreases with an increase in d. On the other hand, in a pneumatically long column (strong decompression under optimal conditions), $\bar{u}_{opt.thin}$ falls with an increase in L and rises with an increase in d. This is in a great contrast with the optimal specific flow rate, $f_{opt,thin}$, which is independent of d, L, and gas decompression. If one were to construct a table of quantities $\bar{u}_{opt.thin}$ for several gases, it would not be as simple as Table 10.3 for $f_{opt,thin}$. The table for $\bar{u}_{opt.thin}$ would have to reflect the dependence of $\bar{u}_{opt.thin}$ not only on the carrier gas but also on a column diameter and length.

Formula for H_{thin} in Eq. (10.119) is significantly more complex than its counterpart in Eq. (10.112). Thus (unlike the function $H_{thin}(f)$ in Eq. (10.112) which is practically indistinguishable, Figure 10.12, from a hyperbola in the (f, H_{thin})-plane, and, which due to a minor approximation, Eq. (10.115), becomes such hyperbola) function $H_{thin}(\bar{u},)$ in Eq. (10.119) can be significantly different from a hyperbola in ($\bar{u}$, H_{thin})-plane.

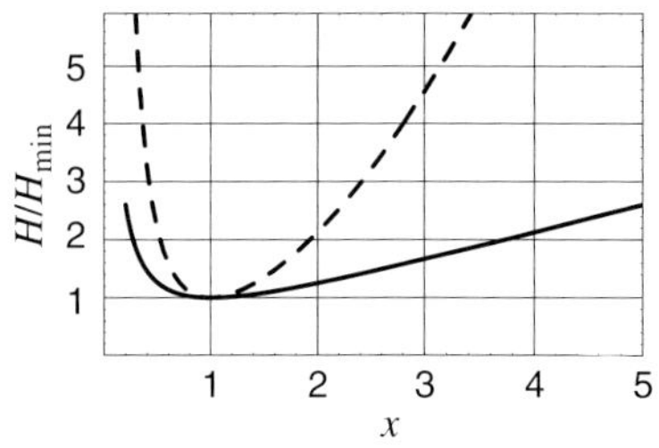

Figure 10.15 Relative plate height, H/H_{min}, in a thin film column as a function of x where (solid line, a hyperbola): $x = f/f_{opt}$ at any decompression, $x = u/u_{opt} = \bar{u}/\bar{u}_{opt}$ at weak decompression, and (dashed line, not a hyperbola) $x = \bar{u}/\bar{u}_{opt}$ at strong decompression. The dashed line shows significantly higher sensitivity of H to departure of x from one. Thus, at $x = 2, H = 1.25H_{min}$ for the solid line and $H = 2.125H_{min}$ for the dashed one.

For weak gas decompression, Eq. (10.119) becomes a hyperbola described in Eq. (10.102) with $c_{u2} = 0$. In a symmetric form, the hyperbola can be expressed as (Figure 10.15)

$$H_{thin} = \frac{H_{min,thin}}{2}\left(\frac{\bar{u}^2_{opt,thin}}{\bar{u}^2} + \frac{\bar{u}}{\bar{u}_{opt,thin}}\right) \quad \text{at} \quad \Delta p \ll p_o \qquad (10.126)$$

where $H_{min,thin}$ is described in Eq. (10.46) and $\bar{u}_{opt,thin}$ is described in Eq. (10.123).

When the decompression is strong, Eq. (10.119) converges to Eq. (10.106) with $c_{u2} = 0$. The latter can be expressed in the following symmetric form (Figure 10.15):

$$H_{thin} = \frac{H_{min,thin}}{2}\left(\frac{\bar{u}^2_{opt,thin}}{\bar{u}^2} + \frac{\bar{u}^2}{\bar{u}^2_{opt,thin}}\right) \quad \text{at} \quad \Delta p \gg p_o \qquad (10.127)$$

where $H_{min,thin}$ is described in Eq. (10.114) and $\bar{u}_{opt,thin}$ is described in Eqs. (10.124) and (10.125). Equation (10.127) describes a hyperbola in $(\bar{u}^2, H_{thin})$-plane. However, it is not a hyperbola in $(\bar{u}, H_{thin})$-plane of Figure 10.15.

The fact that dependence of H_{thin} on $\bar{u}$ and dependence of $\bar{u}_{opt,thin}$ on column dimensions changes with pressure is the source of many complexities associated with using $\bar{u}_{opt}$ as a pneumatic variable. In contrast, H_{thin} is the same hyperbola in (f, H_{thin})-plane (Figure 10.15) at any pressure, and $f_{opt,thin}$ is the same quantity for all column dimensions and all pressures. This is the key advantage of variable f over variable $\bar{u}$.

The pressure-dependent nature of relations between $\bar{u}_{opt}$ and column dimensions significantly complicates the evaluation of the effect of changes in column dimensions on its performance. Thus, although the following example evaluates only one nonoptimal value of $\bar{u}$, it is more complex than Example 10.13 where three nonoptimal values of f were evaluated.

Example 10.14

Consider a pneumatically long column (strong gas decompression under optimal conditions). Suppose that at some original column length, L_{orig}, optimal gas velocity and corresponding minimal plate height and maximal plate number were $\bar{u}_{opt.orig}$, $H_{min,orig}$, and $N_{max,orig}$, respectively. Compare the plate height and the plate number at $\bar{u} = 2\bar{u}_{opt.orig}$ and at arbitrary column length (L) with their original values. The quantity $2\bar{u}_{opt}$ is known from Scott and Hazeldean [135] as optimum practical gas velocity (OPGV). Let us denote $2\bar{u}_{opt}$ and the corresponding H and N as OPGV $= 2\bar{u}_{opt}$, H_{OPGV}, and N_{OPGV}, respectively. (Because $\bar{u}$ is proportional to $f^{1/2}$ (Eq. (7.200)), the case of $\bar{u} =$ OPGV is equivalent to the case of $f = 4f_{opt}$ in Example 10.13.)

The optimal gas velocity, $\bar{u}_{opt}$, is a function, Eq. (10.124), of a column length, L. Therefore, according to Eq. (10.127), changing L while keeping gas velocity ($\bar{u}$) fixed at $\bar{u} =$ OPGV can change H_{OPGV}. As a result, considering the effect of a change in L at OPGV on N_{OPGV} that can be found as $N_{OPGV} = L/H_{OPGV}$, one needs to take into account not only the direct effect of L on N_{OPGV} but also the effect of L on N_{OPGV} through its effect on H_{OPGV}. For that, one needs to know the effect of L on optimal gas velocity, $\bar{u}_{opt}$, at an arbitrary L.

According to Eq. (10.124), quantities $\bar{u}_{opt.orig}$ and $\bar{u}_{opt}$ can be expressed as $\bar{u}_{opt.orig} = X/L_{orig}^{1/2}$ and $\bar{u}_{opt} = X/L^{1/2}$ where quantity X is independent of L. Equation (10.127) yields

$$H_{OPVG} = \frac{H_{min}}{2}\left(\frac{\bar{u}_{opt}^2}{4\bar{u}_{opt.orig}^2} + \frac{4\bar{u}_{opt.orig}^2}{\bar{u}_{opt}^2}\right) = \frac{H_{min}}{2}\left(\frac{L_{orig}}{4L} + \frac{4L}{L_{orig}}\right)$$

$$\frac{N_{OPVG}}{N_{max.orig}} = 2\frac{L}{L_{orig}}\left(\frac{L_{orig}}{4L} + \frac{4L}{L_{orig}}\right)^{-1} = \frac{8(L/L_{orig})^2}{1+16(L/L_{orig})^2} = \begin{cases} 0.47, & \text{at } L = L_{orig} \\ 0.5, & \text{at } L \gg L_{orig} \end{cases}$$

This means that when gas decompression is strong, using OPGV without changing the column length ($L = L_{orig}$) reduces the plate number (N_{OPGV}) to 47% of its maximal value ($N_{max,orig}$) corresponding to $\bar{u}_{opt.orig}$. It further means that no increase in a column length at OPGV can raise N_{OPGV} to half of $N_{max,orig}$.

When N changes from $N_{max,orig}$ to N_{OPGV}, the hold-up time, t_M, changes from $t_{M,opt,orig} = L_{orig}/\bar{u}_{opt.orig}$ to $t_{M,OPGV} = L/(2\bar{u}_{opt.orig})$. Therefore

$$\frac{t_{M,OPGV}}{t_{M,opt,orig}} = \frac{0.5L}{L_{orig}}$$

This means that, when, at $\bar{u} =$ OPGV, column gets longer, t_M increases in proportion with the increase in the column length while having practically no effect on N_{OPGV}. In other words, raising the column length while changing $\bar{u}$ from $\bar{u}_{opt.orig}$ to OPGV is practically harmful.

Combining the last two formulae, one has

$$\frac{N_{OPVG}}{N_{max.orig}} = \frac{32(t_{M,OPVG}/t_{M,opt.orig})^2}{1+64(t_{M,OPVG}/t_{M,opt.orig})^2}$$

This shows that using OPGV under large gas decompression is always harmful to plate number (N) which, at OPGV, is at least two times lower than its original value at $\bar{u}_{\text{opt,orig}}$ and L_{orig}. The loss in N can be even larger if a significant reduction in t_{M} (and, therefore, in the analysis time) by using a shorter column is attempted. More favorable tradeoff between N and t_{M} at large gas decompression is available by avoiding the use of OPGV and by using speed-optimized flow rate (SOF) defined as $\text{SOF} = \sqrt{2}F_{\text{opt}}$ [127, 136]. □

10.5.3
Plate Height and Pressure Drop

At $c_{u2} = 0$, Eq. (10.98) yields for the plate height (H_{thin}) in a thin film column:

$$H_{\text{thin}} = \left(\frac{8b\Omega p_{\text{st}}}{\pi d(p_{\text{i}}^2 - p_{\text{o}}^2)} + \frac{\pi c_1 d(p_{\text{i}}^2 - p_{\text{o}}^2)}{8p_{\text{st}}\Omega}\right) j_G \tag{10.128}$$

At fixed j_{G}, this formula combined with Eqs. (10.39), (10.40), (10.51) and (7.38) yields for optimal pressure drop, $\Delta p_{\text{opt,thin}} = p_{\text{i,opt,thin}} - p_{\text{o}}$:

$$\begin{aligned}\Delta p_{\text{opt,thin}} &= p_{\text{o}}\sqrt{1 + \frac{512\sqrt{3}LD_{\text{pst}}p_{\text{st}}\eta}{d^3 p_{\text{o}}^2 \vartheta_{\text{G1}}}} - p_{\text{o}} \\ &= p_{\text{o}}\sqrt{1 + \frac{512\sqrt{3}\gamma_D^2 D_{\text{g,st}} L p_{\text{st}}\eta \cdot (10^3\varphi)^{0.09}}{d^3 k^{0.1} p_{\text{o}}^2 \vartheta_{\text{G1}}}\left(\frac{T}{T_{\text{st}}}\right)^{\xi - 0.25}} - p_{\text{o}}\end{aligned} \tag{10.129}$$

Due to Eqs. (6.18) and (6.20), this can be also expressed as

$$\Delta p_{\text{opt,thin}} = p_{\text{o}}\sqrt{1 + \frac{120\sqrt{3}\pi L p_{\text{st}}^2 (\gamma_D \lambda_{\text{st}})^2 (10^3\varphi)^{0.09}}{d^3 k^{0.1} p_{\text{o}}^2 \vartheta_{\text{G1}}}\left(\frac{T}{T_{\text{st}}}\right)^{2\xi - 0.25}} - p_{\text{o}} \tag{10.130}$$

indicating that $\Delta p_{\text{opt,thin}}$ increases with the increase in the product $\gamma_D\lambda_{\text{st}}$ where λ_{st} is mean free path of the gas molecules at standard pressure and temperature and γ_D is the empirical parameter of the gas. The quantities γ_D, λ_{st}, and $\gamma_D\lambda_{\text{st}}$ listed in Table 6.12 show that the gases lined-up from the lowest to the highest $\Delta p_{\text{opt,thin}}$ in the same thin film column are nitrogen, argon, hydrogen, and helium. The same line-up can be found in Figure 10.10c where the difference in $\Delta p_{\text{opt,thin}}$ of nitrogen and argon is very small as is the difference in the products $\gamma_D\lambda_{\text{st}}$ for these gases in Table 6.12.

For pneumatically short ($\Delta p_{\text{opt,thin}} \ll p_{\text{o}}$) and long ($\Delta p_{\text{opt,thin}} \gg p_{\text{o}}$) columns, Eqs. (10.129) and (10.130) yield

$$\Delta p_{\text{opt,thin}} = \frac{256\sqrt{3}D_{\text{pst}}Lp_{\text{st}}\eta}{d^3 p_{\text{o}}\vartheta_{\text{G1}}} = \frac{256\sqrt{3}D_{\text{g,st}}Lp_{\text{st}}\gamma_D^2\eta(10^3\varphi)^{0.09}}{d^3 k^{0.1} p_{\text{o}}\vartheta_{\text{G1}}}\left(\frac{T}{T_{\text{st}}}\right)^{\xi-0.25}$$

$$= \frac{60\sqrt{3}\pi L p_{\text{st}}^2(\gamma_D\lambda_{\text{st}})^2(10^3\varphi)^{0.09}}{d^3 k^{0.1} p_{\text{o}}\vartheta_{\text{G1}}}\left(\frac{T}{T_{\text{st}}}\right)^{2\xi-0.25}, \quad \Delta p_{\text{opt,thin}} \ll p_{\text{o}} \tag{10.131}$$

$$\Delta p_{\text{opt,thin}} = 16\sqrt{\frac{2\sqrt{3}D_{\text{pst}}Lp_{\text{st}}\eta}{d^3\vartheta_{\text{G1}}}} = 16\sqrt{\frac{2\sqrt{3}D_{\text{g,st}}Lp_{\text{st}}\gamma_D^2\eta(10^3\varphi)^{0.09}}{d^3 k^{0.1}\vartheta_{\text{G1}}}\left(\frac{T}{T_{\text{st}}}\right)^{\xi-0.25}}$$

$$= 2p_{\text{st}}\gamma_D\lambda_{\text{st}}\sqrt{\frac{10\pi\sqrt{27}L(10^3\varphi)^{0.09}}{d^3 k^{0.1}\vartheta_{\text{G1}}}\left(\frac{T}{T_{\text{st}}}\right)^{2\xi-0.25}}, \quad \Delta p_{\text{opt,thin}} \gg p_{\text{o}} \tag{10.132}$$

10.5.4
Specific Flow Rate as a Pneumatic Variable of Choice

Typically, pneumatic conditions of carrier gas in GC analysis are close to optimal. For this reason, it is desirable to use the pneumatic variable in the plate height formulae that can be directly used as a control parameter for the setting of pneumatic conditions of GC analysis so that the optimum of pneumatic variable becomes a pneumatic setpoint.

The following is the summary of factors that should be considered when choosing pneumatic variable in the plate height formulae. Some of these factors were discussed earlier.

(a) Direct measurement and monitoring GC instrument during analysis.
(b) Simple formula for plate height in a thin film column.
(c) Simple formula for optimal value of pneumatic variable.

Only the pressure is (or can be) directly measured and monitored in contemporary GC instruments. Parameters like average gas velocity ($\bar{u}$) or flow rate (F) are typically calculated from pressure parameters and column dimensions. The latter are typically supplied by the instrument operator. Incorrect column dimensions cause incorrect calculation of parameters like $\bar{u}$ and F. From this perspective, pressure drop (Δp) or other pressure parameters like inlet pressure (p_{i}) and virtual pressure ($\Delta\tilde{p}$, Chapter 7) should be the pneumatic variables of choice.

Equation (10.128) for H_{thin} as a function of pressure is relatively simple (especially if the term $p_{\text{i}}^2 - p_{\text{o}}^2$ is treated as a single variable). However, the dependence of $\Delta p_{\text{opt,thin}}$ on column dimensions (Eqs. (10.129) and (10.130)) is complex. Simpler dependencies of $\Delta p_{\text{opt,thin}}$ on column dimensions depend on the degree of gas decompression in the column (Eqs. (10.131) and (10.132)). This is certainly more complex compared

Table 10.4 Properties of several pneumatic parameters.

Parameter	f	F	$\bar{u}$	Δp
Simple formula for plate height H_{thin}	Yes	Yes		Yes
Optimum is independent of column diameter (d)	Yes			
Optimum is independent of column length (L)	Yes	Yes		
Optimum is independent of decompression	Yes	Yes		
Can be directly measured during analysis				Yes

to complete independence of optimal specific flow rate ($f_{opt,thin}$) on column dimensions at any pressure.

Disadvantages of variable $\bar{u}$ compared to variable f were discussed earlier.

Properties of several pneumatic parameters as variables in expressions for H_{min} and as GC control parameters are listed in Table 10.4.

When it comes to simplicity of formulae for H_{thin} and the formulae for the optimum of the pneumatic parameter, specific flow rate (f) is far ahead of average gas velocity ($\bar{u}$) and pressure. The only parameter that is close to f in that regard is flow rate (F, Eq. (10.48)) – its optimum (F_{opt}) is proportional to column diameter (d). Based on criteria listed in Table 10.4, gas average velocity ($\bar{u}$) is the least suitable pneumatic parameter in GC.

Due to its advantages over other pneumatic parameters, specific flow rate (f) is the pneumatic variable of choice in this book.

10.6 Thick Film Columns

Three levels of stationary phase film thickness can be considered. The film can be relatively thin, intermediate, or thick. All depends on the relationship between resistances to mass transfer in carrier gas and in stationary phase, that is, on the relationship between the c_1-term and the c_{u2}-term in Eqs. (10.38) and (10.99). The latter is more general than the former. The film is considered to be thin if the c_1-term is dominant so that the c_{u2}-term can be ignored. The film is thick in the opposite case when the c_1-term can be ignored compared to the c_{u2}-term. The film thickness is intermediate when neither of the two terms dominates the other.

Thin film columns have been studied earlier. The intermediate film thickness represents the most complex case [61, 127–129, 131, 137]. Only the thick film columns are evaluated here. The following are the reasons behind special attention to the thick film columns.

Thicker film increases the column loadability (sample capacity), which helps to reduce distortion of large peaks [10, 62–65]. It can be shown that the same or better result can be obtained by proportional increase in all dimensions – length, diameter, and film thickness – of a thin film column without transforming it into a thick film one. Better understanding of performance of the thick film column is a step toward quantitative comparison of their performance with the thin film ones.

Several factors contribute to the effect of the film thickness on plate height.

Whether a column does or does not behave as a thin film one depends not on the absolute thickness (d_f) of a stationary phase film, but on the dimensionless film thickness, $\varphi = d_f/d$ (Eq. (2.7)), where d is the column internal diameter. According to Eqs. (10.41), parameter c_{u2} is proportional to φ^2.

Resistance to mass transfer in stationary phase can be significantly different for different solutes. According to Eqs. (10.41) and (10.26), parameter c_{u2} is zero for unretained solutes. It also diminishes for highly retained solutes. Therefore, the c_{u2}-term can be dominant (if at all) mostly for the solutes with moderate retention [61]. This means that if the effect of the c_{u2}-term on H is significant and if it affects $f_{\rm opt}$, then optimal conditions for moderately retained solutes can be significantly different from those for slightly and highly retained ones. This complicates performance evaluation and optimization of the thick film columns. For the solutes eluting during a heating ramp in temperature-programmed analysis, this phenomenon is less pronounced because they all elute with more or less equal retention (Chapter 8). However, accounting for the effect of the film thickness on H in temperature-programmed analysis has its own problems.

The lack of consistent dependence [138–143] of diffusivity (D_S) of variety of solutes in liquid stationary phases on column temperature adds another layer of uncertainty in evaluation of the effect of the film thickness on the column performance.

Comparative nature of this study helps to avoid some of these uncertainties.

The c_{u2}-term in Eq. (10.99) is dominant and, therefore, the column behaves as a thick film one at any f when

$$\frac{48c_{u2}f\Omega}{\pi^2 d^2 p_{\rm st}}\left(\left(\frac{p_{\rm o}^2}{p_{\rm st}^2}+\frac{8f\Omega}{\pi d p_{\rm st}}\right)^{\frac{3}{2}}-\frac{p_{\rm o}^3}{p_{\rm st}^3}\right)^{-1} \gg c_1 \quad \text{(thick film column)} \tag{10.133}$$

Under this condition, Eq. (10.99) combined with Eqs. (10.39) and (10.41) becomes

$$H_{\rm thick} = \frac{\pi d D_{\rm pst}}{2f} j_G + \frac{96 f^2 \varphi^2 \vartheta_{G2}^2 \Omega}{3\pi^2 D_S p_{\rm st}}\left(\left(\frac{p_{\rm o}^2}{p_{\rm st}^2}+\frac{8f\Omega}{\pi d p_{\rm st}}\right)^{\frac{3}{2}}-\frac{p_{\rm o}^3}{p_{\rm st}^3}\right)^{-1} \tag{10.134}$$

The last formula shows that the effect of the film thickness on $H_{\rm thick}$ depends on the degree of gas decompression [61]. It follows from Eq. (7.70) that at weak decompression, Eq. (10.134) can be reduced to the form

$$H_{\rm thick} = \frac{\pi d D_{\rm pst}}{2f} + \frac{8 d p_{\rm st} \vartheta_{G2}^2 \varphi^2}{3\pi D_S p_{\rm o}} f, \quad \text{at} \quad \Delta p \ll p_{\rm o} \tag{10.135}$$

Equation (10.135) is identical to Eq (10.101) with $c_1 = 0$. Optimal specific flow, $f_{\rm opt,thick}$, and the corresponding minimal plate height, $H_{\rm min,thick}$, can be found from Eq. (10.135) as

$$f_{\text{opt,thick}} = \frac{\pi}{4\varphi\vartheta_{G2}}\sqrt{\frac{3D_S D_{\text{pst}} p_o}{p_{\text{st}}}}, \quad \text{at} \quad \Delta p \ll p_o \tag{10.136}$$

$$H_{\text{min,thick}} = 4d\varphi\vartheta_{G2}\sqrt{\frac{D_{\text{pst}} p_{\text{st}}}{3D_S p_o}} \quad \text{at} \quad \Delta p \ll p_o \tag{10.137}$$

These formulae show that in a thick film column with small gas decompression, an increase in the film thickness proportionally reduces optimal gas flow rate and raises minimal plate height.

At strong decompression ($p_i \gg p_o$), $f \gg \pi d p_o^2/(8p_{\text{st}}\Omega)$ as follows from Eq. (7.60). Under this condition, Eq. (10.134) combined with Eq. (7.38) yields

$$H_{\text{thick}} = \frac{9\pi d D_{\text{pst}}}{16f} + \frac{\vartheta_{G2}^2\varphi^2}{4D_S}\sqrt{\frac{d^5 p_{\text{st}}}{\pi L\eta}}\sqrt{f} \quad \text{at} \quad \Delta p \gg p_o \tag{10.138}$$

$$f_{\text{opt,thick}} = \frac{\pi}{d}\left(\frac{9D_S D_{\text{pst}}}{2\vartheta_{G2}^2\varphi^2}\right)^{2/3}\left(\frac{L\eta}{p_{\text{st}}}\right)^{1/3} \quad \text{at} \quad \Delta p \gg p_o \tag{10.139}$$

$$H_{\text{min,thick}} = \frac{d^2}{8}\left(\frac{3^5 D_{\text{pst}} p_{\text{st}} \vartheta_{G2}^4 \varphi^4}{2D_S^2 L\eta}\right)^{1/3} \quad \text{at} \quad \Delta p \gg p_o \tag{10.140}$$

These formulae show that in a thick film column with large gas decompression, an increase in φ more than proportionally reduces the optimal flow rate and raises the minimal plate height. The detrimental effects of the increase in φ can be compensated to some degree by the increase in the column length, L. This is due to the fact that longer columns require higher pressure for the same flow rate. That, in turn, reduces the gas velocity at the beginning of the column, which reduces the detrimental effect of the film thickness in that column segment. Additional discussion on this phenomenon can be found elsewhere [61].

10.7 Temperature-Programmed Analyses

As mentioned earlier, Eq. (10.93) is a recommended and widely accepted definition of plate number (N) in chromatography. Frequently, Eq. (10.93) is treated as the unconditional definition of N. Unfortunately, the definition might be unsuitable [49] for dynamic conditions such as those existing in pressure- and/or temperature-programmed GC.

As shown later, all peaks eluting during a long heating ramp in a temperature-programmed analysis have roughly the same width [144, 145]. On the other hand, the ratio, $t_{R,\text{last}}/t_{R,\text{first}}$, of retention times of the last and the first peak could exceed an order of magnitude. As a result, N in Eq. (10.93) can, for the last peak, be several orders of magnitude larger than it is for the first peak. This means that in dynamic analysis in general and in temperature-programmed GC analysis in particular, N in Eq. (10.93) can no longer be treated as a column parameter (as it is the case with N in a static analysis). Instead, N becomes an attribute of each particular peak – the attribute that strongly depends on the peak retention time. As a result, it becomes unclear what useful purpose can that attribute serve. With quantities t_R and N in it being strong functions of each other, Eq. (10.94) can no longer serve as a means for the peak widths calculation, or, as Giddings suggested [49] (p. 723, column 1), "the use of the latter [formula as a definition of plate number] is incorrect."

A meaningful definition of the apparent plate number (N) in an isobaric temperature-programmed GC analysis has been proposed by Habgood and Harris [146] in 1960 and further elaborated in their 1966 book [59]. Habgood and Harris defined N

$$N = \frac{t_{R,stat}^2}{\sigma^2} \tag{10.141}$$

where $t_{R,stat}$ is the retention time of a solute in a static analysis executed at temperature (T_R) existed at the solute retention time (t_R) in actual temperature-programmed analysis, and σ is actual (measured in actual temperature-programmed analysis) standard deviation of the peak produced by the solute.

Note 10.13

In the original definition, $t_{R,stat}$ has been described as the retention time in isothermal analysis. However, it is clear from the context and from the state-of-the-art at the time that $t_{R,stat}$ meant retention time in isothermal and isobaric, that is, in static analysis. □

Habgood and Harris also provided experimental data confirming insignificant difference between N measured under isothermal and temperature-programmed conditions. This led Giddings to conclude [49] that the definition in question was "certainly the most complete . . . [and that] very little can be added to the conclusions of these two authors since their deductions appear to be theoretically sound and experimentally consistent."

According to Eq. (8.20), $t_{R,stat}$ can be found as

$$t_{R,stat} = \frac{t_{M,R}}{\mu_R} \tag{10.142}$$

where μ_R is solute elution mobility in actual dynamic analysis as well as its mobility in static analysis conducted under the conditions existed at the time t_R in actual dynamic analysis, and $t_{M,R}$ is the hold-up time in static analysis executed under conditions existed at the solute retention time (t_R) in actual dynamic analysis.

Harris – Habgood formula in Eq. (10.141) does not have the shortfalls that Eq. (10.93) has under dynamic conditions. As will be seen shortly, Eq. (10.141) yields N that is almost the same for all solutes in one analysis, and is suitable for the prediction of σ.

In static analysis, quantities $t_{\mathrm{R,stat}}$ and $t_{\mathrm{M,R}}$ are the same as, respectively, t_{R} and t_{M} in that analysis. As a result, Eq. (10.141) can be treated as generalization of the definition in Eq. (10.74) to apparent plate number in any – static and dynamic – chromatographic analysis.

It follows from Eqs. (10.142), (10.78) and (10.81) that adoption of Eq. (10.141) as a general definition of plate number (N) in chromatographic analysis allows one to conclude that in an arbitrary analysis, N as well as apparent plate height (H) and peak width (σ) can be found as

$$N = \frac{t_{\mathrm{M,R}}^2}{\mu_{\mathrm{R}}^2\sigma^2} = \frac{t_{\mathrm{M,R}}^2}{\sigma_{\mathrm{m}}^2} \tag{10.143}$$

$$H = \frac{\sigma^2}{t_{\mathrm{R,stat}}^2}L \tag{10.144}$$

$$H = \frac{\sigma_{\mathrm{m}}^2}{t_{\mathrm{M,R}}^2}L \tag{10.145}$$

$$\sigma = \frac{t_{\mathrm{R,stat}}}{\sqrt{N}} \tag{10.146}$$

$$\sigma = \frac{t_{\mathrm{M,R}}}{\mu_R\sqrt{N}} \tag{10.147}$$

Formulae containing μ_{R} could be more convenient for some applications because they allow one to separate the peak width dependence on a solute dispersion parameters (H and N) from its dependence on the solute mobility μ. In a less specific form, the last formula was used by several authors [147–149] for the estimation of peak widths in temperature-programmed GC analyses.

Equations (10.141) and (10.144) show how H and N can be found from experimental data. However, these formulae leave unanswered the question of prediction of H or N and, therefore, σ from parameters of dynamic analysis. Finding plate height (H_{thin}) and optimal specific flow rate ($f_{\mathrm{opt,thin}}$) in a thin film column in temperature-programmed analysis is the main task for this section.

It is convenient to find H from spatial width ($\tilde{\sigma}$) of eluting solute zone. When conditions are uniform (and possibly dynamic), $\tilde{\sigma}$ can be found from Eq. (10.21) that describes $\tilde{\sigma}^2$ as an additive function of distance z. Equation (10.21) for $\tilde{\sigma}^2$ in uniform (possibly dynamic) analysis is similar in many ways to Eq. (10.69) for σ^2 in static (possibly nonuniform) analysis. Both differential equations have similar structure describing their respective unknown variables $\tilde{\sigma}^2$ and σ^2 as additive functions of z. However, not everything is similar in these equations.

Previously solving Eq. (10.69), we had to deal with only one type of nonuniform conditions – the nonuniformity of gas velocity (u) that was caused by the decompression of ideal gas along the column and described by unique, relatively simple mathematical formula in Eq. (7.67). This led to a closed form solution of Eq. (10.69) and to a closed form expression for plate height (H) and plate number (N) in static analysis.

A general treatment of the uniform (possibly dynamic) conditions described in Eq. (10.21) encounters the reality of more than one way that conditions of a GC analysis could be dynamic. Indeed, temporal changes in GC analysis could come from temperature or from the pressure programming as well as from a combination of both. Generally, each program is a product of imagination of an operator. Furthermore, even the isobaric single-ramp temperature-programmed GC analysis with negligible gas decompression along a column – the simplest practically important type of uniform dynamic conditions – involves a rather complex formula that controls a solute migration. This leads to complex solutions of Eq. (10.21) that are not suitable for practical use. As a result, unlike in the case of exact closed form solution of Eq. (10.69) for static (possibly nonuniform) conditions, one has to accept approximate solutions of Eq. (10.21) for uniform (possibly dynamic) conditions.

10.8 Temperature-Programmed Thin Film Columns

10.8.1 Isobaric Heating Ramp in a Thin Film Columns

It is possible to find a simple and reasonably accurate approximate solution of Eq. (10.21) for highly interactive (highly retained at the beginning of a heating ramp) solutes in isobaric analysis using a thin film column. These conditions are assumed throughout the rest of this section.

It follows from Appendix 10.A.4 combined with Eqs. (10.112), (10.39), (10.51) and (10.40) at $j_G = 1$ that when gas decompression along a thin film column is weak, the plate height (H_{thin}) in static and temperature-programmed isobaric analyses can be expressed as

$$H_{thin} = \frac{B}{f} + C_1 f \quad \text{at} \quad \Delta p \ll p_o \tag{10.148}$$

where

$$B = \gamma_T b = \frac{\pi\gamma_T D_{pst} d}{2} \approx \frac{\pi\gamma_T\gamma_D^2 D_{g.st} d (10^3\varphi)^{0.09}}{2k^{0.1}} \left(\frac{T}{T_{st}}\right)^{\xi-0.25} \tag{10.149}$$

The right-hand side of Eq. (10.149) is valid for $k \geq 0.1$. Other parameters of Eq. (10.148) are

$$C_1 = \frac{d^2 \vartheta_1^2}{48B} \tag{10.150}$$

$$\gamma_T = \begin{cases} 1, & \text{static conditions} \\ 0.82, & \text{isobaric heating ramp} \end{cases} \tag{10.151}$$

and (Eqs. (10.25) and (10.27), Figure 10.2)

$$\vartheta_1 = \vartheta_{G1} = \begin{cases} \sqrt{1+4\omega+6\omega^2} \\ \sqrt{11-16\mu+6\mu^2} \\ \sqrt{1+6k+11k^2}/(1+k) \end{cases} \approx \begin{cases} 1+2.25\omega \\ 3.25-2.25\mu \\ (1+3.25k)/(1+k) \end{cases} \quad \text{(static analysis)} \tag{10.152}$$

$$\vartheta_1 = \sqrt{1+4\bar{\omega}+6\bar{\omega}^2} \quad \text{(isobaric temperature-programmed analysis)} \tag{10.153}$$

where $\bar{\omega}$ is the distance-averaged immobility of a solute in temperature-programmed analysis. The quantity ϑ_1 defined in Eq. (10.152) incorporates quantity ϑ_{G1} (Eq. (10.25)) for static analyses and extends it beyond the static analyses in the form of Eq. (10.153).

To make Eq. (10.148) and the formulae for its components suitable for static conditions and for the heating ramps in temperature-programmed analyses, no distinction is made in notations for quantities f, k, and T. However, it is understood that, in the case of temperature-programmed analysis, these quantities represent their respective parameters at the solute elution time, t_R. In other words, in the case of temperature-programmed analysis, quantities f, k, and T in Eq. (10.148), in its components, and in expressions that follow from these formulae should be interpreted as

$$f = f_R, \quad k = k_R, \quad T = T_R \quad \text{(temperature program)} \tag{10.154}$$

where T_R is a solute elution temperature, k_R is its retention factor at $T = T_R$, and f_R is specific flow rate of carrier gas at $T = T_R$.

For highly interactive solutes eluting during a heating ramp, $\bar{\omega}$ can be found from Eq. (8.122) where r_T is the dimensionless heating rate. Equation (10.153) becomes

$$\vartheta_1 = \sqrt{1+4\frac{1-e^{-r_T}}{r_T}+6\left(\frac{1-e^{-r_T}}{r_T}\right)^2} \approx 1+\frac{9(1-e^{-r_T})}{4r_T} \quad \text{(isobaric heating ramp)} \tag{10.155}$$

The approximation error in the right-hand side of this formula is insignificant at any r_T, Figure 10.16. For moderate heating rates, the formula can be further simplified as, Figure 10.16,

$$\vartheta_1 = 2.4r_T^{-0.2} \quad \text{(isobaric heating ramp, } 0.2 \le r_T \le 2) \tag{10.156}$$

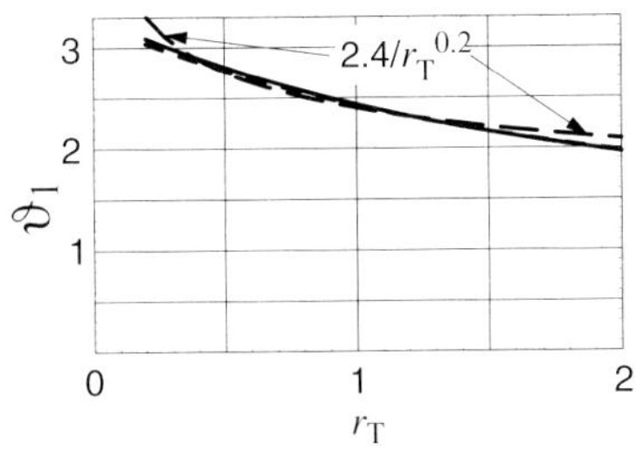

Figure 10.16 Parameter ϑ_1 calculated from Eq. (10.155) (solid line) and Eq. (10.156) (dashed line). The difference between the exact and the approximate values in Eq. (10.156) is barely visible in the graph.

As in the case of only the static conditions, B as a function of a solute diffusivity, D_{pst}, in Eq. (10.149) corresponds to the most accurate description of H_{thin}. Unfortunately, D_{pst} is known for minority of solutes typically analyzed by GC. The right-hand side approximation in Eq. (10.149) removes this uncertainty by expressing D_{pst} via self-diffusivity, $D_{g,st}$, and an empirical parameter, γ_D, of a carrier gas – both can be found from Table 6.12 – and via the solute elution temperature, T.

So far, gas decompression in temperature-programmed analysis has not been considered. The following observations suggest that the gas decompression affects plate height in a typical temperature-programmed GC analysis in approximately the same way as it does the plate height in static analysis.

According to Statement 8.8, gas decompression has no effect on solute elution temperatures and other elution parameters, Figure 8.6b. The same follows from scalability of methods of GC analyses described in Chapter 8 and elsewhere [136, 149–156]. Thus it is possible to construct methods of isobaric GC analyses with arbitrary temperature programs that allow one to reproduce elution temperatures of *all* (not only the highly interactive) solutes regardless of gas decompression. Furthermore, during major portion of migration of a highly interactive solute, the column temperature remains relatively close (Figure 8.4) to the solute elution temperature, T.

Therefore, it is reasonable to assume that during the major portion of a solute migration, gas velocity remains almost static as if the column temperature was fixed at some temperature close to its elution temperature (T). One can conclude that gas decompression should have the same affect on H_{thin} in Eq. (10.148) as it has on H_{thin} in Eqs. (10.112). Equation (10.148) becomes

$$H_{thin} = \left(\frac{B}{f} + C_1 f\right) j_G \qquad (10.157)$$

where, as before, quantities B and C_1 are described in Eqs. (10.149) and (10.150).

The last formula can be also expressed in symmetric forms

$$H_{thin} = \frac{H_{min,thin}}{2}\left(\frac{f_{opt,thin}}{f} + \frac{f}{f_{opt,thin}}\right) \qquad (10.158)$$

$$h_{\text{thin}} = \frac{H_{\text{thin}}}{d} = \frac{h_{\text{min,thin}}}{2}\left(\frac{f_{\text{opt,thin}}}{f} + \frac{f}{f_{\text{opt,thin}}}\right) \tag{10.159}$$

that reproduce Eqs. (10.113) and (10.116), respectively. As follows from Eqs. (10.149) and (10.150), quantities $H_{\text{min,thin}}$, $h_{\text{min,thin}}$, and $f_{\text{opt,thin}}$ in these formulae can be found as

$$H_{\text{min,thin}} = 2\sqrt{BC_1} = \frac{j_G d\vartheta_1}{2\sqrt{3}} = h_{\text{min,thin}}d, \quad h_{\text{min,thin}} = \frac{j_G\vartheta_1}{2\sqrt{3}} \tag{10.160}$$

$$f_{\text{opt,thin}} = \sqrt{\frac{B}{C_1}} = \frac{2\sqrt{3}\pi\gamma_T D_{\text{pst}}}{\vartheta_1} \tag{10.161}$$

$$f_{\text{opt,thin}} = \frac{2\sqrt{3}\pi\gamma_T\gamma_D^2 D_{\text{g,st}}(10^3\varphi)^{0.09}}{k^{0.1}\vartheta_1}\left(\frac{T}{T_{\text{st}}}\right)^{\xi-0.25}, \quad k \geq 0.1 \tag{10.162}$$

The last formulae for H_{thin} and for its components represent the most general description of these quantities in this book. From now on, these formulae are treated as the primary source for H_{thin} and its parameters. The use of notations B, C_1, and ϑ_1 instead of previous b, c_1, and ϑ_{G1}, respectively, that were suitable only for static conditions facilitates the isolation of the final formulae from the intermediate ones.

The formulae lead to several interesting observations regarding highly interactive solutes in a thin film column.

Temperature programming influences the effect of a solute–column interaction on $H_{\text{min,thin}}$, $h_{\text{min,thin}}$, and $f_{\text{opt,thin}}$. In static analysis, parameter ϑ_1 representing the effect of the solute–column interaction in Eqs. (10.160), (10.161) and (10.162) is fixed for each particular solute during the analysis. It can be expressed (Eq. (10.152)) via the solute parameters k, μ, or ω. On the other hand, in the case of temperature programming, ϑ_1 for a given solute is a function (Eq. (10.153)) of the distance-averaged ($\bar{\omega}$) solute immobility (interaction level) that changes during the analysis. Equation (10.160) also shows that other than through its effect on ϑ_1, temperature programming has no effect on $H_{\text{min,thin}}$ and $h_{\text{min,thin}}$.

A typical single-ramp temperature program reduces $f_{\text{opt,thin}}$ by 20% compared to $f_{\text{opt,thin}}$ of the same solute migrating at the same temperature in isothermal analysis. While changing $f_{\text{opt,thin}}$, temperature programming does not change the ratios, Table 10.2, of $f_{\text{opt,thin}}$ for different gases.

Let us consider the evolution of plate height during heating ramp. In doing so, it should be noticed that according to Eqs. (10.147), (10.79) and (10.110), the widths of the peaks are inversely proportional to $N^{1/2}$ and proportional to $H^{1/2}$ and $h^{1/2}$. Therefore, quantities $H^{1/2}$, $h^{1/2}$, and $N^{1/2}$ are more representative than their counterparts H, h, and N, respectively.

According to Eq. (10.162), $f_{\text{opt,thin}}$ is proportional to $T^{\xi-0.25}$. On the other hand, according to Eqs. (7.60), (7.38) and (6.20), f at fixed pressure is proportional to $T^{-\xi}$. Therefore, the ratio $f/f_{\text{opt,thin}}$ is proportional to $T^{0.25-2\xi}$. For helium, hydrogen, and nitrogen, this, due to the data in Table 6.12, becomes

$$f_{\text{opt,thin}} \sim T^{\xi-0.25} \sim T^{0.45} \tag{10.163}$$

$$\frac{f}{f_{\text{opt,thin}}} \sim T^{0.25-2\xi} \sim T^{-1.15} \tag{10.164}$$

These relations, together with Eqs. (10.53) and (10.54) explain earlier found [60] experimental results like $F_{\text{opt,thin}} \sim T^{-0.55}$ and $F/F_{\text{opt,thin}} = f/f_{\text{opt,thin}} \sim T^{-1.1}$.

Equation (10.164) suggests a possibility of a substantial mismatch between optimal and actual flow rates during a heating ramp. While the solutes eluting toward the end of the ramp require higher $f_{\text{opt,thin}}$ than the earlier eluites, the temperature increase reduces the actual f. How large is the impact of the mismatch on $h^{1/2}$?

For a heating ramp, quantities f and $f_{\text{opt,thin}}$ can be expressed as $f = f_{\text{init}}(T_R/T_{\text{init}})^{-\xi}$, $f_{\text{opt,thin}} = f_{\text{opt,thin,init}}\,(T_R/T_{\text{init}})^{\xi-0.25}$, where T_{init} is temperature at the beginning of the ramp and T_R is a solute elution temperature. Substitution of these formulae in Eq. (10.159) yields

$$\frac{h_{\text{thin}}}{h_{\text{min,thin}}} = \frac{1}{2}\left(\frac{(T_R/T_{\text{init}})^{2\xi+0.25}}{X_{\text{init}}} + \frac{X_{\text{init}}}{(T_R/T_{\text{init}})^{2\xi+0.25}}\right), \quad X_{\text{init}} = \frac{f_{\text{init}}}{f_{\text{opt,thin,init}}} \tag{10.165}$$

In the following discussion of Eq. (10.165), subscript "thin" is dropped because only thin film columns are considered. According to Eq. (10.160), quantity $h_{\text{min,thin}}$ is independent of T_R. Therefore, the right-hand side of the last formula is a function of only the ratio T_R/T_{init} and of the relative mismatch, $f_{\text{init}}/f_{\text{opt,init}}$, between f and f_{opt} at the beginning of the ramp, Figure 10.17.

The ratio T_R/T_{init} in Figure 10.17 covers a relatively wide temperature range. For example, if $T_{\text{init}} = 300$ K then T_R changes from 300 to 600 K. Figure 10.17 shows that if, at $T = T_{\text{init}}$, f in a thin film column is not smaller than $f_{\text{opt,init}}$ and not larger than

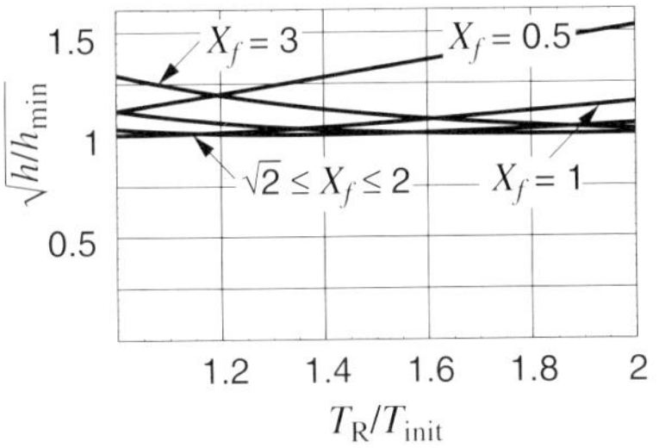

Figure 10.17 Quantity $(h/h_{\text{min}})^{1/2}$ in a thin film column as a function, Eq. (10.165), of relative temperature, T_R/T_{init}, and relative specific flow rate, $X_{\text{init}} = f_{\text{init}}/f_{\text{opt,init}}$, for highly interactive solutes eluting during isobaric heating ramp. T_{init} is a column temperature at the beginning of the ramp, T_R is a solute elution temperature, f_{init} and $f_{\text{opt,init}}$ are actual and optimal specific flow rates of a carrier gas at $T = T_{\text{init}}$.

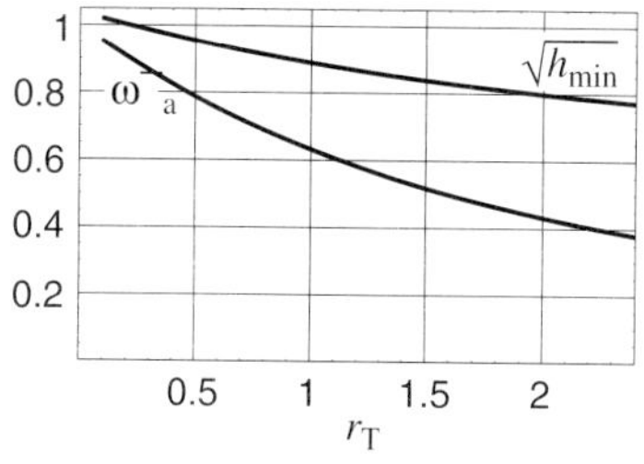

Figure 10.18 Quantities $\bar{\omega}_a$ and $h_{min}^{1/2}$ (Eqs. (8.122) and (10.160)) in a thin film column as functions of dimensionless heating rate (r_T). In the graph, quantity $h_{min}^{1/2}$ corresponds to strong gas decompression ($j_G = 9/8$). When the decompression is weak ($j_G = 1$), $h_{min}^{1/2}$ is about 6% lower at any r_T.

$2f_{opt,init}$ then quantity $(h/h_{min})^{1/2}$ is close to unity for all solutes eluting within a wide temperature range. In other words, if, at the beginning of a heating ramp, specific flow rate, f, is confined within a range $f_{opt} \leq f \leq 2f_{opt}$, then $h^{1/2}$ remains close to $h_{min}^{1/2}$ for all highly interactive solutes eluting during the ramp. The difference between $h^{1/2}$ and $h_{min}^{1/2}$ is especially small when, at the beginning of the ramp, $f \approx 2^{1/2} f_{opt}$.

The quantity h in a thin film column also depends on a heating rate. Indeed, according to Eq. (10.160), minimum ($h_{min,thin}$) in h is a function of quantity ϑ_1 which, in a temperature-programmed analysis, is a function (Eq. (10.153)) of a solute average immobility ($\bar{\omega}_a$). The latter is a function (Eq. (8.122)) of dimensionless heating rate (r_T). Dependence of $\bar{\omega}_a$ and $h_{min}^{1/2}$ on r_T is shown in Figure 10.18. Optimal r_T is typically close to 0.5 (Example 8.13). According to Figure 10.18, $h_{min}^{1/2}$ in this case is close to 1.

Quantities $h^{1/2}$ and $N^{1/2}$ for several dimensionless heating rates are show in Figure 10.19. The flow rate in Figure 10.19 is about 40% higher than optimal.

All in all, the following simple formulae

$$h_{thin} \approx 1, \quad H_{thin} \approx d, \quad N_{thin} \approx L/d = \ell \text{ (optimal or near optimal conditions)} \tag{10.166}$$

appear to be justified approximations for the key solute dispersion parameters under optimal or near optimal static and temperature-programmed conditions.

The relatively tight range of the $H_{min,thin}$ values leads to an important observation regarding the width ($\tilde{\sigma}$) of a solute zone right before its elution. Equations (10.82) and (10.145) yield

$$\tilde{\sigma} = u_o \sigma_m = u_o t_M \sqrt{\frac{H}{L}} = u_o \frac{L}{\bar{u}} \sqrt{\frac{H}{L}} = \frac{L}{j\sqrt{N}} = \frac{\sqrt{HL}}{j} \tag{10.167}$$

where j is James–Martin compressibility factor, Eq. (7.114). Due to Eq. (10.166), this becomes for optimal or near optimal conditions in a thin film column:

$$\tilde{\sigma}_{opt,thin} = \sqrt{\frac{d_c L}{j}} \tag{10.168}$$

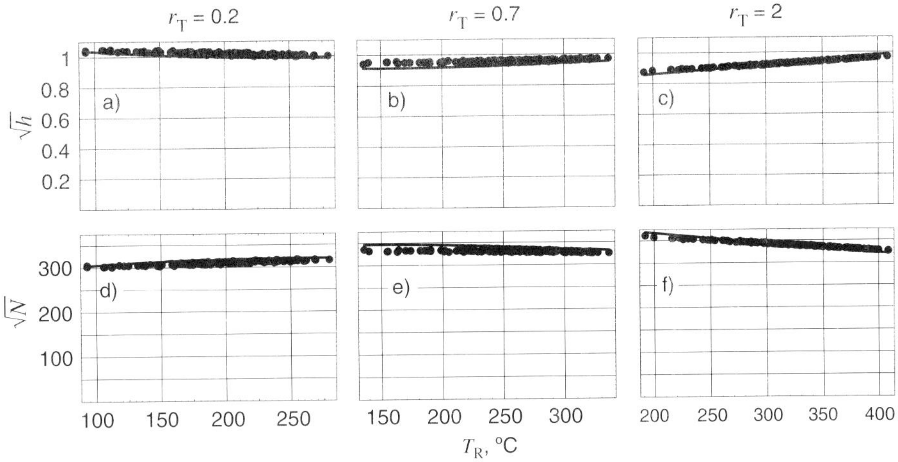

Figure 10.19 Quantities $h^{1/2}$ and $N^{1/2}$ for the pesticides in chromatogram of Figure 8.3. Conditions are the same as in Figures 8.3 and 10.4 except for the dimensionless heating rates, r_T, that were 0.2, 0.7, and 2.0 (at initial hold-up time of 0.32 min, these rates correspond to absolute heating rates, R_T, of 14.67, 51.34, and 146.7 °C/min, respectively). Solid lines represent Eqs. (10.159) and (10.110) with parameters calculated from the method setpoints for each dimensionless heating rate. The dots were calculated from the peak width data in computer-generated chromatograms using Eqs. (10.80), (10.143), and (10.110) and (10.110). Because peak widths are proportional to $h^{1/2}$ and inversely proportional to $N^{1/2}$, these quantities better represent a column performance than their counterparts h and N. In all cases, $h^{1/2} \approx 1$. Therefore, $N^{1/2} \approx (L/d_c)^{1/2}$ (for this column, $(L/d_c)^{1/2} \approx 316$).

$$\tilde{\sigma}_{\text{opt,thin}} = \sqrt{d_c L} \quad \text{at} \quad \Delta p \ll p_o \tag{10.169}$$

Example 10.15

In a column with $L = 60$ m, $d_c = 0.53$ mm under optimal or near optimal conditions, the width, $\tilde{\sigma}_{\text{opt,thin}}$, of *any* solute zone right before its elution is $\tilde{\sigma}_{\text{opt,thin}} = (60\text{ m} \times 0.53\text{ mm})^{1/2} \approx 18$ cm, which is about 0.3% of the column length. It is widely accepted to estimate total space occupied by a Gaussian distribution as 6σ. This means that the total space occupied by *any* Gaussian solute zone along the column in question is about 1 m or about 2% of the column length. □

10.8.2
Critical Length of a Column

A column critical length (L_{crit}) at a given flow of carrier gas is the column length that requires a borderline pressure drop (Chapter 7) $\Delta p = 2p_o$ for that flow. The quantity L_{crit} allows one to recognize the conditions of a column operation (weak or strong decompression) directly from the column dimensions without calculating the column pressure for each particular case.

Due to Eqs. (7.76) and (7.39), *critical length* of a capillary column with a predetermined f can be expressed as

$$L_{\text{crit}} = \frac{3\pi d^3 p_o^2}{256 f p_{\text{st}} \eta} \tag{10.170}$$

For a thin film column at $f = f_{\text{opt}}$, this formula yields after substitution of Eqs. (10.161) and (10.162):

$$L_{\text{crit,opt}} = \frac{\sqrt{3} d^3 p_o^2 \vartheta_1}{512 \gamma_T D_{\text{pst}} p_{\text{st}} \eta} \tag{10.171}$$

$$L_{\text{crit,opt}} = \frac{\sqrt{3} d^3 k^{0.1} \vartheta_1 p_o^2}{512 \gamma_T \gamma_D^2 D_{\text{g,st}} p_{\text{st}} (10^3 \varphi)^{0.09} (T/T_{\text{st}})^{\xi - 0.25} \eta} \quad k \geq 0.1 \tag{10.172}$$

In the latter, quantities $D_{\text{g,st}}$ and η are not independent of each other. They both depend on the carrier gas type. Taking this into account allows one to trace $L_{\text{crit,opt}}$ down to fundamental molecular properties of the gas. Due to Eqs. (6.18) and (6.20), one has

$$L_{\text{crit,opt}} = \frac{0.0046 k^{0.1} (p_o/p_{\text{st}})^2 \vartheta_1}{\gamma_T (10^3 \varphi)^{0.09} (T/T_{\text{st}})^{2\xi - 0.25}} \left(\frac{d}{\gamma_D \lambda_{\text{st}}} \right)^2 d, \quad k \geq 0.1 \tag{10.173}$$

where λ_{st} is the mean free path of gas molecules at standard pressure (p_{st}) and standard temperature (T_{st}); and γ_D is empirical parameter of the gas. Both parameters are listed in Table 6.12. The values of $L_{\text{crit,opt}}$ for several column diameters and carrier gas types are compiled in Table 10.5.

Example 10.16

According to Eq. (10.173) and Table 10.5, a thin film 0.53-mm column of typical length ($L \leq 100$ m) is pneumatically short (has weak gas decompression at optimal flow) with any carrier gas at almost any temperature. On the other hand, a 3-m thin film column with $d = 0.1$ mm and with helium is pneumatically long (has significant gas decompression at optimal flow). □

Table 10.5 Critical lengths, $L_{\text{crit,opt}}$ (in meters), of thin film columns at $f = f_{\text{opt}}$ for several internal diameters (d) and several types of carrier gas at $p_o = 1$ atm, 100 °C, $\varphi = 0.001$, $k = 1$[a)].

d (mm)	0.05	0.1	0.18	0.25	0.32	0.45	0.53
He	0.132	1.05	6.15	16.5	34.5	96	157
H_2	0.23	1.84	10.7	28.8	60.3	168	274
N_2	0.48	3.87	22.5	60.4	127	352	576
Ar	0.43	3.42	19.9	53.4	112	311	509

a) The data are based on Eq. (10.173) and Table 6.12.

Table 10.5 shows that the critical length for all column diameters is the shortest for helium. This implies that helium is the most pressure-demanding of all gases listed in the tables – the fact that can be also observed in Figure 10.10.

10.8.3
Peak Width

Widths of chromatographic peaks directly affect quality of chromatographic separation – the wider are the peaks the smaller is the number of solutes that can be separated during the same time or the larger should be the time required for the separation of the same number of solutes. In addition to that, the width of the narrowest peak in a chromatogram dictates the highest speed requirement for the data acquisition electronics which, in turn, might affect the level of the electronic noise and, therefore, minimal amount or minimal concentration of solutes that the system can detect and quantify.

General formulae for calculation of the peak width (σ) have already been considered. This discussion further explores the dependence of σ on column dimensions and the change of σ with time (t) during a GC analysis. The discussion will continue to focus on thin film columns and on isobaric conditions. A special attention is given to a somewhat surprising but very useful fact [3] that in the case of a strong gas decompression in a thin film column, σ is proportional to the column length and does not depend on the column diameter. This significantly simplifies the prediction of σ.

To find the effects of different factors on σ in both isothermal and temperature-programmed analyses, let us start with the evaluation of unretained width ($\sigma_m = \mu_R\sigma$, Eq. (10.80)) of a peak – the quantity that depends on the presence or absence of temperature programming in a rather minor way.

Due to Eq. (10.145), σ_m can be expressed as a function,

$$\sigma_m = t_M\sqrt{\frac{H}{L}} = \frac{\sqrt{HL}}{\bar{u}} \tag{10.174}$$

of average velocity ($\bar{u}$) of a carrier gas. Due to Eq. (7.178), $\bar{u}$ and, therefore, σ_m can be expressed as a function of a carrier gas specific flow rate (f), Figure 10.20.

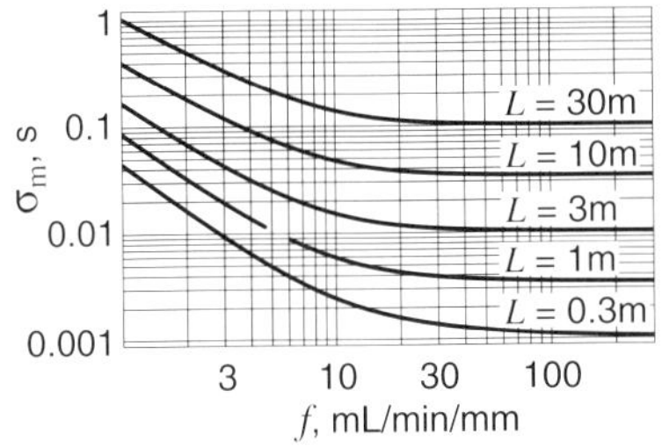

Figure 10.20 Quantity $\sigma_m = \mu_R\sigma$ (Eq. (10.80)) as a function (Eqs. (10.174) and (7.178)) of gas-specific flow rate (f) in L-long column. Conditions: $d = 0.1$ mm, thin film, $\varphi = 0.001$, $p_o = 1$ atm, helium at $T = 100\,^\circ$C, $k = 1$. Each curve has a plateau at high flow rate.

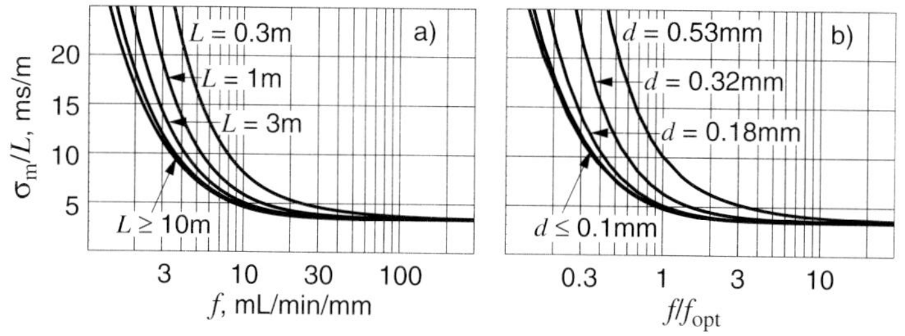

Figure 10.21 Specific unretained peak width (σ_m/L) in thin film columns: (a) as a function of specific flow rate (f) and column length (L); (b) as a function of normalized flow rate ($f/f_{opt,thin}$) and column internal diameters (d). All conditions are the same as in Figure 10.20 except in (b) where $L = 30$ m and d changes as shown in the graph.

The formulae for σ_m as functions of $\bar{u}$, f, or F are somewhat complex. Several interesting insights regarding σ_m can be obtained from the graphs in Figure 10.20 that represent the formulae.

Not surprisingly, quantity σ_m (and therefore σ) declines with an increase in gas flow (and, therefore, increase in gas velocity, pressure, and so forth). Less obvious is the fact that after a certain point, σ_m reaches a plateau and falls no further. The level of the plateau appears to be proportional to the column length (L) suggesting that the plateau for the ratio σ_m/L should be the same for all values of L. This indeed is the case as shown in Figure 10.21a. Not only that, but the level of the plateau in σ_m/L is independent of a column diameter (d) as shown in Figure 10.21b. It is also worth noticing that quantities σ_m/L as a functions of $f/f_{opt,thin}$ in a pneumatically long columns ($d < 0.32$ mm in Figure 10.21b) are nearly the same, and all transition to the plateau at about $f = f_{opt,thin}$. These facts suggest that the plateau in σ_m/L is an important invariant of a carrier gas – its fundamental property that does not depend on the column dimensions, but only on the gas itself.

To find the plateau ($\sigma_{m,min}$) in σ_m, let us express σ_m as a function of inlet pressure (p_i). This appears to be the simplest formula for σ_m as a function of a pneumatic variable. Substitution of Eqs. (10.98) and (7.120) in Eq. (10.174) allows to express σ_m in a thin film column as

$$\sigma_{m,thin} = \sqrt{\sigma^2_{m,B} + \sigma^2_{m,C1}} \tag{10.175}$$

where

$$\sigma^2_{m,B} = \frac{16(p_i^2 + p_o^2)BLp_{st}\Omega^3}{\pi(p_i^2 - p_o^2)^3 d}, \quad \sigma^2_{m,C1} = \frac{\pi(p_i^2 + p_o^2)C_1 dL\Omega}{4(p_i^2 - p_o^2)p_{st}} \tag{10.176}$$

At $f = f_{opt}$, $\sigma_{m,B}$ and $\sigma_{m,C1}$ are equal to each other. Therefore, it follows from the last formulae that $\sigma_{m,B}$ is a dominant component of σ_m in an underoptimized column where $|p_i - p_o| \ll |p_{i,opt} - p_o|$ (or, equivalently, $f \ll f_{opt}$, $\bar{u} \ll \bar{u}_{opt}$, and so forth). On the other hand, $\sigma_{m,C1}$ is a dominant component of σ_m in an overoptimized column where

$|p_i - p_o| \gg |p_{i,opt} - p_o|$ (or, equivalently, $f \gg f_{opt}$, $\bar{u} \gg \bar{u}_{opt}$, and so forth). In the later case, $\sigma_{m,B}$ vanishes and σ_m approaches the plateau (Figure 10.20) where it has its lowest value ($\sigma_{m,min}$) equal to $\sigma_{m,C1}$ at $p_i = \infty$. Combining Eqs. (10.175) and (10.176) at $p_i = \infty$ with Eqs. (10.149), (10.150), (7.38) and (6.17) one has

$$\sigma_{m,min} = \lim_{p_i \to \infty} \sigma_{m,C1} = \sqrt{\frac{\pi C_1 d L \Omega}{4 p_{st}}} = \sqrt{\frac{\eta}{3\gamma_T D_{pst} p_{st}} \vartheta_1 L} \tag{10.177}$$

$$\sigma_{m,min} = \frac{\sqrt{20} k^{0.05} (T/T_{st})^{1/8} \vartheta_1}{3\sqrt{\pi \gamma_T} (10^3 \varphi)^{0.04}} \frac{L}{\gamma_D \upsilon_{st}}, \quad k \geq 0.1 \tag{10.178}$$

where υ_{st} and γ_D (Table 6.12) are, respectively, average molecular speed of the gas at standard temperature and gas-dependent empirical constant.

Equations (10.177) and (10.178) show that for the same stationary phase type, film thickness, and dimensionless heating rate, $\sigma_{m,min}$ is proportional to the ratio, $L/(\gamma_D \upsilon_{st})$, of the column length to the product of gas-dependent quantities γ_D and υ_{st}. The key factor in the ratio is the quantity L/υ_{st} representing average time that a carrier gas molecule would take at the standard temperature to travel along the column if there were no collisions along the way. For helium and hydrogen, quantity $\gamma_D \upsilon_{st}$ is about two times smaller than υ_{st}. Therefore, $L/(\gamma_D \upsilon_{st})$ is about two times larger than L/υ_{st}.

Interestingly, the column diameter does not affect $\sigma_{m,min}$, as follows from Eqs. (10.177) and (10.178), and as shown in Figure 10.21b.

Equations (10.177) and (10.178) have a feature that might look surprising. The quantity σ_m is the width of unretained peak and yet $\sigma_{m,min}$ is a function of a solute retention factor, k. It depends on k via parameter ϑ_1 in Eqs. (10.177) and (10.178), and via a weak function $k^{0.05}$ of k in Eq. (10.178). To explain this phenomenon, recall that σ_m is not the width of actual unretained peak, but rather is unretained equivalent (Eq. (10.80)) of the width (σ) of actual retained peak. An actual peak having actual σ would have width σ_m if it was not retained, but if its dispersion rate (represented by the plate height) was not affected by the absence of the retention.

A plateau level ($\sigma_{m,min}$) of unretained equivalent (σ_m) of a peak can be used for finding the plateau level (σ_{min}) of the width (σ) of the peak itself. For the peaks in a thin film column, quantities σ_{min} can be found as follows.

Peak width (σ) can be found from its unretained equivalent σ_m as $\sigma = \sigma_m/\mu_R$ (Eq. (10.81)), where μ_R is the elution mobility of respective solute. In static analysis, μ_R for a given solute does not change with time and can be expressed via the retention factor (k, Eq. (8.6)). Equations (10.81), (10.178), (8.6) and (10.152) yield for the narrowest width (σ_{min}) of a retained peak:

$$\sigma_{min} = \frac{0.84 (T/T_{st})^{0.13} k^{0.05} (1 + 3.25k)}{(10^3 \varphi)^{0.04}} \times \frac{L}{\gamma_D \upsilon_{st}} \quad \text{(static analysis, } k \geq 0.1\text{)} \tag{10.179}$$

For highly interactive solutes eluting during isobaric heating ramp, μ_R can be found from Eq. (8.115). Equations (10.178), (10.81), (8.115) and (10.155) yield

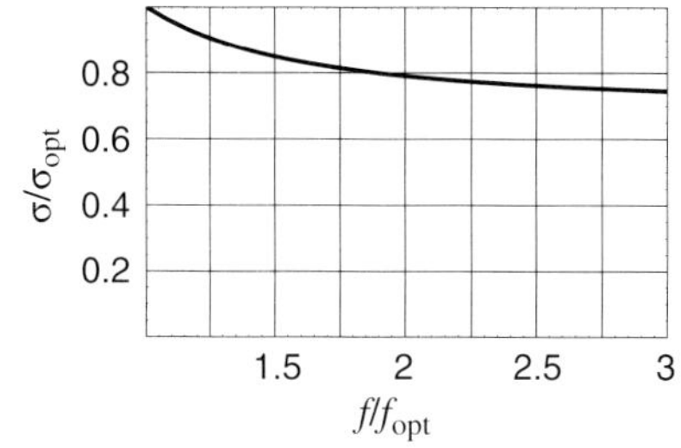

Figure 10.22 When specific flow rate (f) of carrier gas in a thin film column with large gas decompression increases beyond its optimal value (f_{opt}), only a limited reduction in peak width (σ) below its optimal value (σ_{opt}) takes place as follows from Eq. (10.182).

$$\sigma_{min} = \frac{0.23(4/(1-e^{-r_T}) + 9/r_T)(T/T_{st})^{0.13}}{(e^{r_T}-1)^{0.05}(10^3\varphi)^{0.04}} \times \frac{L}{\gamma_D \upsilon_{st}} \quad \text{(isobaric heating ramp)} \tag{10.180}$$

Plateau level (σ_{min}) of the peak width (σ_{thin}) in a thin film column can be used for finding optimal value ($\sigma_{opt,thin}$) of σ_{thin}. A general formula for σ_{thin} at an arbitrary degree of gas decompression is complex. It becomes simpler if the decompression is strong or weak.

At strong decompression, combination of Eq. (10.174) with Eqs. (10.158), (10.160), (10.81), (7.200), and (7.38) yields

$$\sigma_{thin}^2 = \frac{32\pi h_{min,thin} L^2 \eta}{9 f p_{st} \mu_R^2}\left(\frac{f_{opt,thin}}{f} + \frac{f}{f_{opt,thin}}\right) \quad \text{at} \quad \Delta p \gg p_o \tag{10.181}$$

The quantity $\sigma_{opt,thin}$ is σ_{thin} at $f = f_{opt,thin}$. The last formula yields, Figure 10.22,

$$\sigma_{thin} = \sigma_{opt,thin}\sqrt{\frac{1 + (f/f_{opt,thin})^2}{2(f/f_{opt,thin})^2}} \quad \text{at} \quad \Delta p \gg p_o \tag{10.182}$$

When f approaches infinity, σ_{thin} approaches its plateau at σ_{min}. Therefore, it follows from the last formula that

$$\sigma_{opt,thin} = \sqrt{2}\sigma_{min} \quad \text{at} \quad \Delta p \gg p_o \tag{10.183}$$

This means that

Statement 10.9

When gas decompression at optimal flow in a thin film column is strong, an increase in the flow beyond the optimum can reduce the widths of the peaks by no more than 30%.

This means that when gas decompression is strong, there is almost no room for reduction in σ below its optimal value in a thin film column. This known fact [3] provides useful shortcuts to solutions of many problems in chromatography [65, 157, 158].

The inverse peak width 1/σ – the number of peaks per unit of time – is a good measure of speed of analysis [136, 159–161]. Reduction in column diameter is frequently considered to be a sure way to reduce the widths of the peaks, and, therefore, to increase the speed of analysis. Apparently, it is not so when the gas decompression is strong (vacuum outlet operations, analyses with typical 0.25 mm and smaller bore capillary columns, and so forth). In all these cases, the speed of analysis is a function only of a column length and a carrier gas type. However, the number of peaks that a column can separate at a given speed is a different matter.

A closer look at relations between column dimensions and its performance suggests that reduction in a column diameter (*d*), while not reducing the peak width and not increasing the speed of analysis under the optimal pneumatic conditions, increases the column plate number (*N*, Eq. (10.166)) and the hold-up time. As a result, while not increasing the number of peaks that a column can generate during the same time, the analysis utilizing a column with a smaller diameter (and the same length) can last longer and separate more peaks [149]. In other words, reduction in a column diameter while keeping its length fixed increases the column ability to separate complex mixtures, but not the speed of analysis.

To identify the relationship between column dimensions and speed of analysis, it is more appropriate to assume a fixed plate number (*N*) rather than fixed *L*. In order to keep *N* fixed, a reduction in *d* should be accompanied by proportional reduction in *L*, which could lead to proportional reduction in the widths of all peaks [149] and to proportional increase in the speed of analysis without reducing the degree of separation of each pair of peaks.

So far, only pneumatically long columns (strong gas decompression under optimal conditions) were considered. In a thin film column of arbitrary length

$$\sigma_{\mathrm{opt,thin}} = \sigma_{\min}\sqrt{\frac{2(p_{\mathrm{i,opt}}^2 + p_{\mathrm{o}}^2)}{p_{\mathrm{i,opt}}^2 - p_{\mathrm{o}}^2}} \tag{10.184}$$

where $p_{\mathrm{i,opt}}$ is the optimal inlet pressure that, according to Eq. (7.48), could be found as

$$p_{\mathrm{i,opt}} = \sqrt{p_{\mathrm{o}}^2 + \frac{8 f_{\mathrm{opt}} p_{\mathrm{st}} \Omega}{\pi d}} \tag{10.185}$$

When gas decompression is strong ($p_{\mathrm{i,opt}} \gg p_{\mathrm{o}}$), Eq. (10.184) converges to Eq. (10.183).

At weak gas decompression, Eq. (10.184) can be expressed as

$$\sigma_{\mathrm{opt,thin}} = \sigma_{\min}\sqrt{\frac{2p_{\mathrm{o}}}{\Delta p_{\mathrm{opt}}}}, \quad (\Delta p_{\mathrm{opt}} \ll p_{\mathrm{o}}) \tag{10.186}$$

where, according to Eqs. (7.186) and (10.171), optimal pressure drop (Δp_{opt}) can be found as

$$\Delta p_{opt} = p_{i,opt} - p_o = \frac{4 f_{opt} p_{st} \Omega}{\pi d p_o} = \frac{3 L p_o}{2 L_{crit,opt}}, \quad (\Delta p_{opt} \ll p_o) \qquad (10.187)$$

where $L_{crit,opt}$ is the column critical length under optimal conditions (when $\Delta p_{opt} = p_o$).

Typically, reduction in a column length reduces peak width (Figure 10.20). However, the specific peak width, σ/L – peak width per unit of a column length – increases (Figure 10.21a) with the length reductions. When gas decompression in a thin film column is strong, quantity σ_{opt}/L, while staying above σ_{min}/L, asymptotically approaches σ_{min}/L. Things are different under the weak decompression when quantity σ_{opt}/L could be substantially larger than σ_{min}/L. This leaves a room for a substantial reduction in σ and σ_{opt} in columns of the same length.

Gas decompression in any column with its outlet at vacuum ($p_o = 0$) is always strong Statement 7.3. If outlet pressure (p_o) of a column is not zero, significant increase in the speed of analysis can be obtained in some cases by switching to vacuum outlet operations [64, 131, 162–166]. Equations (10.183) and (10.186) suggest that going from nonzero p_o in a thin film column to vacuum at the outlet could reduce σ_{opt} by a factor of $(p_o/\Delta p_{opt})^{1/2}$. When gas decompression is strong, Eq. (10.186) is no longer valid. Instead, a more general formula in Eq. (10.184) should be used to evaluate a potential reduction in $\sigma_{opt,thin}$. The latter suggests that no significant reduction in $\sigma_{opt,thin}$ is possible if $p_{i,opt} \gg p_o$.

Example 10.17

Suppose that at $p_o = 1$ atm, $\Delta p_{opt} = 0.25 p_o$. According to Eq. (10.186), $\sigma_{opt,thin} \approx 2.8 \sigma_{min}$, which is two times larger than $\sigma_{opt,thin}$ in Eq. (10.183). This means that by using vacuum at the column outlet, $\sigma_{opt,thin}$ (as well as the analysis time) can be reduced by a factor of 2 without reducing quality of separation of any pair of peaks. More accurate evaluation based on Eq. (10.184) (at $p_{i,opt} = 1.25$ p_o) suggests that the peak width reduction would be slightly better than 2.

In the case of a thin film column with critical length at optimal flow ($\Delta p_{opt} = p_o$), Eq. (10.186) shows no potential reduction in $\sigma_{opt,thin}$. However, condition $\Delta p_{opt} \ll p_o$ of Eq. (10.186) is not satisfied and Eq. (10.186) might be inaccurate. More accurate formula in Eq. (10.184) suggests that at $p_{i,opt} = 2 p_o$, $\sigma_{opt,thin}$ could be reduced, but only by about 30%. □

For a quick evaluation of a potential peak width reduction, it might be useful to express potential peak width reduction as a function of critical length ($L_{crit,opt}$) at optimal flow. It follows from Eqs. (10.186) and (10.187) that, when $\Delta p_{opt} \ll p_o$ and, therefore, $L \ll L_{crit,opt}$,

$$\sigma_{opt,thin} = \sigma_{min} \sqrt{\frac{4 L_{crit,opt}}{3L}} \quad (L \ll L_{crit,opt}) \qquad (10.188)$$

Switching this column to the vacuum outlet operation reduces the optimal peak widths to the value described in Eq. (10.183) – a factor of $(2L_{\text{crit,opt}}/(3L))^{1/2}$ reduction.

Example 10.18

According to Table 10.5, the critical length ($L_{\text{crit,opt}}$) of 0.32 mm column with hydrogen is about 60 m. As a consequence of Eq. (10.188), the peaks in a 10-m long pneumatically optimized column are about two times narrower in GC-MS (requiring vacuum at the column outlet) than they are at $p_o = 1$ atm. Therefore, while in both cases, plate number is the same, GC-MS analysis with this column requires twice as shorter time than the analysis with $p_o = 1$ atm.

On the other hand, $L_{\text{crit,opt}}$ in 0.1-mm column with hydrogen at $p_o = 1$ atm is about 2 m. Therefore, for a 10-m column, $L \gg L_{\text{crit,opt}}$. As a result, condition $L \ll L_{\text{crit,opt}}$ of Eq. (10.188) is not satisfied. The condition $L \gg L_{\text{crit,opt}}$ is equivalent to the condition $p_{\text{i,opt}} \gg p_o$. It follows from general formula in Eq. (10.184) that, when $p_{\text{i,opt}} \gg p_o$, further reduction of p_o to $p_o = 0$ would not lead to further significant peak width reduction. □

To find how σ changes with time in an isothermal analysis, and with column temperature in a temperature-programmed analysis, let us go back to the general formulae. Equations (10.81), (10.175) and (8.6) allow one to express σ as

$$\sigma = \sqrt{\sigma_B^2 + \sigma_{C1}^2} \tag{10.189}$$

$$\sigma_B = (1 + k_R)\sigma_{m,B}, \quad \sigma_{C1} = (1 + k_R)\sigma_{m,C1} \tag{10.190}$$

where quantities $\sigma_{m,B}$ and $\sigma_{m,C1}$ are described in Eq. (10.176) and k_R is a solute elution retention factor.

In isothermal analysis, k_R of a solute remains fixed during its migration. Therefore, $k_R = k$. According to Eq. (8.22) quantity $1 + k$ could be viewed as a normalized retention time in an isothermal analysis. This, in view of Eqs. (10.189) and (10.190), means that σ would be proportional to time if all solutes had the same σ_m, that is, if σ_m was independent of k. Although this is not exactly the case, it follows from Eq. (10.176) that components $\sigma_{m,B}$ and $\sigma_{m,C1}$ are weak functions of k for all significantly retained peaks. Indeed, according to Eq. (10.176), the only parameters of $\sigma_{m,B}$ and $\sigma_{m,C1}$ that depend on k are, respectively, B and C_1 (Eqs. (10.149) and (10.152), Figure 10.6). Taking into account also Eq. (10.179) for $\sigma_{\min}$, one can conclude that

$$\sigma_B \sim \frac{(1+k)}{k^{0.05}}, \quad \sigma_{C1} \sim \sigma_{\min} \sim (1 + 3.25k)k^{0.05} \tag{10.191}$$

Both components are nearly proportional to $1 + k$ (Figure 10.23a) for significantly retained peaks. There is no surprise that σ_{C1} is proportional to $\sigma_{\min}$. Both the quantities are dominant components of σ in an overoptimized thin film column and $\sigma_{\min}$ is a limit (Eq. (10.177)) of σ_{C1} at infinite gas flow.

The fact that σ_B and σ_{C1} are nearly proportional to $1 + k$ for significantly retained peaks, means that

Statement 10.10

In static analysis using a thin film column, the widths of significantly retained peaks ($k \geq 1$) are nearly proportional to their retention times (t_R) regardless of the degree of a column pneumatic optimization.

This conclusion is in line with Eq. (10.146) where N is a fixed quantity.

What about the changes in σ during a heating ramp?

Highly interactive solutes elute with roughly the same k (Chapter 8). The changes in components σ_B and σ_{C1} of σ (Eqs. (10.189) and (10.190)) occur due to the effect of T on parameters Ω, B, and C_1 of $\sigma_{m,B}$, $\sigma_{m,C1}$ (Eq. (10.176)). The quantity Ω is proportional (Eq. (7.38)) to temperature-dependent gas viscosity while B and C_1 are functions of temperature-dependent solute diffusivities. It follows from Eqs. (7.38) Eqs. (10.149), (10.150), (10.190) and (10.176) that, Figure 10.23b,

$$\sigma_B \sim \sqrt{T^{\xi-0.25}T^{3\xi}} = T^{2\xi-0.125} \approx T^{1.3}, \quad \sigma_{C1} \sim \sqrt{T^{\xi}T^{0.25-\xi}} = T^{1/8} \qquad (10.192)$$

Numerical approximations in these formulae were based on the assumption that $\xi = 0.7$, which is close to the values of ξ of all gases listed in Table 6.12.

Both components, σ_B and σ_{C1}, of σ in a thin film column in isobaric analysis tend to increase with the increase in the solute elution temperature (T). However, there is a strong difference in the rate of the increase. The quantity σ_B (and, therefore, σ in an under-optimized column) can more than double over a wide temperature range, while σ_{C1} (and, therefore, σ in an overoptimized column) just barely change over the same range.

The net effect of components σ_B and σ_{C1} on σ depends on the relative contribution of these components to σ. In an underoptimized column ($f < f_{opt}$), σ_B has a dominant effect on σ, while in an overoptimized column ($f > f_{opt}$) σ_{C1} dominates. In reality, because f_{opt} is a function of a solute retention and both f_{opt} and f are functions of temperature (Eq. (10.152)), the dominance of one component of σ over the other can change during the analysis. This complicates the accurate prediction of evolution of σ during the analysis.

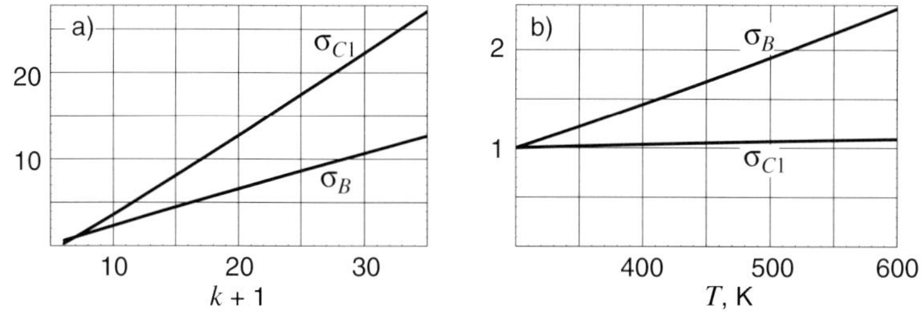

Figure 10.23 Relative changes in peak width components σ_B and σ_{C1} (Eq. (10.189)) as functions (Eqs. (10.191) and (10.192)) of (a) normalized retention time ($k + 1$) in an isothermal analysis, and (b) elution temperatures (T) of highly interactive solutes eluting during a heating ramp in isobaric analysis. Both components are normalized to have relative values of one at $k = 1$ in (a), and at $T = 300$ K in (b).

At optimal or near-optimal conditions, the ratio f/f_{opt} has a minor effect on σ. This allows one to significantly simplify the prediction of σ. To explore this possibility, let us go back to the basic formula (Eq. (10.147)) for σ in static and dynamic analysis.

According to Eqs. (10.147), (8.20) and (10.110), σ in the static analysis is a direct function,

$$\sigma = \frac{t_R}{\sqrt{N}} = \sqrt{\frac{dh}{L}}t_R \quad \text{(static analysis)} \tag{10.193}$$

of a solute retention time (t_R) where d, L, and h are column internal diameter, length, and dimensionless plate height, respectively. At optimal or near-optimal conditions, h can be approximated as h_{min}. Due to Eqs. (10.160) and (10.152), the last formula yields for σ in a thin film column:

$$\sigma_{opt,thin} \approx \frac{\sqrt{dh_{min,thin}}t_R}{\sqrt{L}} \approx 0.97t_M\sqrt{\frac{dj_G(1+k)(0.31+k)}{L}} \quad \text{(static analysis)} \tag{10.194}$$

where k is a solute retention factor and t_M is hold-up time. For significantly retained solutes, the formula can be further simplified as

$$\sigma_{opt,thin} \approx \sqrt{\frac{dj_G}{L}}\, t_R \quad \text{(static analysis, significantly retained solutes)} \tag{10.195}$$

Let us now turn to σ in temperature-programmed analysis. According to Eq. (10.147), $\sigma = t_{M,R}/(\mu_R N^{1/2})$ where $t_{M,R}$ and μ_R are, respectively, hold-up time and solute mobility measured under static conditions existing at the time of the solute elution. In isobaric analysis, $t_{M,R}$ is proportional to gas viscosity (η, Eq. (6.20)) and, therefore, can be described as $t_{M,R} = t_{M,init}(T_R/T_{init})^{\xi}$, where T_{init} and $t_{M,init}$ are initial temperature and hold-up time, respectively, T_R is a solute elution temperature, and ξ is a gas-dependent empirical quantity (Table 6.12). Accounting also for Eqs. (8.132) and (8.115), one can express Eq. (10.147) as

$$\sigma = \frac{t_{M,init}(1-e^{-(T_R-T_{init})/\theta_{T,init}})}{\sqrt{N}(1-e^{-r_T})}\left(\frac{T_R}{T_{init}}\right)^{\xi}, \quad T_R \geq T_{init} + R_T t_{M,init} \tag{10.196}$$

where quantities $\theta_{T,init}$ and r_T are described in Eqs. (8.90) and (8.96).

Similarly to static analysis, quantity $N^{1/2}$ in a thin film column at optimal or near-optimal conditions is a weak function of a solute retention time and, therefore, a weak function of a solute elution temperature. Thus, as shown in Figure 10.17, when specific flow rate (f_{init}) at $T = T_{init}$ is confined within the bounds $f_{opt,init} \leq f_{init} \leq 2f_{opt,init}$, where $f_{opt,init}$ is optimal f at $T = T_{init}$, quantity $h^{1/2}$ in a thin film column is close to its minimum ($h^{1/2}_{min}$). This means that $N^{1/2} \approx \sqrt{L/(dh_{min})}$. Equation (10.196) yields

$$\sigma_{\text{opt,thin}} \approx \sqrt{\frac{dh_{\text{min,thin}}}{L}\frac{t_{\text{M,init}}\left(1-e^{-(T_{\text{R}}-T_{\text{init}})/\theta_{T,\text{init}}}\right)}{1-e^{-r_{\text{T}}}}}\left(\frac{T_{\text{R}}}{T_{\text{init}}}\right)^{\xi}, \quad T_{\text{R}} \geq T_{\text{init}} + R_{\text{T}}t_{\text{M,init}} \tag{10.197}$$

where $h_{\text{min,thin}}$ can be found from Eq. (10.160) combined with Eq. (10.155) or Eq. (10.156). Approximation $h_{\text{min,thin}} = 1$ further simplifies this formula as

$$\sigma_{\text{opt,thin}} \approx \sqrt{\frac{d}{L} \times \frac{t_{\text{M,init}}\left(1-e^{-(T_{\text{R}}-T_{\text{init}})/\theta_{T,\text{init}}}\right)}{1-e^{-r_{\text{T}}}}}\left(\frac{T_{\text{R}}}{T_{\text{init}}}\right)^{\xi}, \quad T_{\text{R}} \geq T_{\text{init}} + R_{\text{T}}t_{\text{M,init}} \tag{10.198}$$

As expected (Eq. (10.166)), the last two formulae yield close results shown in Figure 10.24.

The graphs of $\sigma_{\text{opt,thin}}$ as a function of T_{R} have two distinct segments evident in Figure 10.24.

In the first segment, $\sigma_{\text{opt,thin}}$ rapidly increases with the increase in T_{R}. This segment represents the solutes that are slightly or moderately retained (Figure 5.4) at the beginning of the ramp. The earliest peaks are the sharpest because they represent the solutes that elute with almost no interaction with the column stationary phase. However, the interaction level (ω_{R}) of the eluites rapidly increases with T_{R} approaching the asymptotic level $\omega_{\text{R,a}} = \exp(-r_{\text{T}})$ (Eq. (8.114)) that depends only on the dimensionless heating rate, r_{T}. All highly interactive solutes (those that were highly retained at the beginning of the ramp) elute with $\omega_{\text{R}} = \omega_{\text{R,a}}$. Once the asymptotic level is reached (at about $T_{\text{R}} - T_{\text{init}} \geq 3\theta_{T,\text{init}} \approx 70\,^{\circ}\text{C}$), the asymptotic level ($\sigma_{\text{opt,thin,a}}$) of $\sigma_{\text{opt,thin}}$ becomes proportional to $(T_{\text{R}}/T_{\text{init}})^{\xi}$. Equation (10.198) yields, Figure 10.24:

$$\sigma_{\text{opt,thin,a}} \approx \sqrt{\frac{d}{L}\frac{t_{\text{M,init}}}{1-e^{-r_{\text{T}}}}}\left(\frac{T_{\text{R}}}{T_{\text{init}}}\right)^{\xi} \approx \sqrt{\frac{d}{L}\frac{t_{\text{M,init}}}{1-e^{-r_{\text{T}}}}}\left(\frac{T_{\text{R}}}{T_{\text{init}}}\right)^{0.7}, \quad T_{\text{R}}-T_{\text{init}} \geq 3\theta_{T,\text{init}} \tag{10.199}$$

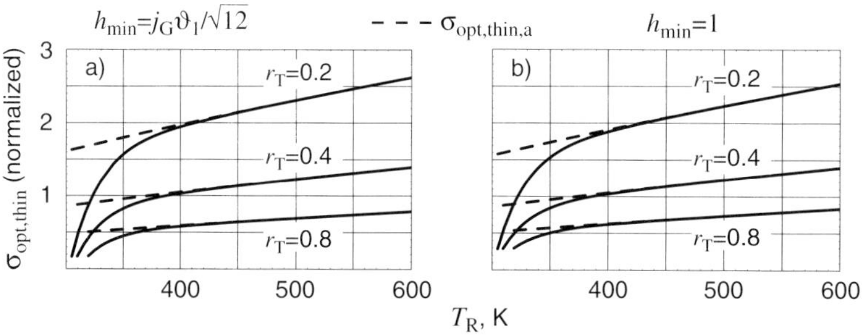

Figure 10.24 Near optimal widths $\sigma_{\text{opt,thin}}$ (solid lines) and their asymptotic levels $\sigma_{\text{opt,thin,a}}$ for peaks generated by solutes eluting from a thin film column during linear heating ramp. T_{R} is a solute elution temperature. (a) Eq. (10.197), (b) Eqs. (10.198) and (10.199). Conditions: $T_{\text{init}} = 300$ K, $\xi = 0.7$. Vertical scales of all curves are normalized by multiplier $1/\sigma_{\text{opt,thin,a}}$ where $\sigma_{\text{opt,thin,a}}$ is found from Eq. (10.199) at $T_{\text{R}} = 100\,^{\circ}$C and $r_{\text{T}} = 0.4$.

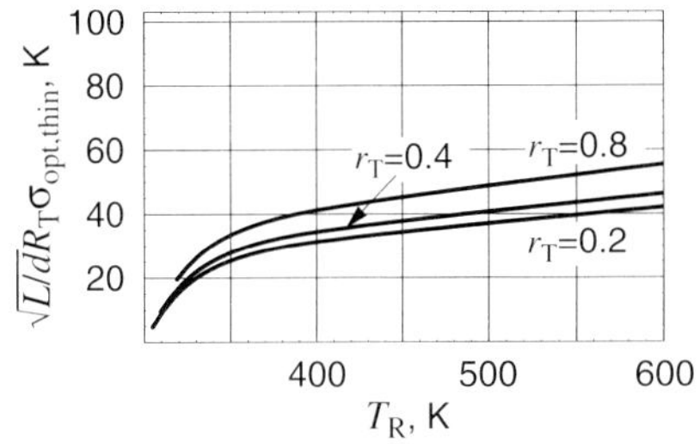

Figure 10.25 Normalized temperature domain widths, $\sqrt{L/d}R_T\sigma_{opt.thin}$, of the peaks in Figure 10.24b. Conditions are the same as in Figure 10.24.

All parameters in this formula are typically known or can be found from the method parameters.

An increase in the heating rate in the analysis covering the same temperature range proportionally reduces the analysis time. Figure 10.24 shows that the increase in the heating rate in the same column with the same carrier gas and pressure makes the peaks sharper. This does not mean, however, that more peaks can be separated. To quantify this fact, it is convenient to represent the entire chromatogram corresponding to a heating ramp in the temperature rather than time domain, that is, as a function of temperature (T) rather than time (t). For the ramp starting at $t = 0$ and $T = T_{init}$, elution temperature, T_R, of a peak can be found as $T_R = T_{init} + R_T t_R$ and its standard deviation in the temperature domain as $R_T\sigma$. For comparison, verticle temperature domain version of Figure 10.24b is shown in Figure 10.25.

Figure 10.25 reveals several interesting facts.

First, the order of the curves in Figure 10.25 is reversed compared to the order in Figure 10.24. An increase in r_T makes peaks narrower in time domain and wider in temperature domain, that is, the faster is the heating of the same column with the same gas and its pressure, the wider is the temperature interval occupied by each peak. This means that an increase in the heating rate reduces the number of peaks that can be separated with the same separation quality.

Second, the spread of the peak widths in temperature domain is relatively tight – much tighter that the spread in time domain. This means that although an increase in the heating rate reduces the quality of separation, the reduction is not very large. It also means that the temperature domain widths are easier to predict that the time domain ones. This leads to the next observation.

Optimal r_T is close to 0.4 (Example 8.13). The average of the flat part of the curve of $\sqrt{L/d}R_T\sigma_{opt.thin}$, at $r_T = 0.4$, can be estimated as 40 °C. As the ratio L/d is a reasonable estimate for a column plate number (N), one can conclude that under optimal or near-optimal conditions, standard deviation ($R_T\sigma$) of a peak in temperature domain can be estimated as 40 °C per square root of N, that is, $R_T\sigma \approx 40\,^\circ C/N^{1/2}$. The inverse of $R_T\sigma$ can be expressed as

$$\frac{1}{R_T\sigma_{\text{opt,thin}}} \approx \frac{0.025N^{1/2}}{°\text{C}} = \frac{2.5N^{1/2}}{100\,°\text{C}} \qquad (10.200)$$

meaning that

Statement 10.11

Under optimal or near-optimal conditions for a linear heating ramp in a thin film column, a region of elution of highly interactive solutes contains about $2.5N^{1/2}$ σ-wide intervals per each 100 °C temperature span of a heating ramp.

As a final look at the quality of peak width evaluation, comparison of theoretical predictions of σ from the basic principles (Eqs. (10.147), (10.97), (10.157) and (8.115)) with experimental or computer simulated data are shown in Figure 10.26. There are several differences between conditions in the graphs. However, one important factor is roughly the same in all three graphs – the normalized heating rate is close to 10 °C per hold-up time in all cases (9.63 °C in (a), 12.7 °C in (b), and 11.7 °C in (c)), which means that all graphs represent similarly scaled heating rates (Section 8.5) [136].

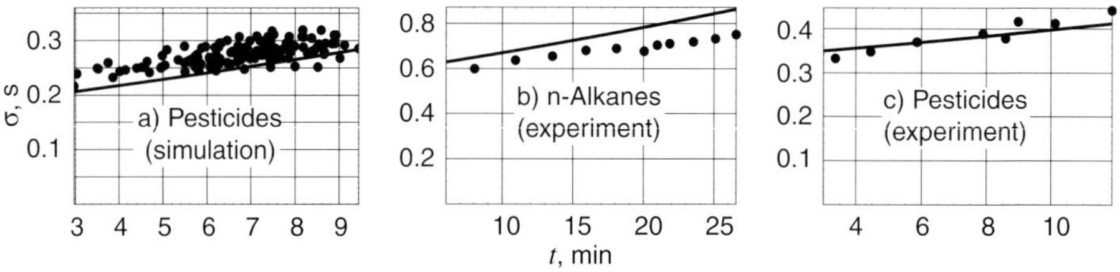

Figure 10.26 Computer-generated (a) and experimental (b, c) peak widths, σ, (the dots) as functions of the peak retention time (t). The pesticides in (a) are the same as in Figure 8.3. Appearing in the order of increase in retention time in (b) are the even numbered n-alkanes C_{12} through C_{30}, and in (c) are the pesticides: [unknown], dichlorvos, vernolate, lindane, chlorpyriphos-methyl, malathion, dieldrin, mirex. Solid lines represent theoretical predictions calculated from Eqs. (10.147), (8.115), (10.97) and (10.157). Conditions for (a) are the same as in Figures 8.3 and 10.4. Conditions for (b) are: 30.3 m × 0.32 mm × 0.25 μ HP-5 column; helium at $p_o = 1$ atm and $p_g = 81.01$ kPa (initial flow 2.56 mL/min, $t_{M,\text{init}} = 1.27$ min); $T = 50\,°\text{C} + (10\,°\text{C/min})t$ ($R_T = 10\,°\text{C/min} = 12.7\,°\text{C}/t_{M,\text{init}}$). Conditions for (c) are: 30 m × 0.32 mm × 0.32 μ, HP-5 column; hydrogen at $p_o = 1.034$ atm and $p_g = 81.01$ kPa (initial flow 5.68 mL/min, $t_{M,\text{init}} = 0.584$ min); $T = 50\,°\text{C} + (20\,°\text{C/min})t$ ($R_T = 20\,°\text{C/min} = 11.7\,°\text{C}/t_{M,\text{init}}$). (The source of departure of the solid line from a reasonable fit to the computer-generated data in (a) is the same as in Figure 8.9a. There could be many reasons for the difference between the solid line and the experimental data in (b). Difference between nominal and actual dimensions of a column, between nominal and actual pressure, and so forth, could be some of them.).

10.9
Packed Columns

Known models of plate height in packed columns are based on empirical theories [10, 72, 132, 167–170] leading to overlapping but not identical results. The most important for comparison of performance of capillary and packed columns are minimal dimensionless plate height, $h_{\mathrm{p,min}}$, and optimal specific flow rate, $f_{\mathrm{opt,p}}$, in a packed column.

It appears to be certain that a somewhat generous estimate of $h_{\mathrm{p,min}}$ for moderately and highly retained solutes in packed columns can be described as [10, 72, 73, 168, 171–174]

$$h_{\mathrm{p,min}} \geq 2 \tag{10.201}$$

For unretained peaks, slightly lower experimental values of $h_{\mathrm{p,min}}$ are also known [72, 172]. On the other hand, some sources indicate that $h_{\mathrm{p,min}}$ for moderately and highly retained solutes is closer to three than to two [171]. The values of $h_{\mathrm{p,min}}$ reaching hundreds have also been reported [174]. Overall, it is probably fair to assume that the estimate in Eq. (10.201) is on the favorable side.

Experimental data for $f_{\mathrm{opt,p}}$ have much narrower distribution compared to the data for $h_{\mathrm{p,min}}$. Nevertheless, because the effect of $f_{\mathrm{opt,p}}$ on column performance is not as straightforward as that of $h_{\mathrm{p,min}}$, it is more difficult to come up with a one-sided estimate for $f_{\mathrm{opt,p}}$. On the one hand, available inlet pressure is one of the key limiting factors for performance of the packed columns [52, 58, 163, 175–179], and the lower is $f_{\mathrm{opt,p}}$ the lower is the pressure required for a given plate number. On the other hand, the higher is $f_{\mathrm{opt,p}}$ the shorter is the analysis at a given plate number. It can be also taken into account that due to eddy diffusion [10, 46, 51, 72, 73] in packed columns, only relatively minor increase in h_{p} can occur when f exceeds $f_{\mathrm{opt,p}}$ several times. According to Giddings [73] (Figure 2.11-1 on page 64), $f_{\mathrm{opt,p}}$ "in a typical efficient chromatographic column" can be estimated as

$$f_{\mathrm{opt},p} \approx D_{\mathrm{pst}} p_{\mathrm{st}} \tag{10.202}$$

where D_{pst} is a solute diffusivity in the carrier gas at standard pressure (p_{st}). In view of these observations and Eq. (10.161), it seems reasonable to use the estimate

$$1/4 \leq f_{\mathrm{opt},p}/f_{\mathrm{opt,thin}} \leq 1 \tag{10.203}$$

for an approximate comparison of performance of the packed columns with the capillary ones.

10.10
Scalability of Peak Widths in Isobaric Analyses

As shown in Section 8.5, GC method translation makes it possible to translate parameters of original isobaric analysis A into parameters of translated isobaric analysis B in such a way that retention time, $t_{R,B}$, of any solute in analysis B relates to retention time, $t_{R,A}$, of the same solute in analysis A as (Eq. (8.59)) $t_{R,B} = t_{R,A}/G$, where speed gain G in analysis B compared to analyses A is the same for all solutes.

In order to predict translated chromatogram from original one, it is necessary to predict not only the effect of the translation on the retention times but also on the peak widths.

Method translation preserves elution temperature, T_R, and elution mobility, μ_R, of each solute. As a result, the ratio σ_B/σ_A of the widths of the peaks corresponding to the same solute in analyses A and B can be found from Eq. (10.147) as

$$\frac{\sigma_B}{\sigma_A} = \frac{t_{M,R,B}}{t_{M,R,A}}\sqrt{\frac{N_A}{N_B}} \tag{10.204}$$

where $t_{M,R}$ is static hold-up time measured at the conditions existing at the time of a solute elution and N is column plate number for the solute.

Due to Eqs. (7.96) and (6.20), Eq. (10.204) can be expressed as

$$\frac{\sigma_B}{\sigma_A} = \frac{t_{M,ref,B}}{t_{M,ref,A}}\left(\frac{T_R}{T_{ref}}\right)^{\xi_B-\xi_A}\sqrt{\frac{N_A}{N_B}} \tag{10.205}$$

where T_{ref} is the predetermined reference temperature somewhere within the temperature range (which is the same for both analyses), $t_{M,ref}$ is hold-up time at $T = T_{ref}$, and ξ is the empirical parameter listed for several carrier gases in Table 6.12. If the carrier gas in analyses A and B is the same, then $\xi_B = \xi_A$. However, even if A and B use different gases, the difference in ξ_B and ξ_A for the gases in Table 6.12 is minor. Suppose that $T_R/T_{ref} = 2$. Then, in the worst case (argon and helium), the effect of the difference in ξ_B and ξ_A on the ratio σ_B/σ_A is under 5%. In the more important case of helium and hydrogen, the effect of the difference on σ_B/σ_A is under 1%. Ignoring the difference between in ξ_B and ξ_A, one can approximate the last formula as

$$\frac{\sigma_B}{\sigma_A} \approx \frac{t_{M,ref,B}}{t_{M,ref,A}}\sqrt{\frac{N_A}{N_B}} = \frac{1}{G}\sqrt{\frac{N_A}{N_B}} = \frac{1}{G}\sqrt{\frac{d_B\, h_B\, L_A}{d_A\, h_A\, L_B}} \tag{10.206}$$

where $G = t_{M,ref,A}/t_{M,ref,B}$ (Eq. (8.57).

For thin film columns, the last formula can be more specific. According to Eqs. (10.110) and (7.96), N can be expressed as $N = L/(hd)$, where d, L, and h are the column internal diameter, length, and dimensionless plate number, respectively. Let f_{ref} and $f_{opt,ref}$ be f and f_{opt} at $T = T_{ref}$. If in analyses A and B the ratio $f_{ref}/f_{opt,ref}$ is the same, then the only source of significant difference in h_A and h_B for the same solute can be the difference in Giddings compressibility factors $j_{G,A}$ and $j_{G,A}$ (Eqs. (10.160) and (10.162)). As a result, Eq. (10.206) can be expressed as

$$\frac{\sigma_{thin,B}}{\sigma_{thin,A}} \approx \frac{1}{G}\sqrt{\frac{d_B\, j_{G,B}\, L_A}{d_A\, j_{G,A}\, L_B}} \approx \frac{1}{G}\sqrt{\frac{d_B\, L_A}{d_A\, L_B}} \quad \left(\text{at} \quad \frac{f_{ref,B}}{f_{opt,ref,B}} = \frac{f_{ref,A}}{f_{opt,ref,A}}\right) \tag{10.207}$$

10.11 Plate Height: Evolution of the Concept

By now, we know that the dispersion of solutes during their migration along the column is the root cause of the peak widths in a chromatogram. The plate height (H) – the rate (or the apparent rate) of the solute dispersion – plays a key role in prediction of the peak width measured as the peak standard deviation (σ).

Not always the peak width was attributed to the solute dispersion. In its early days, theory of chromatography associated the widths of the peaks with the number of imaginary elementary separation stages – theoretical plates – in a column. Each stage was represented by H.E.T.P. – the term that eventually evolved into the term *plate height*.

The subject of this section is evolution of the concept of the plate height from the introduction of H.E.T.P. [180] as a parameter of a distillation column, through adoption of H.E.T.P. [53] as an elementary migration–equilibration stage in a chromatographic column, to the plate height as a measure of a dispersion rate of any statistically large group of objects (not necessarily molecules of a solute) moving with similar velocities along the same path.

This review is not intended to be a complete history of the evolution in question. Rather, it is a brief analysis of the factors that justified the transformation of the original understanding of H.E.T.P. into contemporary understanding of the plate height.

10.11.1 Distillation Columns

The concept of a column plate height came in chromatography from the field of distillation where the prototype of the concept was introduced in 1922 by Peters [180] as H.E.T.P. ("height of equivalent theoretical plate").

As a separation technique, the distillation utilizes different volatility of components of a liquid mixture [181–188]. A key distillation device can be a vertical column (tower) consisting of many vertically cascaded plates (trays). There are many ways to implement the distillation. In one implementation, a binary mixture to be separated is supplied from the top of a (heated if necessary) distillation column and slowly flows down from one plate to another. At the same time, vapor of the mixture moves up the column. During this process, the vapor leaving the top of the column becomes enriched with more volatile component while the liquid leaving the bottom of the column becomes enriched with less volatile component. As a part of the distillation process, the liquid at the bottom of the column can be partially reboiled and sent back up the column. Similarly, a part of the recondensed vapor at the top of the column can be returned down the column. The purity of the end-products is positively affected by relative volatility of the components and by the number of plates.

The primary role of the plates in a distillation column is to provide large contact surface between the liquid and the vapor phases of the mixture that is being separated. Instead of the discrete plates, various types of packing (filling) of the interior of a column can be used to increase the liquid–vapor contact surface within the column.

Peters introduced the H.E.T.P. concept [180] as a metric for comparing a separation performance of a continuous packed ("filled") column with a discrete plate column. He first described "a theoretically perfect plate column" – the one that provides a certain degree of separation of components of binary mixture for a given difference in the component volatility. By testing a real filled or plate column, one can make a conclusion regarding the number of plates that a theoretically perfect plate column with similar performance would have. Thus, from testing of several 28-plate columns with similarly structured plates, Peters found that all performed close to the "theoretically perfect plate column" having 15 plates.

Peters defined H.E.T.P. in a filled column as follows. "If the column tested is a filled column, the number of plates in the calculated theoretical column divided by the height of the filled column . . . gives a measure of the efficiency of the filled column. . . . The reciprocal of this efficiency . . . will be called the 'height of equivalent theoretical plate,' or H.E.T.P." Experimentally H.E.T.P. was "found to vary nearly directly with the mean diameter of the filling pieces." Peters concluded that "It is evident that columns filled with very small pieces can be extremely short to make a given separation."

Filling with discrete pieces or particles was not the only type of packing of distillation columns. Other techniques of increasing the liquid–vapor contact surface, such as placing a spiral structure along a column inner surface and leaving a wide axial space open, were also known [186–188] at the time of introduction and early development of the H.E.T.P. concept. In his 1942 study of "open-tube distillation columns" with parabolic distribution of liquid flow, Westhaver found that (all symbols are the same as in the original text)

$$\text{H.E.T.P.} = \frac{D}{v_a} + \frac{11 r_o^2 v_a}{48D} \quad \text{(open tubular distillation column)} \tag{10.208}$$

where v_a is the radially averaged velocity, r_o the radius, and D the molecular diffusion coefficient, of the vapor stream. From comparison with several experimental H.E.T.P. versus v_a data sets from other sources, Westhaver concluded that his theory correlated very well with experimental data.

An important assumption in the derivation of Eq. (10.208) was "that liquid is not a transfer-limiting factor" meaning in the context of the chapter that the liquid layer flowing down the inner wall was sufficiently *thin* – the factor that is similar to the thin stationary phase film in a capillary (open tubular) chromatographic column. Interestingly, the H.E.T.P. in Eq. (10.208) is two times smaller than $\mathscr{H}$ in Golay formula, Eq. (10.22), for a thin film column ($d_f = 0$) at $k = \infty$.

Note 10.14

The difference between H.E.T.P. in Eq. (10.208) and $\mathscr{H}$ in Eq. (10.22) at $d_f = 0$ and $k = \infty$ has a transparent explanation reaching beyond the needs of this discussion. □

Interestingly, Eq. (10.208) directly relates H.E.T.P. only to one physical property of a mixture – its diffusivity (D) that has no direct relation to volatility of the mixture components. And, like in chromatography, D only affects the optimal value of vapor velocity (v_a) but not the minimal value of H.E.T.P., which is "equal to $0.96r_o$ or, roughly, the tube radius" as Westhaver pointed out analyzing Eq. (10.208).

10.11.2 The First Plate Models in Chromatography

In 1941, Martin and Synge [53] adopted the concept of H.E.T.P. ("height equivalent to one theoretical plate") for chromatography thus giving birth to the theory of the new technique.

Since its original adoption by Martin and Synge, the designation of H.E.T.P. in chromatography was different from what the concept represented in the science of its origin. In distillation, H.E.T.P. was the length of a column segment providing a certain degree of separation of *two* mixed liquids. Martin and Synge, on the other hand, presented H.E.T.P. as a way of explaining the broadening of a single chromatographic peak.

To explain the mechanics of the peak broadening, Martin and Synge suggested to treat a column as a cascade of the H.E.T.P.-long elementary equilibration stages or "'theoretical plates' within each of which perfect equilibrium between the two phases occurs." According to this model, the process of a solute migration along a column can be viewed as a sequence of the two-step transformations. During the first step, the solute fraction contained in the mobile phase of each plate is transferred from preceding to the following plate down the mobile phase flow stream. During the second step, the equilibration of the distribution of the solute between the two phases in each stage takes place. Using the plate model, Martin and Synge explained the experimentally known nearly Gaussian shape of chromatographic peaks.

In 1944, Craig described a 20-stage *extraction apparatus* [189–193], also known as *Craig's machine* [106, 108, 194]. Although the physics of the separation in Craig's machine are different from those in chromatography, a general model describing the operation of the apparatus as the stage-to-stage solute transfer and its distribution between two solvents within each stage is the same as the *Martin-Synge plate model*. As a sign of recognition of Martin and Synge influence on his work, Craig called a stage in his apparatus as a plate. Craig's machine experimentally confirmed the earlier prediction regarding the Gaussian shape of distribution of a solute along the plates traversed by it. With the passage of time, Craig's machine became known not only as a physical multistage separation device, but also as a conceptual equivalent of the Martin–Synge plate model of chromatography. Indeed, as a conceptual model of a multi-stage broadening of a solute zone, Craig's machine is identical to the Martin–Synge plate model.

Each stage in a Craig's extraction apparatus is a separation device. With adoption of the Craig's machine model as a physical implementation of the

Martin–Synge plate model, the emphasis in interpretation of a theoretical plate has shifted from the plate being an elementary equilibration stage to the plate as the elementary separation stage. This interpretation can still be found in contemporary literature.

A significant contribution to the development of the H.E.T.P. concept was made by Glueckauf. In a paper [105] published in 1955, he has shown that the "number of theoretical plates," N, defined as $N = L/H$ where L is the column length and H is "height of theoretical plate," can be found as $N = (t_R/\sigma)^2$. Earlier, this formula was introduced as *Glueckauf formula* (Eq. (10.93)) for apparent plate number in static chromatography (static uniform conditions were implicitly assumed in Glueckauf's and in other studies of that time). In effect, Glueckauf was the first to introduce the concept of the *apparent plate number*, $N = (t_R/\sigma)^2$, as a counterpart to the actual plate number, $N = L/H$.

Although earlier references to Martin and Synge paper are known [7, 189, 195, 196], it appears to have taken more than a decade before H.E.T.P. earned the attention of the key theoreticians in the field. The tide turned in 1955–1956 when three important papers were published [10, 105, 197]. After these publications, the avalanche growth of number of papers on the subject began.

10.11.3
H.E.T.P. and Molecular Diffusion

Distribution of a solute between two phases moving relative to each other is essential for the work of the Martin–Synge and the Craig's machine models of a solute spreading along a column. According to these models, without the distribution, there would be no spreading of a narrow solute zone beyond one plate. It also means that there should be no longitudinal spreading of a solute in an inert tube. These implications contradict to experimental facts exposing substantial shortcomings of the models.

In their 1952 paper, Lapidus and Amundson [198], analyzing longitudinal dispersion ("smearing out") of solutes in ion exchange and chromatographic columns as "diffusion phenomenon," found that although the degree of a solute distribution between two phases in a column affects the dispersion, the key factor in the dispersion is the ratio D/u, where D is the solute diffusivity in a mobile phase and u is the mobile phase velocity. In 1953 and 1954, Taylor, who studied the effect of parabolic velocity profile in laminar flow on longitudinal dispersion of solutes migrating in inert tubes, also found that the solute diffusivity and velocity are the key factors affecting its dispersion [34, 37]. These studies filled a critical void in understanding of the concept of H.E.T.P. altering its perception and facilitating the forthcoming breakthroughs in the theory of H.E.T.P..

Other significant developments of that time that contributed to understanding of a solute dispersion in a chromatographic column were published by Aris [39, 40] in 1956 and 1959. However, these publications appeared or became known too late to have a direct impact on the breakthroughs of 1956–1958.

10.11.4
van Deemter Formula

In a treatment described in a two-paper series published by Klinkenberg and Sjenitzer [197], and by van Deemter et al. [10] in 1956 issue of *Chemical Engineering Science*, Glueckauf results [105] deduced from the Martin–Synge plate model [53] were combined with earlier and more recently appreciated role of diffusion in a solute spreading described by Westhaver [188], Lapidus and Amundson [198], and by Taylor [34, 37]. The result was the celebrated semiempirical formula for H.E.T.P. in a packed column under static uniform conditions. Finally, there was a way to find H.E.T.P. from column and method parameters! The formula, widely known as *van Deemter equation*, can be expressed as [10]

$$H = 2\lambda d_p + \frac{2\gamma D}{u} + \frac{8 d_\mathrm{f}^2 k u}{\pi^2 (1+k)^2 D_\mathrm{S}} \quad \text{(static uniform conditions)} \qquad (10.209)$$

where H is H.E.T.P., D and D_S are diffusivities of a solute in mobile and stationary phases, respectively, d_f is the "effective liquid film thickness," d_p is the diameter of packing particles, k is the retention factor, γ and λ are dimensionless empirical quantities estimated in the original text as $0.5 < \gamma < 1$ and $\lambda \approx 8$, respectively.

Note 10.15

An important term that is proportional to $1/D$ and does not vanish at any k is missing in Eq. (10.209). According to Kieselbach [168], Jones who proposed his own formula for H.E.T.P. in 1956 publicly discussed a correction with van Deemter in 1957. In 1958, Golay [199] spoke about the missing $1/D$ term in Eq. (10.209). A version incorporating his own and van Deemter proposed corrections has been published by Jones [169] in 1961. □

Equation (10.209) has been a significant breakthrough in chromatographic theory. It showed how to find the width (σ) of a chromatographic peak from column and solute parameters. Indeed, once it was known from Eq. (10.209) how to predict H.E.T.P. (H), it became possible to predict the plate number ($N = L/H$) for a given solute and, using Glueckauf formula to find σ of a corresponding peak as $\sigma = t_\mathrm{R}/N^{1/2}$. What is more, Eq. (10.209) shows in a straightforward way that there exists optimal gas velocity minimizing H, and, therefore, maximizing N and minimizing σ for a given t_R. That confirmed that, indeed, H.E.T.P. was one of the key parameters of column performance in chromatography.

The prominent role of a solute diffusivity in Eq. (10.209) was an indication of inconsistency of the existing models of H.E.T.P. with the equation. Not only Eq. (10.209) itself, but also the title of the paper [10] – "Longitudinal Diffusion and Resistance to Mass Transfer as Causes of Nonideality in Chromatography" – suggested that H in Eq. (10.209) had little to do with the stage-by-stage equilibration of the solute zone during its migration along the column.

10.11.5 H.E.T.P. as a Spatial Dispersion Rate

The first clear definition of HETP (Golay's designation form H.E.T.P.) was described by Golay in his famous 1958 lecture [35].

In view of uncertainty of the definition of H.E.T.P., Golay's treatment of a solute zone broadening in a capillary column was brilliantly simple. He was not looking for the value of H.E.T.P., but for the temporal ($\mathscr{D} = d\tilde{\sigma}_z^2/dt$) and spatial ($\mathscr{H} = d\tilde{\sigma}_z^2/dz$) dispersion rates of a solute – the rates of increase in the variance ($\tilde{\sigma}_z^2$) of a solute zone with the increase in its migration time, t, or distance, z, from the inlet. Only after he derived the formulae, Eqs. (10.22) and (10.23), for both rates, Golay suggested that the rate $d\tilde{\sigma}_z^2/dz$ "can be termed as the HETP of the tubular column." The parallel consideration of the two rates $d\tilde{\sigma}_z^2/dt$ and $d\tilde{\sigma}_z^2/dz$ allowed Golay to highlight the fact that $\mathscr{H}$ and, therefore, H.E.T.P. was the solute dispersion rate.

Golay study [35] of the zone dispersion rate was based on implicit assumption of uniform conditions along the column. When conditions are not uniform, the rates $d\tilde{\sigma}_z^2/dt$ and $d\tilde{\sigma}_z^2/dz$ represent not only the solute dispersion – its diffusion-based longitudinal expansion – but the changes (expansion and contraction) of $\tilde{\sigma}_z^2$ due to all mechanisms such as gas decompression in GC and others [44, 45]. Only the dispersion broadens the widths of chromatographic peaks, while other mechanisms typically have negligible effect on the peak widths. It is also important that only the dispersion-based zone broadening depends on the solute and column parameters, while the effects of other mechanisms depend on the width of the zone itself. The definition (Eq. (10.55)) of a spatial dispersion rate ($\mathscr{H}$) of a solute zone that excludes nondispersive zone expansion/contraction mechanisms has been proposed in 1992 [44, 45].

10.11.6 Plate Height

Note 10.16

Recall that the chromatographic concept of H.E.T.P. [53] was adopted from the field of distillation where the concept of H.E.T.P. [180] was based on a clearly defined notion of a "theoretically perfect plate column". Because it was clear how an actually existing plate in a distillation column performed and how it was theoretically expected to perform under ideal conditions, it was clear what the *theoretical plate* and the *height of equivalent theoretical plate* meant. None of that is true of chromatography where there are no actual plates to support the concept of a theoretical plate, and the concept of H.E.T.P.. □

While developing a formula for H.E.T.P., van Deemter *et al.* admitted in their 1956 paper [10] (page 272, column 2) that "H.E.T.P is an empirical quantity and the theory does not deal with the mechanisms which determine it." This was an explanation of the fact (evident from the context of the paper) that the derivation of

H.E.T.P. was not based on the Martin–Singe [53] or the Craig's machine model [106, 108, 189, 191, 194] of a column as a series of (quoting from Martin and Singe [53], page 1359) "'theoretical plates' within each of which perfect equilibrium between the two phases occurs."

A year later, a scientific committee for the nomenclature in GC reported [12] to the *1957 International Symposium on GC* (the first to be held in the US) that "a theoretical plate is an abstract term with no physical significance other than as a measure of the relative variance of a peak." The committee was assembled of the leading authorities (including Martin and Golay) in the field. It is also worth noticing that of the three nomenclature committees (Martin was a member of all three, and chaired two of them) appointed by the first three International meetings on GC held in 1956–1958, neither even mentioned H.E.T.P..

In his 1958 lecture [35] (see also Martin's question and Golay's reply during the discussion [199] of Golay's lecture), Golay explained the reason for identifying the spatial dispersion rate ($\mathscr{H}$) of a solute with H.E.T.P. this way: $\mathscr{H}$ "has the dimension of a length, [and] can be termed as the H.E.T.P. of the tubular column, provided the H.E.T.P. be defined by comparing the column ... not to a Craig's machine, a distinction which increases in importance as [retention factor] k decreases." In other words, H.E.T.P. is the length that is numerically equal to the spatial dispersion rate ($\mathscr{H}$) of a solute. Because so-defined H.E.T.P. is valid at $k = 0$, the concept remains valid outside of chromatography, whereas the Craig's machine model (and the Martin–Synge model) of H.E.T.P. is not. This pinpointed the source of the inconsistency in the concept of H.E.T.P. and the reason for the term H.E.T.P. to be misleading and unjustifiably complex.

The following quotes not only speak about the shortcomings of H.E.T.P., but also explain the transition from the terms *theoretical plate* and H.E.T.P. to simpler terms *plate* and *plate height*.

Giddings and Keller [200], 1961 (page 96): "The current trend is to abandon the postulates that gave birth to the concept of H.E.T.P. and speak of it as a general parameter for the measuring zone spreading, whether spreading is caused by non-equilibrium, molecular diffusion, or eddy diffusion."

Giddings [49], 1962 (page 722, column 2): "Nothing physically resembling a theoretical plate or an equilibration stage can be found in a chromatographic column .... The plate height is no more than a convenient and widely accepted parameter for describing the extent of peak spreading."

Knox and Saleem [52], 1969 (page 615, column 2): "local plate height ... measures the rate of dispersion of a band as it migrates along the column."

Giddings [46] (a comment to derivation of a formula for plate height (H) in his 1991 book, page 97): "The plate model is both cumbersome and inflexible in describing zone spreading in chromatographic columns or other separation systems. Despite the absence of the theoretical plate model in our derivation, H retains the name plate height for historical continuity."

In chromatography, the term *plate* (rather than *theoretical plate*) is as old as the H.E.T.P. concept itself. In their 1941 introductory paper, Martin and Synge [53] used the term "theoretical plate" (with the quotation marks) only once. Apparently, they much preferred the simpler term *plate* (in phrases like "each plate," "the first plate," "number of plates," and so forth), which they used more than a dozen times. Giddings used the phrase "number of plates" at least as early as in 1962 [69].

Littlewood used the term *plate height* in his 1958 lecture [201], but only once, and as a shortcut for the term H.E.T.P. which he used many times. A few years later, plate height and derivative terms like *local plate height, apparent plate height, observed plate height,* and so forth, almost completely replaced the term H.E.T.P. in the papers published by several leading theoreticians in the field [68, 115–117, 167, 168]. Soon after that, the term *plate height* became broadly accepted. Currently, it is recommended by IUPAC [9], although the term H.E.T.P. is still also in use.

In spite of its conflict with reality, the original plate model is still frequently presented as a valid zone broadening model. It is not unusual to find two theories of the zone broadening – the Martin–Singe plate theory of H.E.T.P. typically described via the Craig's machine model and the rate theory utilizing (directly or indirectly) the fact that plate height is the zone dispersion rate. In spite of the irreconcilable conflicts between the two theories, they both are frequently presented as equally valid alternatives.

10.11.7
Local and Apparent Plate Height

This review would be incomplete without mentioning the concepts of *local plate height* ($\mathscr{H}$) and *apparent plate height* (briefly, *plate height*) [46, 52, 59, 67, 68, 73, 100, 115] (H) defined in Eqs. (10.55) and (10.78), respectively. The quantity $\mathscr{H}$ describes spatial dispersion rate of a migrating solute. That rate can change from one location along a column to another. The quantity H, on the other hand, describes what appears to be the solute spatial dispersion rate when the net result of the dispersion is found (Eq. (10.141)) from the measurement of parameters of a peak corresponding to the solute. For practical considerations, only the apparent plate height (H) is important. However, it is the local plate height ($\mathscr{H}$) that allows to express H via column and solute parameters.

Early in the development of plate height theory, Glueckauf [105] and van Deemter *et al.* [10] treated the plate height as the apparent quantity. Golay [35] was the first to treat it as a local quantity. Giddings and coworkers were the first to recognize the essence of the difference between the two approaches and to introduce the concepts of the local [51, 68, 115, 117] and the apparent [68, 115] (measured [67], observed [51, 117]) plate height.

It is probably fair to say that the concept of the apparent plate height is a settled issue. This is not certain yet for the local plate height.

The concept of the local plate height is especially important for nonuniform medium. Until recently, it was generally assumed [46–52, 59] that formula $\mathscr{H} = d\tilde{\sigma}_z^2/dz$ (Eq. (10.21)) represents the definition of the local plate height in a nonuniform medium. However, the definition was not productive and had never been put to work. For example, Giddings formula (Eq. (10.90)) has been originally derived outside of that definition [67, 68]. The definition $\mathscr{H} = d\tilde{\sigma}_z^2/dz$ assigns to $\mathscr{H}$ all zone broadening mechanisms – dispersive and nondispersive (such as the zone broadening due to gas decompression in GC and others [44, 96, 100]). The nondispersive mechanisms significantly affect $d\tilde{\sigma}_z^2/dz$ and can even make it a negative value (Figure 10.7) while having no or little affect on the widths of corresponding peaks. As a result, $\mathscr{H}$ defined as $\mathscr{H} = d\tilde{\sigma}_z^2/dz$ has no effect on the peak widths in some cases, and has no relation to a column performance.

Equation (10.55) that treats the local plate height as only the solute dispersion rate even in a nonuniform medium has been described in 1992 and 1993 [44, 45] along with the differential equation (Eq. (10.61)) allowing one to account for all mechanisms that can change the width of a solute zone and of a corresponding peak. Other equations are described in Appendix 10.A.1. Equation (10.61) was used to solve several theoretical and practical problems including rigorous prove of known from Giddings' [100] *additivity* of temporal equivalents of the variance of a solute zone in a nonuniform medium (Eq. (10.69)), establishing additivity of spatial zones in dynamic uniform medium, derivation (Appendix 10.A.3) of Giddings formula for apparent plate height and Giddings pressure compressibility factor [44], evaluation of the effect of nonuniform film coating on column efficiency [44], evaluation of the effect of nonideal mobile phase compressibility in SFC [96], evaluation of in-column zone focusing by moving thermal gradients in GC, and by moving mobile phase composition gradient in LC [92, 93].

Suitability of Eq. (10.55) as the definition of $\mathscr{H}$ has been also confirmed in recent advanced studies of mathematical moments of chromatographic peaks [98, 99].

10.11.8
A Retrospect

The plate height concept in chromatography went through a tortuous evolution started by the remarkable foresight of Martin and Synge who, during the dark ages of chromatographic theory, foresaw that a length-type parameter is the key to unraveling the mystery of the zone broadening in chromatography. The fact that their vision was based on less than a perfect model (fresh ideas seldom come in a well polished form), makes their foresight even more amazing.

The temporary misunderstanding of the peak broadening model in chromatography was but a natural occurrence for the early stages of a young theory of chromatography. The misunderstanding has been resolved by the early 1960s when it became clear that the solute dispersion is the root cause of the broadening, and the plate height (spatial zone dispersion rate) was its key metric. Unfortunately, the late 1950s and the early 1960s were also the years when a bigger and a longer lasting problem in the plate height theory in GC was solidified. This is the subject of the next section.

10.12 Incorrect Plate Height Theory

It is widely believed that in static (isothermal and isobaric) GC analysis, plate height (H) in packed and capillary columns depends on average velocity ($\bar{u} = L/t_M$) of a carrier gas as, respectively,

$$H = A + \frac{B}{\bar{u}} + C\,\bar{u}, \quad \text{(packed columns)} \tag{10.210}$$

$$H = \frac{B}{\bar{u}} + C\,\bar{u}, \quad \text{(capillary columns)} \tag{10.211}$$

Presumably, quantities A, B, and C are independent of $\bar{u}$ with the implication that optimal gas velocity ($\bar{u}_{opt}$) corresponding to the minimum in H can be found as

$$\bar{u}_{opt} = \sqrt{\frac{B}{C}} \tag{10.212}$$

It is not unusual for the literature presenting Eqs. (10.210)–(10.212) to say very little about quantities A, B, and C (a great degree of uncertainty associated with these formulae is one of the problems of discussing them), and I cannot claim that every reference to these formulae is accompanied by the explicit requirement for A, B, and C to be independent of $\bar{u}$. However, the independence of these quantities from $\bar{u}$ is implied if Eqs. (10.210) and (10.211) are to be meaningful and if Eq. (10.212) to follow from them. Otherwise, if A, B, C are functions of $\bar{u}$ then H in Eqs. (10.210) and (10.211) can be almost any function of $\bar{u}$. It might not have a minimum at all, or, if the minimum exists, there is no reason to expect that it corresponds to $\bar{u}_{opt}$ described in Eq. (10.212).

Example 10.19

Assuming that B and C in the formula $y = B/x + Cx$ could be functions of x, a hyperbola, $y = x^2$ with its minimum, $y_{min} = 0$, at $x_{opt} = 0$ can be expressed as $y = B/x + Cx$ in many ways. One of them is with $B = C = x^3/(1 + x^2)$, which yields $(B/C)^{1/2} = 1$ – a quantity that is different from $x_{opt} = 0$ corresponding to the minimum, $y_{min} = 0$, in $y = x^2$. □

It is assumed throughout the discussion of Eqs. (10.210)–(10.212) that A, B, and C are independent of $\bar{u}$.

Strictly speaking, Eqs. (10.210)–(10.212) are not always incorrect. When decompression of a carrier gas along a column is weak, local velocity (u) of a carrier gas does not change along the column. As a result, $\bar{u} = u$ and, therefore, Eqs. (10.210)–(10.212) can be correct. Thus, when $\bar{u} = u$, the formula in Eq. (10.211) is a compact version of Golay formula in Eq. (10.22). It is important, however, that Eqs. (10.210)–(10.212) can only be correct in a trivial case of weak gas decompression when there is no need in the very concept of average velocity ($\bar{u}$). In all other practically important cases such as all GC-MS analyses (requiring vacuum at the column outlet), capillary columns of 0.32 mm and smaller internal diameters, packed columns – in short, in all nontrivial cases where $\bar{u}$ is different from u – Eqs. (10.210)–(10.212) are not correct.

Typically, Eqs. (10.210) and (10.211) are attributed to van Deemter and/or to Golay. However, neither of them ever derived or endorsed these formulae. Moreover, I am not aware of any rigorous derivation of Eqs. (10.210)–(10.212) or their experimental confirmation for the conditions where the gas decompression is of any significance.

Note 10.17

As described earlier in this chapter, Golay derived [35] Eq. (10.22) for the plate height in a capillary column from Eq. (10.8) of a mass conserving flow under uniform conditions. Not only that, Golay explicitly stated (page 43) that his formulae "are applicable to columns ... in which the input to exit pressure ratio is near unity." Following this comment, Golay proposed a way of using Eq. (10.22) to derive a formula for the plate height "when the pressure variations in a column of uniform cross-section are important" (similar approach [67, 68] was used a year later by Giddings *et al.* to derive Eq. (10.90) for the apparent plate height). Earlier in this chapter, it has been shown that Golay formula in Eq. (10.22) is valid for any degree of gas decompression if u and D are treated as local (coordinate-dependent) and mutually dependent quantities.

The derivation [10, 197] of van Deemter formula (Eq. (10.209)) for the plate height in a packed column is less rigorous than that of Golay formula in Eq. (10.22). However, it is clear from the derivation that the formula does not account for the gas decompression along the column. □

There are many theoretical studies that do not rely on incorrect formulae for H. However, the incorrect formulae in Eqs. (10.210)–(10.212) unquestionably dominate the literature. They are ubiquitous in commercial literature published by GC instrument and column manufacturers. The formulae are also a regular feature of GC textbooks, peer reviewed articles in technical and scientific journals. Far reaching conclusions deduced from the incorrect formulae are known. Several generations of chromatographers learned from the university courses to respect Eqs. (10.210)–(10.212) as a well-established cornerstone of GC theory. All in all, it would not be a big exaggeration to say that among the harmful consequences of Eqs. (10.210)–(10.212) is the fact that they are the major clog in the flow of information between GC science and practice. The lack of the GC method development automation can be also attributed to the faulty plate height theories. A required step in a method development for a contemporary commercial computerized GC system is the interrogation of a human operator by a highly sophisticated computerized system. The purpose of the interrogations is for the computer to find out from a human operator what the very basic carrier gas parameters (velocity, flow rate, pressure, and so forth), and the heating rate in a temperature program should be used for a given column.

Something has to be done to correct the problem.

The logic of science does not require the prove of incorrectness of questionable propositions. Contrary to that, the authors of such propositions are generally expected to justify and verify their validity. Unfortunately, the reality is different from the ideals of dispassionate science. Experience shows that in order to dissuade those who hold their trust in Eqs. (10.210)–(10.212), it is not enough to just present a correct theory, but it is also necessary to explain the errors in Eqs. (10.210)–(10.212) and to expose the roots of these formulae. This is the purpose of the forthcoming discussion.

To simplify the discussion, secondary factors are ignored. Among them is the distinction of the local ($\mathscr{H}$) and the apparent (H) plate height. This also reflects the fact that, although both treatments of the plate height – as an apparent quantity [10, 105] and as a local quantity [35] – were known at the time of emergence of Eqs. (10.210) and (10.211), the distinction has not been clearly recognized.

10.12.1
Conflicts with Reality

A correct formula for the column plate height (H) as a function of $\bar{u}$ at arbitrary degree of gas decompression in a capillary column with arbitrary film thickness is described in Eq. (10.100) (Figures 10.9 and 10.11). Comparison of Eq. (10.211) with Eq. (10.100) immediately indicates that the former does not reflect reality. To simplify the comparison of Eqs. (10.100) and (10.211), only thin film capillary columns are considered through the rest of this section.

An important application of a formula for H is the evaluation of optimal pneumatic conditions corresponding to the minimum in H. Expressing H as a function of $\bar{u}$ allows one to find optimal gas velocity ($\bar{u}_{opt}$). It follows from Eq. (10.100) that $\bar{u}_{opt}$ can be found from Eq. (10.120). According to Eq. (10.120), relation of $\bar{u}_{opt}$ to column dimensions depends on the degree of the gas decompression (Figures 10.9 and 10.11). At weak decompression, Eq. (10.120) converges to Eq. (10.123) which shows that $\bar{u}_{opt}$ is inversely proportional to column internal diameter (d) and does not depend on the column length (L). On the other hand, when the decompression is strong, Eq. (10.120) converges to Eq. (10.124), which shows that $\bar{u}_{opt}$ is proportional to $(d/L)^{1/2}$. Equation (10.211) is not compatible with these facts. Nothing in Eq. (10.212) where B and C are fixed quantities reflects dependence of $\bar{u}_{opt}$ on gas decompression. Similar contradictions exist for the packed columns.

Not only Eq. (10.211) fails to reflect the dependence of relationship between $\bar{u}_{opt}$ and the column dimensions on the degree of the gas decompression, but it also misrepresents the dependence of H on the difference between actual $\bar{u}$ and $\bar{u}_{opt}$. According to Eq. (10.211), a factor of 2 departure of $\bar{u}$ from $\bar{u}_{opt}$ always leads to 25% increase in H (at $\bar{u} = 2\bar{u}_{opt}$, $H = 1.25H_{min}$ where H_{min} is H at $\bar{u} = \bar{u}_{opt}$). In reality, when gas decompression is strong, the twofold departure of $\bar{u}$ from $\bar{u}_{opt}$ more than doubles H. Indeed, according to Eq. (10.127) (Figure 10.15), $\bar{u} = 2\bar{u}_{opt}$ leads to $H = 2.125H_{min}$. Generally, when gas decompression is high, departure of $\bar{u}$ from $\bar{u}_{opt}$ much sharply increases H than it follows from Eq. (10.211). Thus, according to

Eq. (10.211), when $\bar{u}$ is significantly larger than $\bar{u}_{opt}$, H is proportional to $\bar{u}$, that is, at $\bar{u} \gg \bar{u}_{opt}$, $H \sim \bar{u}$. In reality, under the same condition, $H \sim \bar{u}^2$ as follows from Eq. (10.127) and shown in Figure 10.15.

Note 10.18

One of the consequences of the misrepresentation of the dependence of H on $\bar{u}$ is a concept of *optimal practical gas velocity (OPGV)* [135] defined as $\text{OPGV} = 2\bar{u}_{opt}$. Presumably, OPGV provides a better tradeoff between plate number (N) and the analysis time than $\bar{u}_{opt}$ does. The concept of OPGV was based on the following logic.

Duration of static analysis is proportional to hold-up time ($t_M = L/\bar{u}$). Therefore, the effect of a change in $\bar{u}$ and in L on the analysis time is the same as that on t_M. According to Eqs. (10.211) and (10.212), plate height, H_{OPGV}, corresponding to $\bar{u} = \text{OPGV}$ in static analysis is only 25% higher than H_{min} (at $\bar{u} = \bar{u}_{opt}$). Therefore, plate number, N_{OPGV}, is proportionally lower than $N_{max,orig}$ at $\bar{u} = \bar{u}_{opt}$ and at the original column length (L_{orig}). While reducing N, the increase in $\bar{u}$ from $\bar{u}_{opt}$ to OPGV makes t_M 50% shorter. If necessary, column length can be increased by 25% to compensate for the loss in N_{OPGV}. This will also increase t_M by 25% compared to its reduced value. Nevertheless, the net result of the use of OPGV instead of $\bar{u}_{opt}$ would be 37.5% shorter t_M with no loss in N, that is, $N_{OPGV} = N_{max,orig}$. This logic does not work when gas decompression is strong. Thus, according to Eq. (10.127), raising $\bar{u}$ from $\bar{u}_{opt}$ to OPGV raises H by a factor of 2.125 (not by 25% as Eq. (10.211) predicts) causing more than 50% loss in the plate number. To recover from this loss by raising the column length (as the concept of OPGV suggests), it would be necessary to raise the column length by 2.125. This would have two harmful effects. First, it would wipe out the reduction in the analysis time caused by using OPGV in the original column. Second, according to Eq. (10.124), raising L reduces $\bar{u}_{opt}$. This increases the mismatch between OPGV and the new $\bar{u}_{opt}$ and reduces the plate number. Additional details can be found in Example 10.14 and in literature [137]. All in all, contrary to the implications of Eq. (10.211), OPGV approach at strong gas decompression is a loosing proposition. □

The disconnect of Eqs. (10.211) and (10.212) with reality is also a source of frequently sporadic and inconsistent or even outright incorrect recommendations for gas velocity in GC analyses. It is not unusual, for example, to encounter published chromatograms acquired under conditions that were significantly different from the conditions recommended in the same source. Frequently, these recommendations are harmful rather than useful.

Furthermore, correct formulae for $\bar{u}_{opt}$ (Eqs. (10.120), (10.123), and (10.124)) are complex. Much simpler are formulae for optimal flow rate (F_{opt}, Eqs. (10.48) and (10.47)). Nevertheless, although the use of F as a pneumatic variable in equations for H was known [35, 199, 201–203] at least as early as in 1958, and optimal values of F were known since 1999 [127, 136, 158] the overwhelmingly wide use of $\bar{u}$ instead of F can be attributed to oversimplified picture portrayed by Eqs. (10.211) and (10.212). Arguably, that oversimplified picture unjustifiably diverted the attention of chromatographers from correct simple data for F_{opt} to incorrect data for $\bar{u}_{\text{opt}}$.

10.12.2
Origins of Incorrect Formulae

To get to the roots of Eqs. (10.210) and (10.211), and to bring the substance to the story of the emergence of these formulae, the names of several prominent pioneers of GC are mentioned here. It is done with clear understanding that in the development of new ideas, the errors are unavoidable, and that it is much easier to see the past errors than to avoid them at the dawn of a new technique. There is no intention here (and no justification) to negatively judge anyone. The forthcoming story is intended not to be about who did what, but about what happened when, how, and why.

van Deemter, Zuiderweg, and Klinkenberg published their famous formula (Eq. (10.209)) for a column plate height in 1956 [10].

At the *1957 International Symposium on Vapor Phase Chromatography* in London, Keulemans and Kwantes [66] presented a lecture with a detailed analysis of Eq. (10.209). The analysis was based on the following suggestion (page 18). "If we rewrite this equation in a simpler form as $H = A + B/u + Cu \ldots$ we see that it is the equation of a hyperbola with a minimum $A + 2(BC)^{1/2}$ at $u = B/C^{1/2}$ [presumably, this meant $u = (B/C)^{1/2}$]. This means that there is one value of u at which the column is operating under the most efficient gas velocity."

That suggestion is valid as long as Eq. (10.209) is limited to static uniform conditions as it was the case in the original study [10] where, except for the gas velocity (u), all quantities in Eq. (10.209) were treated as fixed parameters independent of u. Unfortunately, Keulemans and Kwantes went beyond uniform conditions. Their aforementioned quotation followed by this text. "It should be observed, however, that owing to the compressibility of the gas phase, [the gas velocity] u is not constant over the column, but gradually increases from inlet to outlet. Hence only a very small section [of a column] can operate at maximum efficiency." This immediately raises two questions. Is Eq. (10.209) valid at strong gas decompression? If the answer is *yes*, then does the form $H = A + B/u + Cu$ make sense?

Note 10.19

Even under static uniform conditions, Eq. (10.209) might not be sufficiently accurate [107, 169]. However, the question before us now is not about the accuracy of Eq. (10.209), but whether or not the template $H = A + B/u + Cu$, when it was proposed, was adequate for representing Eq. (10.209) that the template was intended to represent. An inaccuracy of Eq. (10.209) is irrelevant to that issue. □

Keulemans–Kwantes equation (as it was called in the early years of chromatography [51, 117, 201, 204]), that is, the *Keulemans–Kwantes form* of van Deemter formula, in Eq. (10.209) can be expressed as

$$H = A + B/u + Cu \tag{10.213}$$

$$A = 2\lambda d_p, \quad B = 2\gamma D, \quad C = \frac{8d_f^2 k}{\pi^2(1+k)^2 D_S} \tag{10.214}$$

Assume that outlet pressure is fixed. Strong gas decompression along a column is a result of a large difference between inlet and outlet pressure. In that case, somewhere in the middle of a column, not only the local gas velocity (u) but local solute diffusivity (D) is a function of local pressure (p) at that location, that is, $u = u(p)$ and $D = D(p)$. Excluding p from this system, one can express D as a function of u, that is, $D = D(u)$. This implies that other than at the column outlet, B in Eqs. (10.213) and (10.214) is a function of u. This also suggests that in the presence of a significant gas decompression, Eq. (10.213) does not actually mean what it implies according to existing conventions. Indeed, if Eq. (10.213) is to mean something, it should be assumed (Example 10.19) that all its parameters A, B, and C are independent of u. This is certainly not the case at strong gas decompression. And, certainly, at strong gas decompression, quantity $(B/C)^{1/2}$ is different from the local gas velocity (u_{opt}) minimizing the plate height at that location.

One can conclude that at strong gas decompression, Eq. (10.213) is not a proper template for expressing van Deemter formula in Eq. (10.209). For similar reasons, Eq. (10.213) is not a proper template for Golay formula (Eq. (10.24)) for H (see also Note 10.4).

Due to its simplicity, Eq. (10.213) was quickly accepted by leading theoreticians [115, 117, 135, 169, 201, 205–210] as a template for van Deemter and Golay formulae. The fact that Eq. (10.213) was valid only when the gas decompression was weak went almost unnoticed.

The unconditional acceptance of Eq. (10.213) by some workers left unnoticed one nuisance. This was the fact that in the presence of a significant gas decompression, the local gas velocity (u) was not a single variable, but a set of variables that were different at different coordinates along the column.

As far as I could find, the resolution of the problem in the form of Eq. (10.210) should be attributed to Purnell.

In their 1958 lecture [205], Bohemen and Purnell stated that "the rate theory of van Deemter, Zuiderweg and Klinkenberg ... leads to an equation relating the height equivalent to a theoretical plate (H) with the average superficial linear gas velocity (u) which ... for convenience of discussion ... may be written more simply" as Eq. (10.213). It is apparent from the lecture [205] context that the term "average superficial linear gas velocity" has the same meaning as the conventional term *average gas velocity* does.

In the paper [211] published next year, Purnell was even more specific. Referring to Eq. (10.213), he suggested that "van Deemter *et al.* have obtained an expression" where "u represents the average carrier-gas velocity." The transformation of van Deemter and Golay formulae into Eqs. (10.210) and (10.211) was complete when, in the lecture at the *1960 Symposium*, Purnell and Quinn [208] replaced symbol u in Eq. (10.213) with symbol $\bar{u}$, which represented "the compressibility-averaged carrier gas velocity."

In the aforementioned publications [205, 208, 211], no explanation was given to the following inevitable question. At strong gas decompression, quantity u in van Deemter and Golay formulae can only be treated as local velocity of a carrier gas. Keulemans and Kwantes also clearly recognized and treated u in Eq. (10.213) as the local velocity. Why then "the compressibility-averaged carrier gas velocity" ($\bar{u}$) replaced the local carrier gas velocity (u) in Eq. (10.213)? Possible answers to this question will be explored shortly.

Referring to the familiar paper [10], Purnell and Quinn identified Eq. (10.210) as "the van Deemter equation." In his 1962 book [108], Purnell also stated that Eq. (10.210) was "equation of van Deemter, Zuiderweg and Klinkenberg." However, Eq. (10.210) is fundamentally different from Eq. (10.209) known from van Deemter *et al.* and should not be attributed to these workers. For the same reason, Eq. (10.211) cannot be attributed to Golay as it is customarily done.

It appears from the publications of that time that the treatment of quantity u in Eqs. (10.210) and (10.211) as the average velocity of a carrier gas was inevitable. Purnell was the first in it, but other workers adopted the same approach independently and only slightly later. Not always, however, the average gas velocity was designated by symbol $\bar{u}$. In some cases, symbol u in Keulemans–Kwantes equation (Eq. (10.213)) was treated as the average gas velocity just blurring the distinction between Eq. (10.213) and, eventually, Eqs. (10.210) and (10.211) on the one hand, and the true van Deemter and Golay formulae (Eqs. (10.209) and (10.22)) on the other. This added the confusion to what could have been clearly identifiable errors, and further complicated the subject.

10.12.3
What Stimulated Adoption of Incorrect Formulae?

Effect of gas velocity on a column plate height was quantified by van Deemter *et al.* [10] and by Golay [35] in the second half of 1950s causing an avalanche of further theoretical and experimental research. The easiest way to find the gas velocity in an L-long column was to inject an unretained solute (like methane), measure its retention time, t_M, and find the gas velocity as L/t_M.

During those early years of development of GC theory, GC-MS with its requirement for vacuum at the column outlet was years away [212]. Carrier gas was typically nitrogen (requiring lower optimal pressure than that for helium or hydrogen, Figure 10.10c) and, in 1960 – the year of introduction of Eq. (10.210) – exotic experiments of Desty *et al.* [213] with high-resolution capillary column (270 m × 0.15 mm, nitrogen at 150 psi ≈ 1000 kPa) were yet to be reported a year later.

At that state-of-the-art, the distinction between different gas velocities (average, local, outlet, and so forth) was practically insignificant and not imperative. As a result, quantity L/t_M – later to be named as *average* velocity and designated as $\bar{u}$ – might have been not only the most convenient, but also a sufficiently accurate measure of the gas velocity. This did not justify the arbitrary modification of theoretical formulae. Nevertheless, the modifications were made and quickly accepted by chromatographers.

There are probably three main factors that brought to life and kept alive the unsubstantiated formulae in Eqs. (10.210) and (10.211). First, Eqs. (10.210) and (10.211) express H via $\bar{u}$ that is easy to measure in laboratory conditions. Second, the formulae portrayed a conveniently simple and familiar picture of (actually more complex, Eqs. (10.100) and (10.106)) dependence of H on $\bar{u}$. Third, the formulae were incorrectly attributed to van Deemter and to Golay who were widely (and justifiably) respected as some of the most prominent founders of GC theory.

It can be also mentioned that Golay's own terminology might have added to the confusion. In his early publications, Golay referred to quantity u in Eq. (10.22) as to "average velocity of carrier gas" [35], "average carrier gas velocity" [214], "average gas velocity" [215], and so forth. However, starting with his original definition [35] in 1958, Golay always used the term *average velocity* to mean cross-sectional average of a parabolic gas velocity profile. At that time, the meanings of the term average velocity has not been established in GC. Thus the concept of the *time-averaged velocity* defined as $\bar{u} = L/t_M$ was introduced by Bohemen and Purnell [205] (also in 1958) as the "average superficial linear gas velocity." Two years later, at 1960 symposium, Purnell and Quinn [208] referred to the same concept as to the "compressibility-averaged carrier gas velocity." Only later, the current meaning of the term average velocity or *average linear velocity* with the meaning of time-averaged velocity denoted as $\bar{u}$ was firmly established and confirmed by IUPAC [9]. However, its dual meaning during the early days of GC might have added to the confusions in the plate height theory.

10.12.4 Immediate Objections and Retractions

Speed of analysis was the main focus of theoretical lectures presented at *1960 European Symposium on Gas Chromatography.* Aforementioned Purnell and Quinn lecture [208] on the subject has been preceded by the lectures of Scott and Hazeldean [135] and by Desty and Goldup [207]. All three were based on Eq. (10.213) where u was explicitly or implicitly designated as the average carrier gas velocity. Similar approach was used in the paper [177] published by Knox in 1961.

It did not take long time to realize the incorrectness of that treatment.

At the next symposium in June 1961, Desty, Goldup, and Swanton reported their results on fast, high-resolution GC [213] utilizing small bore (0.035 mm) or very long (270 m × 0.15 mm) capillary columns. In the report, referring to their own and their peers previous assumption [135, 177, 207, 208], the authors recognized (page 114) that the assumption "is only approximately true providing the pressure drop along the column is small. Where large pressures are encountered, as in long columns, or columns having small bores this assumption may lead to erroneous conclusions."

Almost simultaneously with Desty and Goldup retraction, stronger and more specific objections were published by Ayers *et al.* [216] in July 1961.

Note 10.20

To graphically illustrate the validity of Eq. (10.210), Bohemen and Purnell earlier suggested [205] that quantity A in Eq. (10.210) can be ignored without significant sacrifice in the accuracy. With that assumption, it follows from Eq. (10.211), as Bohemen and Purnell pointed out, that the product $\bar{u}$ should be a linear function, $H\bar{u} = B + C\bar{u}^2$, of $\bar{u}^2$ (symbol u was used to denote the average gas velocity in the original text). Experimental plots (at unspecified gas decompression) confirmed this linearity. □

Objecting to Bohemen and Purnell conclusions, Ayers *et al.* wrote (page 988, column 2): "Bohemen and Purnell . . . earlier employed plots of $H\bar{u}$ *vs.* $\bar{u}^2$, where $\bar{u}$ is the average gas velocity, to represent the relationship of plate height to carrier gas velocity. While plots of this type are satisfactory for short columns having low pressure drops ($\bar{u} \approx u_o$), marked deviations from linearity occur if the column pressure drop is high."

Furthermore, in their comments to Eq. (10.210) in Purnell and Quinn study [208], Ayers *et al.* suggested (page 989) that "The method employed by Purnell and Quinn . . . for the estimation of optimum conditions for high speed separations is based upon the assumption that H is a unique function of $\bar{u}$, independent of the length of the column. This assumption is not correct for the conditions we have described. Hence the method which they have proposed is not generally valid."

Also there were clarifications. In their 1969 paper [52], Knox and Saleem suggested (page 614, column 1) that previous [135, 208] "derivations assumed that the carrier was incompressible and so applied only to liquid or low-pressure-drop gas chromatography."

More recently, the disagreement of Eq. (10.211) with experimental data was reported by Grob and Tschour [217, 218] and others [128, 129, 136].

Although the first retractions, objections, and clarifications were rather timely, they were unable to prevent the wide spread of Eqs. (10.210) and (10.211) that was taking place in a rapidly growing field of GC, especially in a laboratory and engineering environment.

During the last two decades, capillary columns became the columns of choice for majority of GC analyses, and GC-MS with its strong gas decompression became the ubiquitous technique. This and more recent development of comprehensive two-dimensional GC (GC × GC and GC × GC-MS) [219] relying on small bore secondary columns [220–228] exposed practical aspects of the problems with Eq. (10.211) and the urgency of clarifications attempted here.

10.12.5
Other Track

The wide spread and the persistence of the incorrect formulae for the column plate height in GC cannot be explained by the absence of the knowledge regarding the correct dependence of the plate height on the column pneumatic conditions.

A formula,

$$\mathscr{H} = A + \frac{B}{pu} + C_1 pu + C_2 u \tag{10.215}$$

correctly representing (at $A = 0$) Golay formula (Eq. (10.22)) in a compact form (leading through a different normalization directly to Eq. (10.38) – the basis for all compact forms in this book) structurally similar to Eq. (10.213) has been described by Giddings *et al.* in 1959 [67, 68]. In Eq. (10.215), quantities p and u are local (coordinate-dependent) gas pressure and velocity. The product pu (designated in this book as the energy flux, $J = pu$) is independent of coordinates. The key difference of Eq. (10.215) from its incorrect counterpart in Eq. (10.213) is that quantities A, B, C_1, and C_2 in Eq. (10.215) are independent of variables u and pu. Equation (10.215) is valid for uniform and nonuniform conditions.

From Eq. (10.215), Giddings *et al.* derived [67, 68] Giddings formula (Eq. (10.90)) for apparent plate height (H) in GC. The latter became the root of all known solutions for H in the presence of gas decompression in GC.

In this book, a single theory was used to derive the previously known and the new formulae for H. All formulae have their roots in Eq. (10.22) expressed in the form of Eq. (10.38) which directly follows from Eq. (10.215) with $A = 0$ and with transformation of variable J into variable f according to Eq. (7.58).

Appendix 10.A

10.A.1 Solute Velocity in Nonuniform Medium

Apparent (net) velocity (v_{app}) of a solute migrating in nonuniform (coordinate-dependent) medium can be found [44] from Eqs. (10.57) and (10.9) as

$$v_{app} = v + \frac{\partial D_{eff}}{\partial z} = \mu u + \frac{\partial D_{eff}}{\partial z} \tag{A1}$$

where u is the gas velocity, μ is the solute mobility, and D_{eff} is its effective diffusivity. Equation (A1) shows that the gradient, $\partial D_{eff}/\partial z$, of D_{eff} appears as a component – the diffusion component – of v_{app}.

Effective diffusivity [44, 45], D_{eff} (effective diffusion coefficient [46], virtual coefficient of diffusion [41], dynamic diffusion constant [43]) of a solute in a carrier gas stream at some coordinate z accounts for all diffusive zone broadening effects in a column (Chapter 10).

In capillary column, D_{eff} can be found (Eqs. (10.14) and (10.23)) as [44, 45] $D_{eff} = \mu u \mathscr{H}/2$ and, therefore, can be estimated as

$$D_{eff} \approx \frac{\mu u d}{2} \tag{A2}$$

where d is the column internal diameter.

The contribution of $\partial D_{eff}/\partial z$ to apparent solute velocity (ν_{app}) in Eq. (A1) becomes negligible when

$$\left| \frac{1}{\mu u} \frac{\partial D_{eff}}{\partial z} \right| \ll 1 \tag{A3}$$

It follows from Eqs. (A2) and (7.67) that

$$\frac{1}{\mu u} \frac{\partial D_{eff}}{\partial z} \approx \frac{1}{4(1 - z/L_{ext})\ell_{ext}} \tag{A4}$$

where quantities L_{ext} and ℓ_{ext} are defined in Eqs. (7.64) and (7.104).

Two extremes – the weak and the strong decompression of a carrier gas along a column – lead to different results in the evaluation of Eq. (A4).

In the case of a weak decompression where inlet pressure, p_i, is only slightly larger than outlet pressure, p_o, Eqs. (7.64) and (7.104) together with the fact that z in Eq. (A4) changes from 0 to L yield the following relations for the components of Eq. (A4): $L_{ext} \gg L$, $\ell_{ext} \gg \ell$, and $|z/L_{ext}| \ll 1$ where L and $\ell = L/d$ are, respectively, the length and dimensionless length of a column. Due to these inequalities, Eq. (A4) yields

$$\left| \frac{1}{\mu u} \frac{\partial D_{eff}}{\partial z} \right| \ll \frac{1}{4\ell} \tag{A5}$$

indicating that in a capillary column with any practically useful dimensionless length (even probably as small as 10) and with weak decompression, the diffusivity component ($\partial D_{eff}/\partial z$) in Eq. (A1) can be ignored.

The margins are not so favorable when gas decompression along a column is strong.

An extreme case of a strong decompressing is the one of the outlet at vacuum ($p_o = 0$). In that case, according to Eqs. (7.64) and (7.104), $L_{ext} = L$ and $\ell_{ext} = \ell$. Equation (A4) becomes

$$\frac{1}{\mu u} \frac{\partial D_{eff}}{\partial z} \approx \frac{1}{4(1 - z/L)\ell} \tag{A6}$$

For small values of z/L, that is, for a region near a column inlet, Eq. (A6) remains valid for all practical values of ℓ. However, when z/L approaches unity at a column outlet, the right-hand side of Eq. (A6) approaches infinity indicating that $\partial D_{eff}/\partial z$ in Eq. (A1) can become a dominant component of a solute velocity. The net outcome depends on the relative size of a column fraction, adjacent to its outlet, where the dominance of $\partial D_{eff}/\partial z$ exists. Assuming that the breakeven condition is the one where the right-hand side of Eq. (A6) is equal to 1, one has for the near-outlet fraction,

$1 - z/L$, of a column with unfavorable conditions:

$$1 - z/L \leq 1/(4\ell) \tag{A7}$$

This inequality suggests that for a column with 10^3 and larger dimensionless length, that is, when

$$\ell = L/d \geq 10^3 \tag{A8}$$

the near-outlet tip of a column with unfavorable conditions does not exceed 0.025% of a total length. For example, in a relatively unfavorable case of $L = 0.5$ m and $d = 0.5$ mm (a short column with large diameter), it would be only 0.5-mm-long outlet tip where $\partial D_{\mathrm{eff}}/\partial z$ might dominate the net solute velocity.

One can conclude that for typical gas flow rates and for not very short (Eq. (A8)) capillary columns, component $\partial D_{\mathrm{eff}}/\partial z$ in Eq. (A1) can be ignored so that the assumption $v_{\mathrm{app}} = v$ is a very accurate approximation. It should be mentioned, however, that there are practically useful applications [229–231], where relatively short columns ($L = 0.5$ m, $d_c = 0.53$ mm, $\ell \approx 10^3$) and well above optimal gas flow rates are used in vacuum outlet operations. In such cases, a column plate height can be much larger than internal diameter, d. As a result, D_{eff} becomes proportionally larger than its estimate in Eq. (A2). And so does the gradient, $\partial D_{\mathrm{eff}}/\partial z$, in Eq. (A1), making the conditions for ignoring $\partial D_{\mathrm{eff}}/\partial z$ as a component of v_{app} proportionally less favorable.

10.A.2 Formation of a Solute Zone

Equation (10.61) can be expressed in several equivalent forms.

In addition to quantities $\mathcal{D}$ and $\mathcal{H}$ described in the main text, let us introduce quantities

$$\mathcal{Q} = \mathcal{H}/v = \mathcal{D}/v^2, \quad \mathcal{W} = \mathcal{Q}/v = \mathcal{H}/v^2 = \mathcal{D}/v^3 \tag{A9}$$

This together with Eq. (10.66) for σ_z allows one to write Eq. (10.61) in the following equivalent forms:

$$\frac{d\tilde{\sigma}_z^2}{dt} = \mathcal{D}(z(t), t) + 2\tilde{\sigma}_z^2 \left.\frac{\partial v(z, t)}{\partial z}\right|_{z=z(t)} \tag{A10}$$

$$\frac{d\tilde{\sigma}_z^2}{dz} = \mathcal{H}(z, t(z)) + \left.\frac{2\tilde{\sigma}_z^2}{v(z, t(z))}\frac{\partial v(z, t)}{\partial z}\right|_{t=t(z)} \tag{A11}$$

$$\frac{d\sigma_z^2}{dt} = \mathcal{Q}(z(t), t) - \left.\frac{2\sigma_z^2}{v(z(t), t)}\frac{\partial v(z, t)}{\partial t}\right|_{z=z(t)} \tag{A12}$$

$$\frac{d\sigma_z^2}{dz} = \mathcal{W}(z, t(z)) - \left.\frac{2\sigma_z^2}{v^2(z, t(z))}\frac{\partial v(z, t)}{\partial t}\right|_{t=t(z)} \tag{A13}$$

In this collection of ordinary differential equations, Eq. (A11) reproduces previously known [44, 45] Eq. (10.61). Equation (A10) follows from Eq. (A11) due to Eqs. (8.16) and (10.23). Transition to Eq. (A12) is based on Eq. (4.12) with $y = v$ in combination with relations $dz = v dt$ and $\tilde{\sigma}_z = v\sigma_z$ described in Eqs. (8.16) and (A9), respectively. These allow one to transform $d\tilde{\sigma}_z^2/dz$ as

$$\frac{d\tilde{\sigma}_z^2}{dz} = \frac{dt}{dz}\frac{d}{dt}(v^2\sigma_z^2) = \frac{1}{v}\left(v^2\frac{d\sigma_z^2}{dt} + \sigma_t^2\frac{dv^2}{dt}\right) = v\frac{d\sigma_z^2}{dt} + 2\sigma_z^2\left(v\frac{\partial v}{\partial z} + \frac{\partial v}{\partial t}\right) \qquad \text{(A14)}$$

Substitution of this expression in Eq. (A11) yields Eq. (A12). Equation (A13) follows from Eq. (A12) due to the relation $dz = v dt$.

It follows from Eqs. (A10) and (A11) that dispersion rates $\mathcal{D}$ and $\mathcal{H}$ are the rates (temporal and spatial, respectively) of growth of $\tilde{\sigma}_z^2$ in uniform (static and dynamic) medium where $\partial v/\partial z = 0$. As quantity $\tilde{\sigma}_z^2$ is measured in units of length^2, the rates $\mathcal{D}$ and $\mathcal{H}$ are measured in units of length^2/time and length^2/length, respectively. The latter can be (and conventionally is) reduced to the units of length.

Equations (A12) and (A13) are complimentary to Eqs. (A10) and (A11) in many ways. It follows from Eqs. (A12) and (A13) that $\mathcal{Q}$ and $\mathcal{W}$ are the rates (temporal and spatial, respectively) of growth of σ_z^2 in static (uniform and nonuniform) medium where $\partial v/\partial t = 0$. As the quantity σ_z^2 is measured in units of time^2, the rates $\mathcal{Q}$ and $\mathcal{W}$ are measured in units of time^2/time and time^2/length. The former can be reduced to units of time and is a prototype of the plate duration [137] – a column parameter known from Purnell [64, 108, 123, 127, 131, 136, 137, 208, 211].

Solute velocity (v) in Eqs. (A10)–(A13) is a product $v = \mu u$ (Eq. (8.2)) of a column fluid (mobile phase) velocity (u) and a solute mobility (μ). In gradient elution LC, $\partial\mu/\partial z \neq 0$. On the other hand, in GC with uniform heating of a uniform column, $\partial\mu/\partial z = 0$. This allows for further simplifications. Equations (A10) and (A11) containing the derivative $\partial v/\partial z$ become

$$\frac{d\tilde{\sigma}_z^2}{dt} = \mathcal{D}(z(t), t) + 2\mu(t)\tilde{\sigma}_z^2\frac{\partial u(z, t)}{\partial z}\bigg|_{z=z(t)} \qquad \text{(A15)}$$

$$\frac{d\tilde{\sigma}_z^2}{dz} = \mathcal{H}(z, t(z)) + \frac{2\tilde{\sigma}_z^2}{u(z, t(z))}\frac{\partial u(z, t)}{\partial z}\bigg|_{t=t(z)} \qquad \text{(A16)}$$

It has been pointed out in the main text that, typically, plate height ($\mathcal{H}$) is roughly the same for all solutes and remains more or less fixed during a typical temperature-programmed GC analysis. This, in view of Eq. (8.2), implies that quantities $\mathcal{D}$, $\mathcal{Q}$, and $\mathcal{W}$ defined in Eq. (A9) are strong functions of a solute mobility (μ) that changes within several orders of magnitude from solute to solute, and for a given solute in temperature-programmed analysis – from one temperature to another. This prevents treatment of quantities $\mathcal{D}$, $\mathcal{Q}$, and $\mathcal{W}$ as a column parameter and makes plate height the parameter of the choice in the studies of solute dispersion and peak broadening in chromatography.

10.A.3 Spatial Width of Eluting Zone in Static GC

Spatial width ($\tilde{\sigma}$) of a solute zone at a column outlet can be found from Eq. (10.62) which, after substitution of Eqs. (10.31), (10.43), (7.67), (7.64) and (7.50), becomes

$$\frac{d\tilde{\sigma}_z^2}{dz} = \mathcal{H}_{\text{thin}} + \frac{c_{u2}(p_i^2 - p_o^2)}{2\Omega\sqrt{p_i^2 - (p_i^2 - p_o^2)z/L}} + \frac{(p_i^2 - p_o^2)\tilde{\sigma}_z^2}{p_i^2 L - (p_i^2 - p_o^2)z} \tag{A17}$$

where, in static analysis, all parameters are independent of z. For $z = L$, the last equation yields

$$\tilde{\sigma}^2 = L\left(\frac{\mathcal{H}_{\text{thin}}(p_i^2 + p_o^2)}{2p_o^2} + \frac{c_{u2}(p_i^3 - p_o^3)}{3p_o^2\Omega}\right) + \frac{p_i^2}{p_o^2}\tilde{\sigma}_i^2 \tag{A18}$$

where $\tilde{\sigma}_i$ is the initial spatial width of a zone – the spatial standard deviation of the zone concentration immediately after its placement in a column inlet. Due to Eqs. (7.114), (7.120) and (7.9), the last formula can be expressed as

$$\tilde{\sigma}^2 = \frac{L}{j^2}\left(\frac{\mathcal{H}_{\text{thin}} 9(P^2 - 1)^2(P^2 + 1)}{8(P^3 - 1)^2} + c_{u2}\bar{u}\right) + P^2\tilde{\sigma}_i^2 \tag{A19}$$

10.A.4 Plate Height in Temperature-Programmed Analysis

Only thin film columns with weak decompression of carrier gas are considered in this section.

It follows from Eq. (10.21) that for arbitrary dynamic conditions, spatial variance, $\tilde{\sigma}_{\text{thin}}^2$, of eluting solute zone can be found as

$$\tilde{\sigma}_{\text{thin}}^2 = \int_0^L \mathcal{H}_{\text{thin}} dz = L\int_0^1 \mathcal{H}_{\text{thin}} d\zeta \tag{A20}$$

where $\mathcal{H}_{\text{thin}}$ is the local plate height (Eq. (10.42)), L is the column length, and $\zeta = z/L$ is the dimensionless distance along the column.

According to Eqs. (10.42) and (10.50), $\mathcal{H}_{\text{thin}}$ can be expressed as

$$\mathcal{H}_{\text{thin}} = \left(\frac{T}{T_{\text{st}}}\right)^{1+\xi}\frac{X}{f} + \left(\frac{T_{\text{st}}}{T}\right)^{1+\xi}\frac{d^2\vartheta_{G1}^2 f}{48X} \quad \text{at} \quad \Delta p \ll p_o \tag{A21}$$

$$X = \frac{\pi\gamma_D^2 D_{g.\text{st}} d(10^3\varphi)^{0.09}}{2}\left(\frac{T_{\text{st}}}{T_{\text{char}}}\right)^{1.25} \tag{A22}$$

where f is the carrier gas specific flow rate.

Instead of quantity $\tilde{\sigma}_{\text{thin}}$ in Eq. (A20), it is more convenient to deal with quantity $\sigma_{\text{m.thin}} = \tilde{\sigma}/u_{\text{oR}}$ (Eq. (10.82)) where u_{oR} is the outlet velocity of a carrier gas at retention time t_R. Substitution of Eqs. (A20) and (A21) in Eq. (10.82) yields

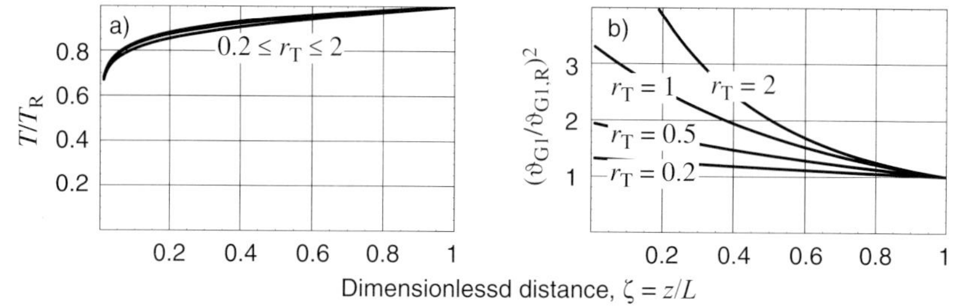

Figure 10.A.1 Column temperature (T, Eqs. (8.112) and (5.46)) (a), and quantity ϑ^2_{G1} (Eq. (10.25) combined with Eqs. (8.109)–(8.111) 110, 111)) (b) vs. distance (z) of a solute from inlet. T_R and $\vartheta_{G1,R}$ are, respectively, T and ϑ_{G1} at the solute retention time; r_T is dimensionless heating rate (Eq. (8.95)). Conditions: thin film, helium at constant pressure, small gas decompression, high retention at the beginning of heating ramp, $T_{char} = 600$ K, $\varphi = 0.001$ (variation of T_{char} and φ within 300 K $\leq T_{char} \leq$ 600 K and $0.0001 \leq \varphi \leq 0.01$ has a negligible effect on the graphs).

$$\sigma^2_{m,thin} = \frac{t^2_{MR}}{L}\int_0^1\left(\left(\frac{T}{T_{st}}\right)^{1+\xi}\frac{X}{f} + \left(\frac{T_{st}}{T}\right)^{1+\xi}\frac{d^2\vartheta^2_{G1}f}{48X}\right)d\zeta \quad \text{at} \quad \Delta p \ll p_o \qquad \text{(A23)}$$

where $t_{M,R}$ is a static hold-up time measured at the conditions existed at $t = t_R$.

When column temperature (T) changes during solute migration along the column, not only T in Eq. (A23) but also ϑ_{G1} and f that change with T become functions of dimensionless distance ζ. For highly interactive solutes – those that are highly retained at the beginning of a heating ramp – the dependence of T and ϑ_{G1} on ζ is shown in Figures 10.A.1 and 10.A.2a, respectively.

Figure 10.A.1a shows that T is a weak functions of z and almost independent of dimensionless heating rate, r_T. The same is true for specific flow rate (f) which, at constant pressure, is inversely proportional to T^{ξ} ($\xi \approx 0.7$, Table 6.12). This indicates that $T^{1+\xi}$ and f in Eq. (A23) can, without adding significant errors, be replaced with their distance-averaged values.

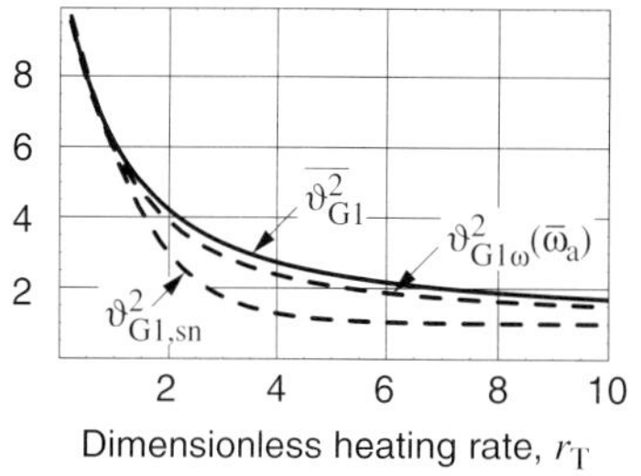

Figure 10.A.2 Average ($\overline{\vartheta^2_{G1}}$, Eq. A26, solid line) of ϑ^2_{G1} as a function of dimensionless heating rate (r_T, Eq. (8.95)). Dashed lines show two approximations to the solid line: $\overline{\vartheta^2_{G1}} \approx \vartheta^2_{G1,sn}$ where $\vartheta_{G1,sn}$ is ϑ_{G1} at $T = T_{sn}$, and as $\overline{\vartheta^2_{G1}} \approx \vartheta^2_{G1\omega}(\bar{\omega}_a)$ where function $\vartheta_{G1\omega}(\cdot)$ is described in Eq. (10.30), and ($\bar{\omega}_a$, Eq. (8.122)) is the average of ω_a over the column length. All conditions are the same as in Figure 10.A.1.

Let T_{sn} be the average of T. Accounting in Eq. (8.112) for Eq. (5.46) and integration of the result shows that (Figure 10.A.1a) for any carrier gas and film thickness, T_{sn} can be estimated as

$$T_{sn} = 0.92T_R \tag{A24}$$

Giddings, recognizing the importance of the temperature T_{sn}, pointed out [232] that (page 64, different symbol for T_{sn} was used in the original text) "the distance average of any slowly varying function of T is the value acquired at temperature T_{sn}," and, "for many purposes ... a programmed temperature process may be considered equivalent to an isothermal process, providing the latter is run at T_{sn}." Giddings called T_{sn} as *significant temperature*. The same term is adopted here.

Let f_{sn} be the value of f at $T = T_{sn}$. Replacing in Eq. (A23) quantities T and f with T_{sn} and f_{sn}, respectively, leads to expression

$$\sigma^2_{m,thin} = \frac{t^2_{MR}}{L}\left(\left(\frac{T_{sn}}{T_{st}}\right)^{1+\xi}\frac{X}{f_{sn}} + \left(\frac{T_{st}}{T_{sn}}\right)^{1+\xi}\frac{d^2\overline{\vartheta^2_{G1}}f_{sn}}{48X}\right) \quad \text{at} \quad \Delta p \ll p_o \tag{A25}$$

where

$$\overline{\vartheta^2_{G1}} = \int_0^1 \vartheta^2_{G1}\,d\zeta \tag{A26}$$

is the distance-averaged ϑ^2_{G1}.

Comparison of quantity in parentheses of Eq. (A25) with Eq. (A21) indicates that the former can be interpreted as the plate height,

$$\mathcal{H}_{sn,thin} = \left(\frac{T_{sn}}{T_{st}}\right)^{1+\xi}\frac{X}{f_{sn}} + \left(\frac{T_{st}}{T_{sn}}\right)^{1+\xi}\frac{d^2\overline{\vartheta^2_{G1}}f_{sn}}{48X} \quad \text{at} \quad \Delta p \ll p_o \tag{A27}$$

at static conditions with $T = T_{sn}$. The only component in this formula that is not consistent with this interpretation is quantity $\overline{\vartheta^2_{G1}}$, which is the average of ϑ^2_{G1} rather than its value ($\vartheta^2_{G1,sn}$) at $T = T_{sn}$. As shown in Figure 10.A.2, there is a significant difference between $\overline{\vartheta^2_{G1}}$ and $\vartheta^2_{G1,sn}$. As a result, replacement of $\overline{\vartheta^2_{G1}}$ with $\vartheta^2_{G1,sn}$ can increase an error in Eq. (A27).

An important difference between Eqs. (A23) and (A27) is that all components of the latter are no longer functions of changing T. Therefore, $\mathcal{H}_{sn,thin}$ can be viewed as temperature-independent plate height $H_{sn,thin}$. Furthermore, accounting for Eq. (A22), one can conclude that all but quantities f_{sn}, T_{char}, T_{sn}, and $\overline{\vartheta^2_{G1}}$ in Eq. (A27) are *a priori* known *fixed* quantities that are the same for all solutes. Let us take a closer look at the unknown quantities.

The quantity T_{char} is a solute parameter that can be treated as a known quantity or can be expressed via the solute retention and elution temperature as shown later. That leaves three unknown quantities – f_{sn}, T_{sn}, and $\overline{\vartheta^2_{G1}}$ – of which f_{sn} is a direct function of T_{sn}. Although they are unknown parameters, they can be expressed via the known ones.

According to Eq. (A24), the significant temperature, T_{sn}, is a function of T_R. The latter can be found from Eq. (8.117) where r_T (Eq. (8.95)) is the dimensionless heating rate and θ_{char} is the characteristic thermal constant of a solute. Like T_{char}, quantity θ_{char} can be treated as a solute known parameter. General trend of dependence of θ_{char} on T_{char} is described in Eq. (8.15).

The quantity, f_{sn} – specific flow rate at $T = T_{sn}$ – can also be expressed as a function of T_R. Indeed, being inversely proportional to gas viscosity (η, Eq. (7.61)) f_{sn} can be expressed, according to Eq. (6.20) and the data for ξ in Table 6.12, as

$$f_{sn} = (T_R/T_{sn})^{\xi} f_R = f_R/0.92^{\xi} \approx f_R/0.95 \tag{A28}$$

where f_R is f at $T = T_R$.

Finally, according to Figure 10.A.2, a good approximation for $\overline{\vartheta^2_{G1}}$ is quantity $\vartheta^2_{G1\omega}(\bar{\omega}_a)$ where function $\vartheta_{G1\omega}(\cdot)$ is described in Eq. (10.30) and $\bar{\omega}_a$ is the distance averaged ω_a described in Eq. (8.122) as a function of dimensionless heating rate (r_T) – a known parameter for each analysis.

Substitution of Eqs. (A24), (A28) and (A22) in Eq. (A27) along with the replacement of $\overline{\vartheta^2_{G1}}$ with $\vartheta^2_{G1\omega}(\bar{\omega}_a)$ and approximation $\xi \approx 0.7$ based on the data in Table 6.12, allows one to express Eq. (A27) as

$$H_{thin} = \frac{B}{f_R} + C_1 f_R \quad \text{at} \quad \Delta p \ll p_o \tag{A29}$$

where

$$B = \frac{\pi\gamma_T\gamma_D^2 D_{g.st} d(10^3\varphi)^{0.09}}{2}\left(\frac{T_R}{T_{st}}\right)^{1+\xi}\left(\frac{T_{char}}{T_{st}}\right)^{-1.25}, \quad C_1 = \frac{d^2\vartheta^2_{G1\omega}(\bar{\omega}_a)}{48B} \tag{A30}$$

$$\gamma_T = 0.92^{1+2\xi} \approx 0.82 \tag{A31}$$

The quantity B in Eq. (A30) allows one to find H_{thin} for a solute with known characteristic parameters T_{char} and θ_{char} of its interaction with the column. In that case, the solute elution temperature (T_R) can be found from Eq. (8.117). In studies of a column performance, it is more convenient to deal with a solute eluting at a given temperature (T_R) rather than with a known solute with (presumably) known T_{char} and θ_{char}. In that case, T_R in Eq. (A30) can be treated as an independent variable. A corresponding T_{char} can be found from Eq. (5.52). Equation (A30) becomes

$$B = \frac{\pi\gamma_T\gamma_D^2 D_{g.st} d(10^3\varphi)^{0.09}}{2k^{0.1}}\left(\frac{T_R}{T_{st}}\right)^{\xi-0.25}, \quad (k \geq 0.1) \tag{A32}$$

where k can be found from Eqs. (8.119) and (8.116). For highly interactive solutes – the ones with the large initial retention (k_{init}) – Eq. (8.119) converges to Eq. (8.116).

References

1 Blumberg, L.M. and Klee, M.S. (2001) *J. Chromatogr. A*, **933**, 1–11.
2 Blumberg, L.M. (1984) *Anal. Chem.*, **56**, 1726–1729.
3 Blumberg, L.M. and Berger, T.A. (1993) *Anal. Chem.*, **65**, 2686–2689.
4 Bracewell, R.N. (1986) *The Fourier Transform and Its Application*, 2nd edn Revised, McGraw-Hill Book Company, New York.
5 Dose, E.V. (1987) *Anal. Chem.*, **59**, 2414–2419.
6 Yan, B., Zhao, J., Brown, J.S., Blackwell, J., and Carr, P.W. (2000) *Anal. Chem.*, **72**, 1253–1262.
7 James, A.T. and Martin, A.J.P. (1952) *Analyst*, **77**, 915–932.
8 Littlewood, A.B. (1970) *Gas Chromatography: Principles, Techniques, and Applications*, 2nd edn, Academic Press, New York.
9 Ettre, L.S. (1993) *Pure Appl. Chem.*, **65**, 819–872.
10 van Deemter, J.J.J., Zuiderweg, F.J., and Klinkenberg, A. (1956) *Chem. Eng. Sci.*, **5**, 271–289.
11 Desty, D.H., Glueckauf, E., James, A.T., Keulemans, A.I.M., Martin, A.J.P., and Phillips, C.S.G. (1957) *Vapor Phase Chromatography* (eds D.H. Destyand C.L.A. Harbourn), Academic Press, New York, pp. xi–xiii.
12 Jones, W.L., Dal Nogare, S., Desty, D.H., Golay, M.J.E., Keulemans, A.I.M., Martin, A.J.P., Ober, S.S., Phillips, C.S.G., Thoburn, J., and Williams, E. (1958) *Gas Chromatography* (eds V.J. Coates, H.J., Noebelsand I.S. Fagerson), Academic Press, New York, pp. 315–317.
13 Martin, A.J.P., Ambrose, D., Brandt, W.W., Keulemans, A.I.M., Kieselbach, R., Phillips, C.S.G., and Stross, F.H. (1958) *Gas Chromatography 1958* (ed. D.H. Desty), Academic Press, New York, p. xi.
14 Johnson, H.W. and Stross, F.H. (1958) *Anal. Chem.*, **30**, 1586–1589.
15 Jones, W.L. and Kieselbach, R. (1958) *Anal. Chem.*, **30**, 1590–1592.
16 Ambrose, D., James, A.T., Keulemans, A.I.M., Kováts, E., Röck, H., Rouit, F., and Stross, F.H. (1960) *Pure Appl. Chem.*, **1**, 177–186.
17 Ambrose, D., James, A.T., Keulemans, A.I.M., Kováts, E., Röck, H., Rouit, F., and Stross, F.H. (1960) *Gas Chromatography 1960* (ed. R.P.W. Scott), Butterworth, Washington, pp. 423–432.
18 Ettre, L.S. (2008) *LC-GC N. Amer.*, **26**, 48–60.
19 Müller, R. (1950) Application of the chromatographic method for the separation and determination of very small quantities of gases. Ph.D. Thesis, University of Innsbruck.
20 Kaiser, R.E. (1962) *Z. Anal. Chem.*, **189**, 1–14.
21 Kaiser, R.E. and Rieder, R.I. (1975) *Chromatographia*, **8**, 491–498.
22 Fourier, J.B.J. (1822) *Théorie Analytique De La Chaleur.* Chez Firmin Didot, père et fils, Paris.
23 Fourier, J.B.J. (1878) *The Analytical Theory of Heat* (Translated With Notes), The University Press, Cambridge.
24 Fourier, J.B.J. (2003) *The Analytical Theory of Heat* (Translated With Notes), Dover Publications, Mineola.
25 Fick, A. (1855) *Ann D. Phys. U. Chem.*, **94**, 59–86.
26 Fick, A. (1855) *Phil. Mag. J. Sci.*, **10**, 30–39.
27 Crank, J. (1989) *The Mathematics of Diffusion*, 2nd edn, Clarendon Press, Oxford.
28 Brown, R. (1828) *Ann D. Phys. U. Chem.*, **14**, 294–313.
29 Brown, R. (1828) *Phil. Mag.*, **4**, 161.
30 Einstein, A. (1905) *Annalen Der Physik*, **17**, 549–560.
31 Einstein, A. (1956) *Investigations on the Theory of the Brownian Movement*, Dover Publications., New York.
32 Zauderer, E. (1989) *Partial Differential Equations of Applied Mathematics*, 2nd edn, John Wiley & Sons, Inc., New York.
33 Risken, H. (1989) *The Fokker-Plank Equation: Methods of Solution and Applications*, 2nd edn, Springer, Berlin.
34 Taylor, G. (1953) *Proc. R. Soc. London, A*, **219**, 186–203.

35 Golay, M.J.E. (1958) *Gas Chromatography 1958* (ed. D.H. Desty), Academic Press, New York, pp. 36–55.
36 Korn, G.A. and Korn, T.M. (1968) *Mathematical Handbook for Scientists and Engineers*, McGraw-Hill Book Company, New York.
37 Taylor, G. (1954) *Proc. R. Soc. London, A*, **225**, 473–477.
38 Westhaver, J.W. (1947) *J. Res. Nat. Bur. Stand.*, **38**, 169–183.
39 Aris, R. (1956) *Proc. R. Soc. London, A*, **235**, 67–77.
40 Aris, R. (1959) *Proc. R. Soc. London, A*, **252**, 538–559.
41 Taylor, G. (1954) *Proc. R. Soc. London, A*, **223**, 446–468.
42 Golay, M.J.E. and Atwood, J.G. (1979) *J. Chromatogr.*, **186**, 353–370.
43 Golay, M.J.E. (1980) *J. Chromatogr.*, **196**, 349–354.
44 Blumberg, L.M. and Berger, T.A. (1992) *J. Chromatogr.*, **596**, 1–13.
45 Blumberg, L.M. (1993) *J. Chromatogr.*, **637**, 119–128.
46 Giddings, J.C. (1991) *Unified Separation Science*, John Wiley & Sons, Inc., New York.
47 Golay, M.J.E. (1958) *Nature*, **182**, 1146–1147.
48 Golay, M.J.E. (1961) *Gas Chromatography* (eds H.J. Noebels, R.F. Wall, and N. Brenner), Academic Press, New York, pp. 11–19.
49 Giddings, J.C. (1962) *Anal. Chem.*, **34**, 722–725.
50 Giddings, J.C. (1964) *J. Gas Chromatogr.*, 2, 167–169.
51 DeFord, D.D. (1963) *Gas Chromatography* (ed. L. Fowler), Academic Press, New York, pp. 23–31.
52 Knox, J.H. and Saleem, M. (1969) *J. Chromatogr. Sci.*, **7**, 614–622.
53 Martin, A.J.P. and Synge, R.L.M. (1941) *Biochem. J.*, **35**, 1358–1368.
54 Bayer, E. (1961) *Gas Chromatography*, Elsevier, Amsterdam.
55 Smolková-Keulemansová, E. (2000) *J. High Resolut. Chromatogr.*, **23**, 497–501.
56 Ettre, L.S. (1987) *J. High Resolut. Chromatogr.*, **10**, 221–230.
57 Ettre, L.S. (2002) *Milestones in the Evolution of Chromatography*, ChromSource, Inc., Franklin, TN.
58 Golay, M.J.E. (1958) *Gas Chromatography* (eds V.J. Coates, H.J. Noebels, and I.S. Fagerson), Academic Press, New York, pp. 1–13.
59 Harris, W.E. and Habgood, H.W. (1966) *Programmed Temperature Gas Chromatography*, John Wiley & Sons, Inc., New York.
60 Blumberg, L.M., Wilson, W.H., and Klee, M.S. (1999) *J. Chromatogr. A*, **842/1-2**, 15–28.
61 Blumberg, L.M. (1999) *J. High Resolut. Chromatogr.*, **22**, 501–508.
62 Ettre, L.S. (1984) *Chromatographia*, **18**, 477–488.
63 Ettre, L.S. (1985) *J. High Resolut. Chromatogr.*, **8**, 497–503.
64 Cramers, C.A. and Leclercq, P.A. (1988) *Crit. Rev. Anal. Chem.*, **20**, 117–147.
65 Blumberg, L.M. (2003) *J. Chromatogr. A*, **985**, 29–38.
66 Keulemans, A.I.M. and Kwantes, A. (1957) *Vapor Phase Chromatography* (ed. D.H. Desty), Academic Press, New York, pp. 15–29.
67 Stewart, G.H., Seager, S.L., and Giddings, J.C. (1959) *Anal. Chem.*, **31**, 1738.
68 Giddings, J.C., Seager, S.L., Stucki, L.R., and Stewart, G.H. (1960) *Anal. Chem.*, **32**, 867–870.
69 Giddings, J.C. (1962) *Anal. Chem.*, **34**, 314–319.
70 DeFord, D.D., Loyd, R.J., and Ayers, B.O. (1963) *Anal. Chem.*, **35**, 426–429.
71 Blumberg, L.M. (1999) *J. High Resolut. Chromatogr.*, **22**, 213–216.
72 Giddings, J.C. (1963) *Anal. Chem.*, **35**, 1338–1341.
73 Giddings, J.C. (1965) *Dynamics of Chromatography*, Marcel Dekker, New York.
74 Fuller, E.N., Ensley, K., and Giddings, J.C. (1969) *J. Phys. Chem.*, **73**, 3679–3685.
75 Fuller, E.N. and Giddings, J.C. (1965) *J. Gas Chromatogr.*, **3**, 222–227.
76 Fuller, E.N., Schettler, P.D., and Giddings, J.C. (1966) *Ind. Eng. Chem.*, **58**, 19–27.
77 Maynard, V.R. and Grushka, E. (1975) *Advances in Gas Chromatography*, vol. **12**

(eds J.C. Giddings, E. Grushka, R.A. Keller, and J. Cazes), Marcel Dekker, New York, pp. 99–140.
78 Rubey, W.A. (1991) *J. High Resolut. Chromatogr.*, **14**, 542–548.
79 Zhang, Y.J., Wang, G.M., and Qian, R. (1990) *J. Chromatogr.*, **521**, 71–87.
80 Griffiths, J., James, D., and Phillips, C.S.G. (1952) *Analyst*, **77**, 897–904.
81 Drew, C.M. and McNesby, J.R. (1957) *Vapour Phase Chromatography* (ed. D.H. Desty), Academic Press, New York, pp. 213–221.
82 Martin, A.J.P., Bennett, C.E., and Martinez, F.W. (1961) *Gas Chromatography* (eds H.J. Noebels, R.F. Wall, and N. Brenner), Academic Press, New York, pp. 363–374.
83 Ettre, L.S. (2003) *LC-GC*, **21**, 144–149, 167.
84 Zhukhovitskii, A.A., Zolotareva, O.V., Sokolov, V.A., and Turkel'taub, N.M. (1951) *Doklady Akademii Nauk S.S.S.R.*, **77**, 435–438.
85 Ettre, L.S. (2000) *LC-GC*, **18**, 1148–1155.
86 Ohline, R.W. and DeFord, D.D. (1963) *Anal. Chem.*, **35**, 227–234.
87 Golay, M.J.E. (1962) *Gas Chromatography* (eds N. Brenner, J.E. Callen, and M.D. Weiss), Academic Press, New York, pp. xi–xv.
88 Rubey, W.A. (1992) *J. High Resolut. Chromatogr.*, **15**, 795–799.
89 Phillips, J.B. and Jain, V. (1995) *J. Chromatogr. Sci.*, **33**, 541–550.
90 Jain, V. and Phillips, J.B. (1995) *J. Chromatogr. Sci.*, **33**, 601–605.
91 Berger, T.A. and Wilson, W.H. (1993) *Anal. Chem.*, **65**, 1451–1455.
92 Blumberg, L.M. (1992) *Anal. Chem.*, **64**, 2459–2460.
93 Blumberg, L.M. (1994) *Chromatographia*, **39**, 719–728.
94 Blumberg, L.M. (1995) *Chromatographia*, **40**, 218.
95 Blumberg, L.M. (1997) *J. Chromatogr. Sci.*, **35**, 451–454.
96 Blumberg, L.M. (1993) *J. High Resolut. Chromatogr.*, **16**, 31–38.
97 Levich, V.G. (1962) *Physicochemical Hydrodynamics*, Prentice-Hall, Inc., Englewood Cliffs, NJ.
98 Lan, K. and Jorgenson, J.W. (2000) *Anal. Chem.*, **72**, 1555–1563.
99 Lan, K. and Jorgenson, J.W. (2001) *J. Chromatogr. A*, **905**, 47–57.
100 Giddings, J.C. (1963) *Anal. Chem.*, **35**, 353–356.
101 Kučera, E. (1965) *J. Chromatogr.*, **19**, 237–248.
102 Grubner, O. (1968) *Advances in Chromatography*, vol. **6** (eds J.C. Giddingsand R.A. Keller), Marcel Dekker, New York, pp. 173–209.
103 Grushka, E. (1972) *J. Phys. Chem.*, **76**, 2586–2593.
104 Jönsson, J.A. (1987) *Chromatographic Theory and Basic Principles* (ed. J.A. Jönsson), Marcel Dekker, New York, pp. 27–102.
105 Glueckauf, E. (1955) *Trans. Faraday Soc.*, **51**, 34–44.
106 Keulemans, A.I.M. (1959) *Gas Chromatography*, 2nd edn, Reinhold Publishing Corp., New York.
107 Dal Nogare, S. and Juvet, R.S. (1962) *Gas-Liquid Chromatography. Theory and Practice*, John Wiley & Sons, Inc., New York.
108 Purnell, J.H. (1962) *Gas Chromatography*, John Wiley & Sons, Inc., New York.
109 Lee, M.L., Yang, F.J., and Bartle, K.D. (1984) *Open Tubular Gas Chromatography*, John Wiley & Sons, Inc., New York.
110 Guiochon, G. and Guillemin, C.L. (1988) *Quantitative Gas Chromatography for Laboratory Analysis and On-Line Control*, Elsevier, Amsterdam.
111 Ettre, L.S. and Hinshaw, J.V. (1993) *Basic Relations of Gas Chromatography*, Advanstar, Cleveland, OH.
112 Grob, R.L. (1995) *Modern Practice of Gas Chromatography*, 3rd edn, John Wiley & Sons, Inc., New York.
113 Jennings, W., Mittlefehldt, E., and Stremple, P. (1997) *Analytical Gas Chromatography*, 2nd edn, Academic Press, San Diego.
114 Poole, C.F. (2003) *The Essence of Chromatography*, Elsevier, Amsterdam.
115 Giddings, J.C. (1960) *Anal. Chem.*, **32**, 1707–1711.
116 Habgood, H.W. and Harris, W.E. (1960) *Anal. Chem.*, **32**, 1206.
117 Stewart, G.H. (1960) *Anal. Chem.*, **32**, 1205.

118 Golay, M.J.E. (1963) *Nature*, **199**, 776.
119 Sternberg, J.C. (1966) *Advances in Chromatography*, vol. **2** (eds J.C. Giddingsand R.A. Keller), Marcel Dekker, New York, pp. 205–270.
120 Gaspar, G., Annino, R., Vidal-Madjar, C., and Guiochon, G. (1978) *Anal. Chem.*, **50**, 1512–1518.
121 Gaspar, G., Vidal-Madjar, C., and Guiochon, G. (1982) *Chromatographia*, **15**, 125–132.
122 Gaspar, G. (1991) *J. Chromatogr.*, **556**, 331–351.
123 Gaspar, G. (1992) *J. High Resolut. Chromatogr.*, **15**, 295–301.
124 Gaspar, G. (1994) *Sep. Times*, **8**, 8–11.
125 Spangler, G.E. (1998) *Anal. Chem.*, **70**, 4805–4816.
126 Reidy, S.M., Lambertus, G.R., Reece, J., and Sacks, R. (2006) *Anal. Chem.*, **78**, 2623–2630.
127 Blumberg, L.M. (1999) *J. High Resolut. Chromatogr.*, **22**, 403–413.
128 Blumberg, L.M. (1997) *J. High Resolut. Chromatogr.*, **20**, 597–604.
129 Blumberg, L.M. (1997) *J. High Resolut. Chromatogr.*, **20**, 704.
130 Ingraham, D.F., Shoemaker, C.F., and Jennings, W. (1982) *J. High Resolut. Chromatogr.*, **5**, 227–235.
131 Leclercq, P.A. and Cramers, C.A. (1985) *J. High Resolut. Chromatogr.*, **8**, 764–771.
132 Knox, J.H. and Scott, H.P. (1983) *J. Chromatogr.*, **282**, 297–313.
133 Knox in, J.H., Brown, P.R., and Grushka, E. (1998) *Advances in Chromatography*, vol. **38** Marcel Dekker, New York, pp. 1–49.
134 Knox, J.H. (1999) *J. Chromatogr. A*, **831**, 3–15.
135 Scott, R.P.W. and Hazeldean, G.S.F. (1960) *Gas Chromatography 1960* (ed. R.P.W. Scott), Butterworth, Washington, pp. 144–161.
136 Klee, M.S. and Blumberg, L.M. (2002) *J. Chromatogr. Sci.*, **40**, 234–247.
137 Blumberg, L.M. (1997) *J. High Resolut. Chromatogr.*, **20**, 679–687.
138 Butler, L. and Hawkes, S.J. (1972) *J. Chromatogr. Sci.*, **10**, 518–523.
139 Kong, J.M. and Hawkes, S.J. (1976) *J. Chromatogr. Sci.*, **14**, 279–287.
140 Millen, W. and Hawkes, S.J. (1977) *J. Chromatogr. Sci.*, **15**, 148–150.
141 Cramers, C.A., van Tilburg, C.E., Schutjes, C.P.M., Rijks, J.A., Rutten, G.A., and de Nijs, R. (1983) *J. Chromatogr.*, **279**, 83–89.
142 Steenackers, D. and Sandra, P. (1995) *J. High Resolut. Chromatogr.*, **18**, 77–82.
143 Vetter, W., Luckas, B., and Mohnke, M. (1996) *J. Microcolumn. Sep.*, **8**, 183–188.
144 Blumberg, L.M. and Klee, M.S. (2000) *Anal. Chem.*, **72**, 4080–4089.
145 Blumberg, L.M. and Klee, M.S. (2001) *J. Chromatogr. A*, **918/1**, 113–120.
146 Habgood, H.W. and Harris, W.E. (1960) *Anal. Chem.*, **32**, 450–453.
147 Bautz, D.E., Dolan, J.W., Raddatz, L.R., and Snyder, L.R. (1990) *Anal. Chem.*, **62**, 1560–1567.
148 Bautz, D.E., Dolan, J.W., and Snyder, L.R. (1991) *J. Chromatogr.*, **541**, 1–19.
149 Blumberg, L.M., and Klee, M.S. (1998) *Anal. Chem.*, **70**, 3828–3839.
150 Snyder, W.D. and Blumberg, L.M. (1992) Fourteenth International Symposium on Capillary Chromatography, Proceedings of the Baltimore, MD, USA, May 25–29, 1992, ISCC92, Baltimore, USA (eds P. Sandraand M.L. Lee), pp. 28–38.
151 Quimby, B.D., Giarrocco, V., and Klee, M.S. (1995) Speed Improvement in Detailed Hydrocarbon Analysis of Gasoline Using 100 μm Capillary Column, Application Note 228-294, Hewlett-Packard Co., Wilmington, DE, Catalog number (43) 5963-5190E.
152 David, F., Gere, D.R., Scanlan, F., and Sandra, P. (1999) *J. Chromatogr. A*, **842**, 309–319.
153 Quimby, B.D., Blumberg, L.M., and Klee, M.S. (1999) Pittcon'99. Book of Abstracts, Orange County Convention Center, Orlando, FL, March 7–12, 1999, p. 1271.
154 Sandra, P. and David, F. (2002) *J. Chromatogr. Sci.*, **40**, 248–253.
155 Reiner, G.A. (2003) 26th International Symposium on Capillary Chromatography & Electrophoresis, The Orleans Hotel & Casino, Las Vegas, Nevada, May 18–22, 2003, Elsevier Internet Conference Publication Services, www.elsubmit.com/esubmit/cce2003/.
156 Blumberg, L.M.(21 October 2003) Method Translation in Gas Chromatography, USA Patent 6,634,211.

157 Blumberg, L.M. (2002) Proceedings of the 25th International Symposium on Capillary Chromatography (CD ROM), Palazzo dei Congressi, Riva del Garda, Italy, May 13–17, 2002, I.O.P.M.S., Kortrijk, Belgium (ed. P. Sandra).
158 Blumberg, L.M., David, F., Klee, M.S., and Sandra, P. (2008) *J. Chromatogr. A*, **1188**, 2–16.
159 Blumberg, L.M. and Klee, M.S. (1998) 20th International Symposium on Capillary Chromatography (CD ROM), Proceedings of the Palazzo dei Congressi, Riva del Garda, Italy, May 26–29, 1998, I.O.P.M.S., Kortrijk, Belgium (eds P. Sandraand A.J. Rackstraw).
160 Magni, P., Facchitti, R., Cavagnino, D., and Trestianu, S. (2002) Proceedings of the 25th International Symposium on Capillary Chromatography (CD ROM), Palazzo dei Congressi, Riva del Garda, Italy, May 13-17, 2002, I.O.P.M.S., Kortrijk, Belgium (ed. P. Sandra).
161 Eiceman, G.A., Gardea-Torresdey, J., Overton, E.B., Carney, K., and Dorman, F. (2004) *Anal. Chem.*, **76**, 3387–3394.
162 Cramers, C.A., Scherpenzeel, G.J., and Leclercq, P.A. (1981) *J. Chromatogr.*, **203**, 207–216.
163 Cramers, C.A. and Leclercq, P.A. (1999) *J. Chromatogr. A*, **842**, 3–13.
164 Smith, H.L., Zellers, E.T., and Sacks, R. (1999) *Anal. Chem.*, **71**, 1610–1616.
165 de Zeeuw, J., Peene, J., Janssen, H.-G., and Lou, X. (2000) *J. High Resolut. Chromatogr.*, **23**, 677–680.
166 Whiting, J.J., Lu, C.-J., Zellers, E.T., and Sacks, R. (2001) *Anal. Chem.*, **73**, 4668–4675.
167 Kieselbach, R. (1960) *Anal. Chem.*, **32**, 880–881.
168 Kieselbach, R. (1961) *Anal. Chem.*, **33**, 23–28.
169 Jones, W.L. (1961) *Anal. Chem.*, **33**, 829–832.
170 Kieselbach, R. (1962) *Gas Chromatography* (eds N. Brenner, J.E. Callen, and M.D. Weiss), Academic Press, New York, pp. 139–148.
171 Dal Nogare, S. and Chiu, J. (1962) *Anal. Chem.*, **34**, 890–896.
172 Kieselbach, R. (1963) *Anal. Chem.*, **35**, 1342–1345.
173 Myers, M.N. and Giddings, J.C. (1965) *Anal. Chem.*, **37**, 1453–1457.
174 Myers, M.N. and Giddings, J.C. (1966) *Anal. Chem.*, **38**, 294–297.
175 Golay, M.J.E. (1957) *Nature (London)*, **180**, 435–436.
176 Knox, J.H. (1960) *Gas Chromatography 1960* (ed. R.P.W. Scott), Butterworth, Washington, pp. 195–197.
177 Knox, J.H. (1961) *Journal of the Chemical Society*, 433–441.
178 Giddings, J.C. (1965) *J. Chromatogr.*, **18**, 221–225.
179 Jennings, W. (2006) *LC-GC N. Amer.*, **24**, 448–457.
180 Peters, W.A. (1922) *J. Ind. Eng. Chem.*, **14**, 476–479.
181 Forbes, R.J. (1948) *Short History of the Art of Distillation*, Brill, Leiden.
182 Billet, R. (1979) *Distillation Engineering*, Chemical Publishing Co., New York.
183 Kister, H.Z. (1992) *Distillation Design*, McGraw-Hill, New York.
184 Stichlmair, J.G. and Fair, J.R. (1998) *Distillation Principles and Practices*, Wiley-VCH, New York.
185 Perry, R.H., Green, d.W., and Maloney, J.O. (1984) *Perry's Chemical Engineer's Handbook*, 6th edn, McGraw-Hill Book Company, New York.
186 Podbielniak, W.J. (1931) *Ind. Eng. Chem. Anal. Ed.*, **3**, 177–188.
187 Bragg, L.B. (1939) *Ind. Eng. Chem. Anal. Ed.*, **11**, 283–287.
188 Westhaver, J.W. (1942) *Ind. Eng. Chem.*, **34**, 126–130.
189 Craig, L.C. (1944) *J. Biol. Chem.*, **155**, 519–534.
190 Craig, L.C. and Post, O. (1949) *Anal. Chem.*, **21**, 500–504.
191 Craig, L.C. and Craig, D. (1950) *Techniques of Organic Chemistry*, vol. **III**, Interscience, New York.
192 Grob, R.L. (1977) *Modern Practice of Gas Chromatography* (ed. R.L. Grob), John Wiley & Sons, Inc., New York, pp. 39–112.
193 Grob, R.L. (1995) *Modern Practice of Gas Chromatography* (ed. R.L. Grob), John Wiley & Sons, Inc., New York, pp. 51–121.
194 Poppe, H. (1997) *J. Chromatogr. A*, **778**, 3–21.
195 Mayer, S.W. and Tompkins, E.R. (1947) *J. Am. Chem. Soc.*, **69**, 2866–2874.

196 James, A.T. and Martin, A.J.P. (1952) *Biochem. J.*, **50**, 679–690.

197 Klinkenberg, A. and Sjenitzer, F. (1956) *Chem. Eng. Sci.*, **5**, 258–270.

198 Lapidus, L. and Amundson, N.R. (1952) *J. Phys. Chem.*, **56**, 984–988.

199 Golay, M.J.E. (1958) *Gas Chromatography 1958* (ed. D.H. Desty), Academic Press, New York, pp. 62–68.

200 Giddings, J.C. and Keller, R.A. (1961) *Chromatography* (ed. E. Heftmann), Reinhold Publishing Corp., New York, pp. 92–111.

201 Littlewood, A.B. (1958) *Gas Chromatography 1958* (ed. D.H. Desty), Academic Press, New York, pp. 23–35.

202 Golay, M.J.E. (1958) *Gas Chromatography 1958* (ed. D.H. Desty), Academic Press, New York, pp. 53–55.

203 Bethea, R.M. and Wheelock, T.D. (1961) *Gas Chromatography* (eds H.J. Noebels, R.F. Wall, and N. Brenner), Academic Press, New York, pp. 1–10.

204 Dal Nogare, S. (1963) *Gas Chromatography* (ed. L. Fowler), Academic Press, New York, pp. 1–22.

205 Bohemen, J. and Purnell, J.H. (1958) *Gas Chromatography 1958* (ed. D.H. Desty), Academic Press, New York, pp. 6–17.

206 Khan, M.A. (1960) *Gas Chromatography 1960* (ed. R.P.W. Scott), Butterworth, Washington, pp. 181–182.

207 Desty, D.H. and Goldup, A. (1960) *Gas Chromatography 1960* (ed. R.P.W. Scott), Butterworth, Washington, pp. 162–183.

208 Purnell, J.H., Quinn in, C.P., and Scott, R.P.W. (1960) *Gas Chromatography 1960*, Butterworth, Washington, pp. 184–198.

209 Kaiser, R.E. (1960) *Chromatographie in Der Gasphase. Band II Kapillar-Chromatographie*, Bibliographisches Institut, Mannheim.

210 Kaiser, R.E. (1963) *Gas Phase Chromatography. Volume II Capillary Chromatography*, Washington, Butterworths.

211 Purnell, J.H. (1959) *Ann. NY Acad. Sci.*, **72**, 592–605.

212 Scott, R.P.W. (1979) *75 Years of Chromatography - A Historical Dialogue* (eds L.S. Ettreand A. Zlatkis), Elsevier, Amsterdam, pp. 397–404.

213 Desty, D.H., Goldup, A., Swanton, W.T., Brenner, N., Callen, J.E., and Weiss, M.D. (1962) *Gas Chromatography*, Academic Press, New York, pp. 105–135.

214 Golay, M.J.E. (1963) *Nature*, **199**, 370–371.

215 Golay, M.J.E. (1968) *Anal. Chem.*, **40**, 382–384.

216 Ayers, B.O., Loyd, R.J., and DeFord, D.D. (1961) *Anal. Chem.*, **33**, 986–991.

217 Grob, K. and Tschour, R. (1990) *J. High Resolut. Chromatogr.*, **13**, 193–194.

218 Grob, K. (1994) *J. High Resolut. Chromatogr.*, **17**, 556.

219 Liu, Z. and Phillips, J.B. (1991) *J. Chromatogr. Sci.*, **29**, 227–231.

220 Dallüge, J., van Stee, L.L.P., Xu, X., Williams, J., Beens, J., Vreuls, R.J.J., and Brinkman, U.A.Th. (2002) *J. Chromatogr. A*, **974**, 169–184.

221 Marriott, P.J., Dunn, M., Shellie, R., and Morrison, P.D. (2003) *Anal. Chem.*, **75**, 5532–5538.

222 Mondello, L., Casilli, A., Tranchida, P.Q., Dugo, P., and Dugo, G. (1019) *J. Chromatogr. A*, **2003**, 187–196.

223 Ryan, D., Shellie, R., Tranchida, P., Casilli, A., Mondello, L., and Marriott, P. (1054) *J. Chromatogr. A*, **2004**, 57–65.

224 Focant, J.-F., Sjödin, A., Turner, W.E., and Patterson, D.G. (2004) *Anal. Chem.*, **76**, 6313–6320.

225 Beens, J., Janssen, H.-G., Adahchour, M., and Brinkman, U.A.Th. (2005) *J. Chromatogr. A*, **1086**, 141–150.

226 Harynuk, J. and Górecki, T. (2007) *Am. Lab. News*, **39**, 36–39.

227 David, F., Tienpont, B., and Sandra, P. (2008) *J. Sep. Sci.*, **31**, 3395–3403.

228 Tranchida, P.Q., Purcaro, G., Conte, L., Dugo, P., Dugo, G., and Mondello, L. (2009) *Anal. Chem.*, **81**, 8529–8537.

229 Dagan, S. and Amirav, A. (1994) *Int. J. Mass Spectrom. Ion Processes*, **133**, 187–210.

230 Dagan, S. and Amirav, A. (1996) *J. Am. Soc. Mass Spectrom.*, **7**, 737–752.

231 Shahar, T., Dagan, S., and Amirav, A. (1998) *J. Am. Soc. Mass Spectrom.*, **9**, 628–637.

232 Giddings, J.C. (1962) *Gas Chromatography* (eds N. Brenner, J.E. Callen, and M.D. Weiss), Academic Press, New York, pp. 57–77.

Index

Temperature-Programmed Gas Chromatography. Leonid M. Blumberg

ISBN: 978-3-527-32642-6

u

v